HIGHWAY INVESTMENT in developing countries

HIGHWAY INVESTMENT

in developing countries

Proceedings of the conference 'Criteria for planning highway investment in developing countries', sponsored by the Commission of the European Communities and the Institution of Civil Engineers, organized by the Institution of Civil Engineers and held at the Café Royal Conference Centre, London, on 17–18 May 1982

THOMAS TELFORD LTD,
London, 1983

Conference sponsored jointly by The Institution of Civil Engineers and the Commission of the European Communities.

ORGANIZING COMMITTEE
J. N. Bulman
Dr J. C. R. Latchford
S. Thomas
J. Tresidder

Published for The Institution of Civil Engineers by Thomas Telford Ltd, PO Box 101, 26–34 Old Street, London EC1P 1JH.

BRITISH LIBRARY CATALOGUING IN PUBLICATION DATA
Institution of Civil Engineers
Highway investment in developing countries
1. Underdeveloped areas—Road construction—Congresses 2. Underdeveloped areas—Roads—Maintenance and repair—Congresses
1. Title
625. 7 TE153

ISBN: 0 7277 0163 0

Printed in Great Britain by The Thetford Press Limited, Thetford, Norfolk

Contents

Foreword

J. N. BULMAN, Overseas Unit, Transport and Road Research Laboratory

Road transport is the dominant mode of transport in virtually all developing countries and is likely to remain so for the foreseeable future. The dependence of road transport on fossil fuel, the real cost of which has increased dramatically over the past decade, has meant that an ever-increasing proportion of the scarce resources of many developing countries has to be expended on the provision and operation of road transport. One important consequence of this is that the validity of the criteria used for appraising investment in roads needs to be critically examined. These criteria must take better account of national, economic and social realities, and must be appropriate to the assessment of investment aimed at improving the efficiency of operation and the maintenance of existing road transport systems, as well as to additions to the road infrastructure.

The papers presented to this conference emphasize these points. The need for improving the cost-effectiveness of the design and construction of roads is put forward in several papers and in the discussion. This requires not only a better understanding of the factors influencing the deterioration of roads in the tropics, but also much-improved means of achieving the desired quality of construction. All too often the standard of construction actually achieved in practice falls far short of that assumed by designers, with the consequence that premature failures occur and resources are wasted.

The need for greatly improving the cost-effectiveness of road maintenance is also widely recognized, but it is very difficult to achieve because resource constraints are reinforced by a combination of institutional and staffing constraints in many countries. One possible solution to the road-maintenance problem put forward at the conference is to undertake more of the work by contract rather than by direct labour. The use of new forms of contract for both construction and maintenance offers a means of providing improved incentives for efficient operation, whilst minimizing some of the drawbacks of traditional methods of contract.

The urgent need to control the overloading of commercial vehicles attracted the attention of several contributors. The excessive damage done to road pavements by the heavy axle loads of such vehicles is evident in many countries, but an effective solution to the problem remains elusive. In the absence of adequate axle-load controls the life of road pavements is impossible to predict, which makes nonsense of economic appraisals made on the assumption that axle loads do not exceed legal maximum limits.

Unusually for such a conference, the subject of the cost of road accidents attracted considerable attention, illustrating a growing awareness of the significant economic cost of road accidents in many developing countries, and the need for road planners and designers to pay more attention to road-safety considerations in planning highway projects.

Together with many other issues, these points are explored in depth in these Papers, which collectively provide a valuable source of information for those concerned with highway investment in developing countries.

Opening address

A. AUCLERT, Directorate General for Development, Commission of the European Communities

Although I am an administrator, not an engineer, during my career in the French civil service in West Africa and in the European Communities since 1958, I have had to travel many times on the dusty roads of Upper Volta, Niger, and Mali, I have had to repair many wooden bridges which are destroyed or flooded each year, I have had to finance many new roads from the resources of the five European development funds of the Communities, so that I feel myself deeply and personally involved in the road business in Africa.

Taking for example only one European Development Fund, EDF IV of the Lomé 1 Convention, we have realized more than 3000 km of new roads in Africa, south of the Sahara. I think sometimes of the old dream of Sir Cecil Rhodes, who wanted to build a single route linking Cairo to Capetown. With just one EDF at his disposal, he could have seen his famous dream realized.

Having read with much interest the excellent Papers, I should like to present some comments on three main topics: the so-called dilemma between roads and agriculture, the great problem of the economic viability of road projects, and the question of road maintenance.

It has now become the fashion to juxtapose, in the selection of priorities for development, the development of rural production to the improvement of communications. The development plans of developing countries are frequently judged, in terms of quality, by their emphasis on the rural sector, compared to the modest share of resources devoted to communications, and especially to roads. This is a false dilemma. A Finance Minister once said: 'Give me good politics, I'll give you good finance.' There is the same relationship between agriculture and roads, between the traffic generated by rural development and the means of communication for supporting that traffic. After the independence of his country, the former President of Ghana, Doctor Kwame Nkrumah, once asked the famous economist Professor Arthur Lewis, 'What should I do to develop industry in my country?' . Professor Lewis replied, 'Develop first your agriculture!' It is the same for roads and agriculture: 'Give me good agriculture, I'll give you good roads.'

Most of the Papers at this conference deal with the difficult problem of the economic viability of roads - a subject which has many complexities.

Firstly, the calculations of the rate of return of a road must be very accurate, so wide is the range of benefits which can be derived from a new road, or even from an improved road. From the excellent paper by Mr Jarvis concerning road-user charges comes this quotation with which I agree entirely: 'Road user charges generate income that generally accrues to the government, which disburses money for the maintenance, policing and signing of the road system and for the construction of new or improved roads.' There are fixed charges (those levied on a vehicle, either only once or annually) and variable charges (those that are related to the amount of use of the vehicle, such as taxes on the sale of fuel and lubricants, and tools). During my career in the French civil service, I had occasion to make a detailed calculation of the profits accruing to the government of Niger, from a major cost-equalizing operation in transport from Eastern Niger to the port of Dahomey. I demonstrated that, due to the importance of taxes levied on road traffic, and on the import of the lorries, spare parts, tyres, and fuel, an apparently artificial operation based on budgetary subsidies was, on balance, highly beneficial to that budget. My calculations led the French aid administration to finance the first bridge built on the River Niger, at Malanville, on the border between Niger and Dahomey.

Secondly, we have to be very careful when we speak in terms of generating new traffic on new roads or improved roads. The stimulating paper by Messrs Grieveson and Winpenny is very clear on this point: 'It is not denied that roads can generate their own traffic, but it is futile to expect sizeable "developmental" effects in remote regions with few resources and population. One of the most persistent myths about roads is that virtually any road can justify itself because of the traffic it generates. Evaluation studies do not support this general view; some roads have witnessed dramatic traffic build-up, but others have had disappointing results. Also, as road building in Scotland demonstrates, roads can depopulate as well as develop.'

For my own part I would be a little more optimistic. In Morocco, Field-Marshal Lyautey designed a very generous road network, anticipating from the very beginning the development of Morocco and the discovery of large phosphate deposits. Maybe you know the reply he made to the wise men who asked him 'What if phosphates had not been discovered?' Field-Marshal Lyautey's reply was straightforward: 'Gentlemen, one always finds phosphates!'

Thirdly, it is quite obvious that in

appraising the viability of roads in developing countries we have to take account of many factors other than the simple rate of return. I have been very interested, in this respect, by the Paper produced by Mr Smith concerning the choice of appraisal techniques when resources are limited. This Paper underlines the need to incorporate social and other criteria into assessment frameworks and to include social, environmental and aesthetic factors in addition to economic and operational criteria, despite the difficulties in quantifying many of the non-economic benefits. It quotes the example of rural road programmes in the Malay peninsula, where the objectives of the programmes were defined as raising the productivity and incomes of the rural population by providing greater access for existing inhabitants and improving marketing and distribution outlets for rural produce and by providing the necessary physical infrastructure to meet the security needs of the country.

Maybe it is true that the road in itself does not generate the traffic. But it has a kind of catalytic effect on traffic, mobilizing potentialities and efforts. When I was a district commissioner in Upper Volta, one of my tasks was collecting taxes in the villages. In collecting them, I had to explain what the money collected would be used for. My speeches to the population always emphasized road improvements, the effects of which would be that the lorry could come right to the village, carrying goods to be sold and carrying back local products, and so regularizing the level of prices. Such language, factual and concrete, was always understood.

Finally, I find myself in complete agreement with the illuminating paper by Messrs Cornwell and Thomson about the development of priorities for rural roads, when it states that 'Rural road projects, by their nature, can give rise to relatively important economic and social development in the areas they serve and a large proportion, if not all, of the traffic on the new or improved roads, may be associated with new development. Road user savings are not, in this case, a reliable measure of project benefits, and attention must be devoted to the underlying activities which create the associated development and its benefits.

On the question of road maintenance, everyone is, I think, in agreement with the Authors of Paper 2, when they note, evidently with regret, that 'it is all too easy to underestimate maintenance costs on the existing network and devote too many resources to new infrastructure, with adverse long-term results.'

The European Development Fund faces such a problem, which is becoming more and more acute, due to the budgetary problems of developing countries. This is why it has become a general policy to insert in the special conditions of financing agreements for new road projects a clause stipulating that 'The government undertakes to set aside annually in its budget the funds necessary for maintenance of the road and to notify the Commission when it draws up its annual budget of the amounts allocated to the maintenance of all the roads financed by Community aid.'

Obviously, we have to take care that such a beautiful clause should not remain wishful thinking! During my career in the French civil service in West Africa, I suggested the creation of road funds, set apart from the current budget, through which expenditure for road maintenance could be financed by additional taxes on fuel consumption.

This is probably a correct and a safe solution. Otherwise, we would be rapidly in the situation that we faced, for example, in Chad, with the Fort Lamy-Massaguet road that we financed three times: in the first phase, as an earth road; in a second phase, as an asphalted road because the government had not made the necessary efforts to maintain the earth road; and in a third phase, to reconstruct the asphalted road, which had been no more properly maintained than the former earth road.

In conclusion, I should like to be rather impertinent and tell you three brief stories, drawn from my field experience of African roads. In 1950, as district commissioner in Upper Volta, I was on mission to a very remote village of the district, helping to collect the taxes, when I met a poor peasant. He had only one goat, and had not paid his tax for that goat, which was only five CFA francs, the present equivalent of one penny. Duly summoned by me, he wanted to pay on the spot. I refused, and took him in my car to the capital of the district where he paid at the counter of my financial agent and afterwards he returned to his village on foot. Was it inhuman? The poor peasant was, I saw it on his face, enthusiastic about his trip, during which he had had the occasion to visit parents and friends, to have personal contact with a remote administration, and to see how life was outside his village. The African is a great traveller, and this man had discovered the virtue of communication. I think that in both English and French, we have a double meaning for the word 'communication.' As well as the physical means of communication (like a road) it means also the function of communicating.

In Ivory Coast, in recent years, the government has built a great new highway between Abidjan and N'Douci, 140 km long. The highway was built by a Swiss contractor, financed by Swiss banks, and designed generously with technical specifications to be 'like a European highway.' You can drive on it at maximum speed, there are telephones every kilometre, and you can enjoy seeing panels indicating 'frequent fog.' Do not smile! In this part of the Ivory Coast, the climate is so humid that at dawn, when the ground literally transpires the humidity of the night, you can actually have fog.

In Guinea in 1952, we inaugurated in the presence of the French Overseas Minister, a new road, well asphalted, that had been financed by French aid. The Minister, having driven himself along this road, became very angry because he had not seen any traffic at all. The Governor, to facilitate the Minister's drive during the inauguration ceremony, had in fact prohibited all traffic on the road! So you will appreciate how very careful we have to be in projecting traffic for new roads, such projections usually being one of the main criteria for planning highway investment in developing countries, the very subject of these proceedings.

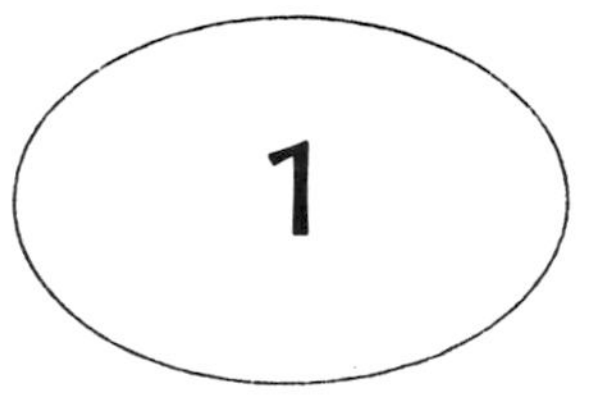

Economic appraisal of the required transport infrastructure for land-locked countries and external financing constraints

P. CABANIUS and A. PUSAR, United Nations Conference on Trade and Development, Geneva

UNCTAD'S concern with the presently used investment criteria for appraising new or up-grading existing roads stems from its activities related to the land-locked developing countries. Over the past three years, ever since the creation of a Transit Transport Task Force in Geneva, great emphasis was given to the transit transport problems being faced by the 21 land-locked countries in Africa, Asia and Latin America.

BACKGROUND

During the seventies, there were a number of UNCTAD and United Nations resolutions adopted concerning the mobilization of international assistance for least developed and particularly for the land-locked developing countries to help them overcome their transport and communications problems. However, assistance which was provided to the least developed countries was often dampened by events such as the sharp increase in fuel prices, significant fluctuations in the prices of raw materials which these countries were exporting, often coupled with increases in the price of finished imported goods, political upheavals, etc. In all, relative economic position the least developed and land-locked developing countries has deteriorated rather than improved during this period.

Most Least Developed Countries have a surplus of labour, so it is not surprising that they have favoured labour intensive practices and methods where and whenever possible, which has resulted in poor operating performance and low technical standards because of unskilled supervision and inadequate technical direction, leading to a need for rising maintenance requirements. Maintenance work itself suffers the same weakness, coupled with the additional burden of inadequate supplies of spare parts and materials, plus inefficient budget and managements support. Thus, in spite of large capital investments, Transport sector expansion and improvements have been slow in being realised.

Of some 31 countries presently considered as least developed (few others are being considered to be added to the list) by the United Nations on the basis of their very low per capita income, very low literacy rate and very low level of industrialization, 15 are land-locked. Their lack of territorial access to the sea, aggravated by remoteness and isolation from world markets and heavy costs of international transport services. This lack of access has acted as a major impediment to the improvement in their foreign trade and it appears to be one of the major causes of the relative poverty of land-locked countries and a serious constraint to their further economic and social development. These difficulties range from the inadequacy of physical facilities along the transit routes and in seaports to complications in the commercial and legal aspects of transitting through a foreign country.

Most Least Developed Countries are in sub-regional groupings and have neighbours suffering similar economic and social dis-abilities. Transport developments in all are limited by the small size and underdevelopment of domestic markets, scarcities of resources for investment, and the necessity of competing with long-established strong transport organizations from the developed market economies. Then again, during the past decade, the opportunities for regional and sub-regional co-operation have been lessened by a rising tide of nationalism, political upheavals and disruptions, as well as growing competition between neighbouring states for access tô limited aid resources and to world export markets.

In fact the problems of land-locked developing countries have been the subject of research by UNCTAD since 1970. A variety of basic documents have been prepared stressing an "integrated planning approach" which can be considered as a recommended transport strategy for land-locked developing countries. It aims at promoting and consolidating cooperative arrangements between land-locked countries and their transit neighbours. Such arrangements include the promotion of joint ventures in the field of transit transport, the simplification and standardization of procedures and formalities, the facilitation of the clearance of goods, the facilitation of road and rail traffic across national borders and the establishment of institutional mechanisms for regular review of transit transport issues.

In this regard, in almost all the land-locked sub-regions of the developing world, UNCTAD has embarked upon a series of detailed costing studies that hopefully will lead to the mobilization of international financial and technical support for improvements in transit infrastructure in both land-locked and transit countries. The costing studies should provide decision-makers with a clear blue-print of what policy changes or investment decisions are likely to have highest pay-offs. They also

Highway investment in developing countries. Thomas Telford Ltd, London, 1983, 5–7

should give a clear guide to the priorities for more detailed follow-up on specific improvements and form concrete basis for attracting needed assistance from the international community.

These studies, however, have already identified difficulties and inconsistencies in the application of traditional economic feasibility criteria when it is applied to transport project selection in the land-locked developing countries. A typical difficulty arises when, on the one hand, the traffic volume to and from the land-locked developing country is relatively low, and on the other hand when the transit route for this traffic is of very low priority to the transit country itself thus reflecting lack of maintenance and improvements.

Although the right for land-locked countries to be "unlocked" is in principle accepted, its application is not yet fully understood. The "right" has implications on transport infrastructure and technically speaking on construction, design, maintenance organization, documentation, etc.

UNCTAD feels, therefore, that project appraisal criteria must recognize the double aim of opening up land-locked countries and ensuring their economic integration within their subregion as well as establishing conditions for their further development.

Consequently, the key consideration is how to finance the transport physical investments, their maintenance costs, the non-physical activities and their corresponding costs within a framework that goes beyond the usual concern for national borders. It calls in question the very principle of determining proper responsibility and solvency. New financial appraisal methods have to be designed and, in this regard, it would be most advisable to work at the subregional level by bringing together countries with shared or closely related economic and commercial interests.

This concern regarding the applicability of traditional project appraisal criteria is neither new nor is it limited to UNCTAD. Recently, the need for review of criteria for investment and financing of road projects has been raised at the IVth African Highway Conference held in Nairobi (20-25 January 1980) then again at the IXth IRF World Meeting in Stockholm (1-5 June, 1981), it is now being discussed here at this Conference and, no doubt, this matter will continue to be debated in the future.

We have no quick answers or solutions to offer. We know there is a need for more research, however, as a result of our investigations from headquarters in Geneve and our presence in the field where we are providing technical assistance in transit transport to the land-locked developing countries and their transit neighbours, we have identified a number of issues for consideration of this meeting.

SECTORAL PLANNING

In a simplified form, the planning process within any economy consists of three principal levels: inter-sectoral (including inter-regional), sectoral, and project level. For the planning to be successful there must be horizontal and vertical interplay between all these levels forming a harmonious politico-socio-economic matrix. This is a complex but a necessary process, which, however, is beyond the scope of this paper. The important thing to remember is that the traditional project evaluation criteria is usually concerned with the "project" level assigning limited interest not only to sectoral but also to intersectoral levels as well as to politico-social factors which are difficult to measure in monetary terms but which play a crucial role in economic project appraisal techniques. The matrix is further complicated by short-term and long-term considerations which result from different national objectives over time. The politicians without exception are primarily concerned with the short-term "project" issues which does not help the analysts, for after all, it is their primary duty to demonstrate the different costs for alternative investment projects based on which the politician might take the necessary decision.

Very often it is taken almost for granted that over long land distances rail transport is cheaper than road. This may or may not be true depending on a variety of issues: availability of the railway, type and value of the goods to be transported, the quality of service provided by each mode, the secondary benefits generated, the developmental impetus provided, as well as issues such as pollution, social disruption, incidence of cost and benefits distribution and so on. The most important link between different sectors of national economy is served by roads rather than by any other mode and at the same time it is the road that has the greatest impact on opening up regions and generating development. Naturally, this cannot be separated form macro and micro considerations, from origins and destinations of imports and exports, from supply and demand of materials, from prices, wages, levels of employment and distances to markets. The point to remember, therefore, is that there is a definite inter-relationship between all sectors of economy, be it agriculture, industry, etc. and transport.

ECONOMIC APPRAISAL OF PROJECTS

Transport as a sector unlike other sectors of economy, is a service industry and as such does not possess a single pure concept of demand. It is a sector which has complex blends of multiple demands interlinked with supply. It is wrongful therefore to apply the traditional economic evaluation criteria to transport projects without accounting for variables such as land-use, population trends, migration tendencies to urban centers, location and availability of arable land and a score of other factors which in one way or another affect the transport investment decision. But above all the criteria has to account for a multiplicity of social indicators and include socio-economic trends and assumptions which has impact on change of social values, for, after all, transport is a communications sector.

When investment projects are being evaluated in the transport sector there are two special aspects requiring attention. The first concerns the significant indirect benefits and the second concerns the "lumpiness" of transportation investments. The indirect benefits include such factors as additional opportunities for internal trade, increased potential for export, increased mobility of population and goods, better prospects for communications, better defence and education opportunities, and so on. The inherent "lumpiness" in the investment of transportation projects results from the fact that by nature transportation projects involve long term commitment. Thus initial investment can only be justified on the basis of many years of obligation allowing for the fact that during the initial years there will be a surplus capacity and in excess of the immediate demand.

The evaluation process is further complicated by the fact that the concept of transport demand can never be defined with absolute certainty particularly given its interaction between economic and social factors. Then again we have to distinguish between need and demand for they both play an important role in project analysis especially in developing countries. It should also be borne in mind that demand can be manipulated and influenced.

The application of the Cost-Benefit Analysis (CBA) or the Multiple-criteria Analysis (MCA) to transport projects in developing countries have its inherent weaknesses for neither account for structural effects. In most countries society is committed to economic development and improvement of quality of life, improvement in balance of payments situation including foreign exchange, amelioration of land-use, abolishment of unemployment, factors which are difficult to incorporate ad-valorem in project analysis.

The choice of realistic alternatives poses another problem in project evaluation. Very often the definition of a problem is vague and the objectives are not properly defined although everyone appreciates the need to account for external effects which affect the community, spatial planning and land-use policies, and cultural impacts. In the decision making process, one of the major weaknesses is the lack of facts, lack of data, lack of statistics. This is one of the reasons why post-project evaluations usually are non-existent, and thus the planners can never learn from past mistakes.

Other weaknesses of the present evaluation techniques stem from failure to properly account for:

- the allocation of costs and benefits between different groups of society.
- attribution of weights and values to factors such as noise, pollution, time, etc.
- financial effects on different parties as a result of the project.
- effects of regulatory issues which affect the project, such as working rules, safety requirements, schedules of operation, controls, subsidies, and so on.

FINANCING

The arrangements of financing transportation projects in the land-locked developing countries pose another major problem. It is commonplace for a land-locked country to desire amelioration of transportation facilities in the transit country. This may involve better maintenance or upgrading of road surface, improvement of port facilities, and so on. Improvement of transport facilities, however, requires investments, involves operating expenses, and eventually the question arises on who is to pay for all this.

The financial institutions, including the World Bank and the African Development Bank, as well as majority of individual country donors have lending policies which do not permit them to lend to land-locked countries for facilities outside of their national borders. The onus of improved facilities rests, therefore, with the transit country. The problems of financing are further complicated by the fact that in majority of cases neither land-locked nor transit country can really afford additional debts even on soft-loan basis.

Even if the investment and recurrent cost problems being faced by land-locked and transit countries could be resolved, there is still the question of allocation and sharing of costs and benefits. One school of thought recommends that investments in transportation projects should be shared between the countries according to benefits that each country reaps. Very often, though, the magnitude of costs involved is in excess of what a developing land-locked country can afford, not to question the validity itself of the above costs and benefits sharing formula.

CONCLUSIONS

There is a pressing need to clarify the concepts of economic evaluation criteria presently used for investment projects in developing countries. All indications are that more just project evaluation criteria needs to be developed which *inter alia* would account for: income distribution, trade balance and foreign exchange issues, the "need" versus the "demand" for services, land use, population growth and migration aspects, improvement in quality of life, creation of employment, and so on.

On the financing side, given the maze of different objectives of financial institutions and donor governments, there is a need for a special effort to assist the land-locked developing countries to meet not only their required transport investment but also the annual maintenance and recurrent costs.

2

Assessing road investment strategies in Argentina

C. J. HOLLAND, BSc, DipTP, MRTPI, MCIT, MICE, R. Travers Morgan International Ltd, London, T. J. POWELL, MA, Coopers and Lybrand Associates, and R. M. DUNMORE, MA, MSc, DIC, R. Travers Morgan International Ltd, London

As part of the National Transport Plan of Argentina a need arose to calculate the future capital and maintenance expenditure associated with various policies for developing the road network. A number of programs were written which used the available data on the network to assess the opportunities for upgrading of each link. Taking into account regional traffic growth and network wear and tear, it was possible to select a pattern of future road investment compatible with a given policy or budget.

INTRODUCTION

1. Between 1979 and 1981 Travers Morgan and Coopers and Lybrand, in consortium, were engaged in the first phase of the National Transport Plan for the Government of Argentina supported by the World Bank. This paper describes work carried out at the end of this period to assist with the planning of national expenditure on highway construction and maintenance. A second phase of the study, which is expected to include development of this work, has now begun.

2. The client, the Secretariat of State for Transport and Public Works, supervised the six contracts into which the study was divided. Four of these (the 'modal contracts') were responsible for the data collected on each main mode of transport in Argentina: road, rail, aviation and water (canal, river and coastal shipping) transport. In each case the work carried out included two elements: the development of a data base describing the transport network for the mode, and the analysis of the structure of costs for transport by the mode. These data were then used in a further contract (No 6) to develop the main computer model of the national transport system.

3. We were responsible, in the fifth contract, for the elements of evaluation and policy formation in the study. The level at which output was required included:-

- appropriate national levels of spending on new transport facilities.
- relative priorities for expenditure on each mode of transport.
- taxation policies for the transport sector.
- regulations and tariffs for public transport operators.
- the extent and development of the networks for each mode.

4. Given this level of approach, and the need to study all the different modes within the same framework, the level of detail of the modelling of highways was necessarily limited. Although general differences in costs and link characteristics in different situations could be allowed for, it would have been impractical, as well as inappropriate, to survey every link in the national network in the level of detail customary for evaluation in the UK using, for example, the COBA program. For this reason the characteristics of individual links in the road network were proxied by variables which could more easily be determined, such as type of terrain and climatic region. Studies of individual schemes in greater depth would take place at a later stage in the analysis.

The Existing Network

5. Argentina is approximately twelve times larger than Britain but its population is only half as great. Further, one third of the population lives in greater Buenos Aires, and the degree of urbanisation in the other big cities is such that only 21% of the national population is considered to be rural. Rural settlements are small and widely spaced, and the road network throughout most of the interior is coarse and lightly trafficked.

6. The country extends over 3,500 km north to south and includes a wide range of climate and conditions, from high-altitude desert in the north-west and sub-tropical jungle in the north-east to the barren tip of Tierra del Fuego. Argentina also has territorial interest in Antarctica, although no road network is required there! Two other features present special problems. The western border is bounded by the Andes, which reach almost 7,000 m in places and are crossed by several high passes into Chile. Secondly, the flat alluvial plain of the pampas which extends inland over 500 km from Buenos Aires,which is virtually free of gravel and rock, and prone to wind erosion in times of drought and flooding in times of heavy rain.

7. Because of the variety of conditions arising within the country, road-building is rarely a straight-forward matter, and the approach to the problem must vary considerably in different regions. A major task of the roads contract was therefore to identify the differing costs of construction, maintenance and operation on roads of many types in widely scattered locations.

8. The road network is relatively dense around Buenos Aires and in its pampas hinterland, but further from the capital becomes progressively more sparse and of lesser standard. Argentina had at one stage a most extensive railway network, which, like the road network, radiated outwards from Buenos Aires and to a lesser extent from the other ports of Bahia Blanca and Rosario. Although many of the outlying parts of the rail network became too lightly-trafficked to survive, the principal lines alone still amount to some 20,000 route-km. Low investment and poor maintenance underlie slow improvement in the services offered. Road services have competitive advantages and, for example, more reliable journey times by road are attractive for the conveyance of livestock or 'difficult' products such as tobacco or cement (which must be kept dry).

9. As a result of road and pipeline competition over the last 15 years, rail share of ton-km hauled in Argentina was halved, and road now accounts for about 85% of the road/rail total; while the absolute increase in road hauled goods has been about 70%. In the future the railway's concentration on block trains is likely to retain their market share on the trunk corridors; nonetheless road goods traffic will continue to increase.

10. In the passenger market (excluding urban services), the last 15 years have seen a 70% growth in passenger-km by bus and a 110% growth in travel by car. Air traffic has grown sixfold to a 6% market share and rail traffic, due largely to closures of lightly used services, has declined by a third to a 7% share.

11. The inter-urban road network is almost exclusively single carriageway except along the Santa Fe - Rosario - Buenos Aires - La Plata corridor, and in large cities where urban motorways are being built. Not all of the network designated 'national' and shown in Fig. 1 (currently some 37,000 km) is paved, although the current policy is to bring it up to paved standard within a few years. Particularly in border areas with Chile, Bolivia and Paraguay many roads are of compacted 'earth' construction, or are 'improved' by the laying of a gravel base. As a result roads may be impassable for long periods in rainy weather, and even where major works have been carried out, sizeable structures have been washed out by flooding. A single closure in such a coarse network can result in detours of 100 km or more as few alternative roads are available.

The Assessment of New Works

12. Because of the size of the country, the on-site costs of internally-produced materials (including gravel, cement and bitumen) can vary widely, as does the scale of works needed for construction of roads of equivalent standard. The current program of construction is aimed largely at improving existing links to a higher standard and to a lesser extent at the provision of new links, but with the wide variations in costs in different locations, the development of 'rules-of-thumb' on what levels of provision are justified by what flows is extremely difficult.

13. Rather than attempt to derive general rules we therefore decided to examine the entire network and treat each link on its merits. It was felt that this approach would provide insights into the type of development of the network that was most beneficial, and also give an indication of the likely nationwide expenditure entailed by following any particular policy.

Development of the program

14. Contract 6 of the study developed a transport model which was able to consider all the stages of traffic generation, distribution, modal split and assignment. The model was designed principally for testing different policies on tariffs and taxation, and the extent to which these would alter the distribution of traffic and revenues. It was also able to estimate the effects of different policies on network development in the steady state: ie to assess the annual costs (including annualised capital costs) and benefits of the travel patterns resulting from given investment and tariffs.

15. The model was not intended to select a network directly given a budget and a brief to carry out those schemes which produced the highest rate of return. Network changes could only be made by manually coding the new link data, and their capital costs could also be only obtained by manual calculation.

16. We therefore developed a program closely related to certain parts of the main model, but with a different function. The basic purpose of the program was to examine each link of the road network, assess whether any new work was worthwhile, and then output both a new network and economic data on the cost of the proposed works and their expected rate-of-return. Later refinements allowed for the deterioration of roads with time and with the growth in traffic flows. The algorithm for selecting link improvements could be set so as to concentrate on pursuing only schemes in line with a particular policy, or to use a minimum rate-of-return on the capital expenditure. It was thought that a program of this type would be able to provide information on such questions as:-

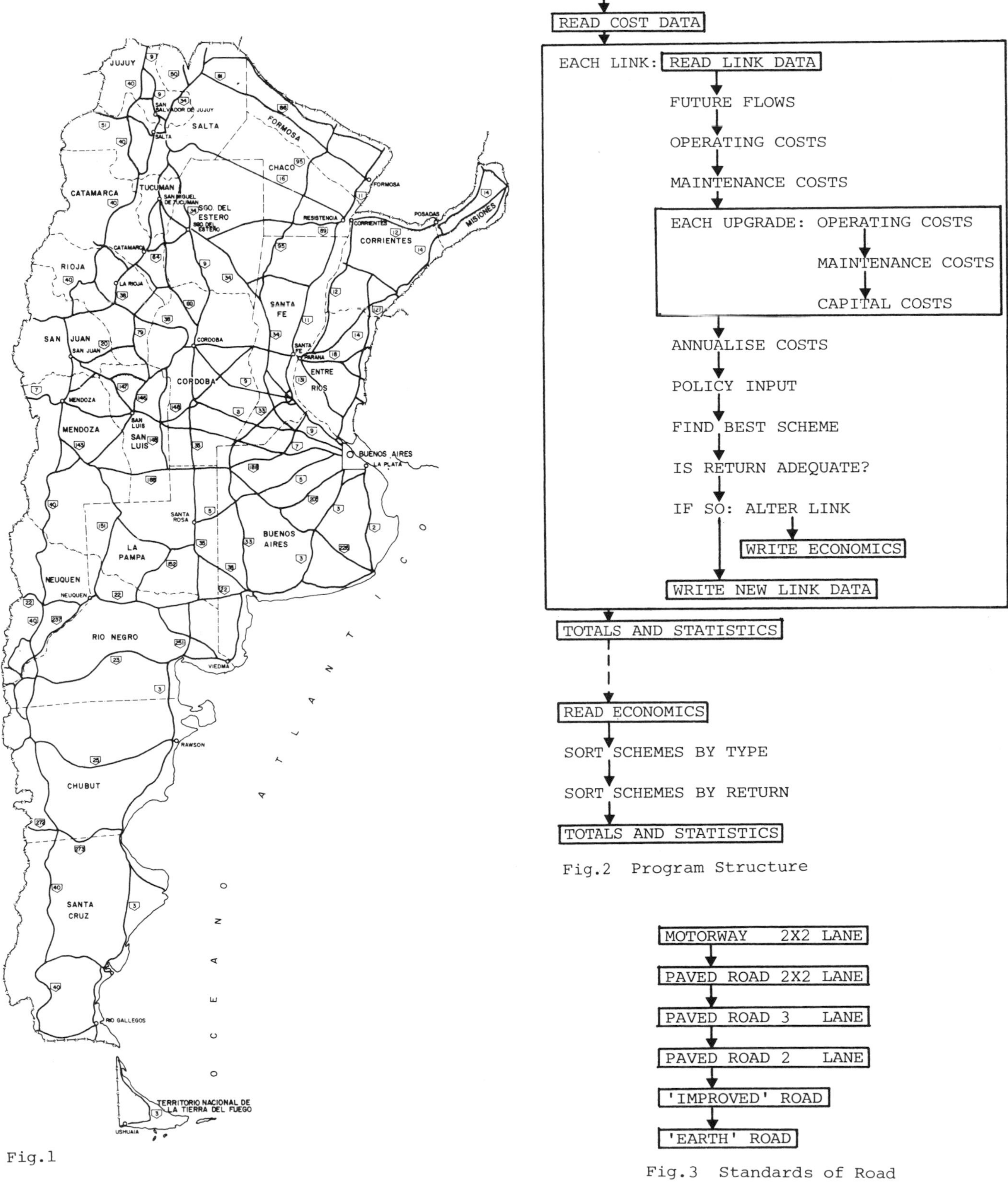

Fig.1

Fig.2 Program Structure

Fig.3 Standards of Road

(i) given a policy,

what will be the capital costs of its implementation?

what will the benefits and rate-of-return be?

where will the activity and benefits be distributed?

(ii) given a budget,

what amount should be allocated to maintenance?

what new schemes could be carried out?

what will the benefits and rate-of-return be?

where will the activity and benefits be distributed?

(iii) on a given link or corridor,

what standard should be provided?

how much will this cost?

how does it compare with other schemes elsewhere?

(iv) for the whole network,

what will its future condition be?

what will it cost to maintain?

17. As well as calculating the capital costs of new construction, the program was also designed to assess the maintenance costs of the network as it developed. This was considered to be particularly important since it is all too easy to underestimate maintenance costs on the existing network and devote too many resources to new infrastructure, with adverse long-term results.

18. No attempt was made to predict the effects of completely new links, as this would have required modelling of modal split and assignment. The program was restricted to assessing the way in which the existing network could best be improved.

Data Sources

19. The bulk of the input data required by the program was already available from the work associated with the main model. Three input data files were required by the program:

- traffic growth rates, province by province
- road network link data
- a cost data file containing all the parameters used to calculate road construction and maintenance costs and all elements of vehicle operating costs.

20. The provincial growth rates table was a simple array of the growth rates for each type of vehicle - car, bus and truck - currently observed in each of the 24 provinces.

21. Road data was provided by the National Highways Department, responsible for the 37,000 km of national network, and by the provincial road departments, responsible for the other roads in the modelled network. (The total modelled length was approximately 40,000 km.) To take into account the variations between roads in different areas, a total of 174 different categories of road were defined. This corresponded roughly to 7 standards of road (ranging from earth road to motorway) x 3 conditions x 7-10 environments. The 'environments' were combinations of terrain, climate and location (urban and rural) which determined vehicle speeds and the costs of land acquisition and road construction and maintenance. The average link in the network was approximately 70 km long and it was judged that over such lengths broad average costs per km based on terrain and climate would suffice. Also coded was the length of the link, its average daily flows of cars, buses and trucks, node numbers in the network and the national route number (if any).

22. Cost data included a variety of different items, which for convenience were read from the same source in a standardised way by a number of programs. Road construction costs had been estimated by the National Highways Department as a function of terrain, climate, and whether urban or rural for each standard of road; these were further subdivided into costs of land acquisition, earthworks, pavement and surfacing. Road maintenance costs were estimated in a similar way: these also varied with the condition in which the road was kept.

23. Cost data for vehicle operations were provided for the three classes of vehicle modelled: cars, buses and trucks. The costs were subdivided into tyre costs (which can be extremely high on poor roads), fuel costs and maintenance. Operating costs and vehicle speed were dependent on the road condition and the type of terrain. Occupants' values-of-time were also input.

Program Structure

24. The program was devised so that it worked in 5-year steps, using as input the network from a base year (initially 1980) and outputting a new network 5 years thence. Fig.2 shows the basis of the program structure.

25. The direct inputs to the program consisted of the three data files. Additional inputs such as changes in policy on what types of scheme would be considered were input indirectly by modifying the program.

26. Calculations began with reading in a link from the base (1980) network, and factoring its flow to correspond to 5 years growth at the appropriate provincial rate. The link condition

was then altered to represent the deterioration expected over a 5-year period. The vehicle operating costs and annual maintenance costs on the link in this projected do-nothing situation were calculated using the cost input data.

27. The next step was to enumerate the various possible methods of investing in the link. Fig. 3 shows the standards of road considered in the model, from earth roads up to 2-lane motorway, which at present is very rare outside the large cities.

28. If no investment was made in a link it would gradually deteriorate according to the use experienced as the surface and pavement wore out at the end of their design lives. The lowest investment option for each link was therefore to incur adequate expenditure to keep it in good condition. (In some cases it would be found that even this level of expenditure could not be justified in financial terms.) The remaining investment possibilities were to improve the road's standard, from 'earth' road to 'improved' or paved, or to increase the number of lanes. The principal benefits in the former case were increased safe speeds and reduced operating costs, and in the latter time-savings resulting from increased capacity on congested links.

29. At this point a policy input could be made to predetermine which type of scheme would be included and excluded from the examination. Examples of such policies would be to deny the upgrading of 'earth' roads to 'improved earth' status, instead paving them directly; or the conversion of all 3-lane roads into dual carriageways.

30. For each possible scheme on the link, the program then calculated the annual operating and maintenance costs in the same way as in the do-nothing situation. Then the scheme capital costs were calculated.

31. All capital costs were converted back to an annual equivalent cost (known as CAE), which was a fixed fraction of the total cost depending on the design life. By this means it was possible to compare schemes with different lifetimes on an equal basis. It was assumed that all the capital costs would be paid off retrospectively over the design life at an interest rate of 10% and the CAE was the annual repayment required. For example, for land acquisition, an asset with an infinite life, the CAE was 10% of the acquisition cost. For road surfacing with a design life of 5 years, the CAE was 26% of the capital cost.

32. An allowance was also made for the residual value of any materials in the old road which could be reused at the end of their design life. As an example, land taken to build a single carriageway could still be used as part of a dual carriageway along the same route and offset against the cost of a complete new dual carriageway. In contrast, the road surfacing material was assumed not to be economically reusable in reconstruction of the road. The capital cost calculated for new schemes was therefore rarely the cost of building a new road from scratch, but was reduced by allowances (mainly for land and earthworks) from what could be saved from the existing link.

33. To compare the various options on a given link, a 'rate-of-return' was calculated as the ratio of the annual cost savings (road maintenance and vehicle maintenance and operation) to the annualised capital cost. (If this was exactly 1.0 the internal rate-of-return of the scheme was 10%.) Once all the possible options for a given link had been costed and their benefits calculated, the one with the highest rate-of-return was identified. If there was a higher standard (and more expensive) option giving a lower overall rate-of-return, it might still be accepted if the rate-of-return on the marginal expenditure exceeded the test rate.

34. The 'best scheme' was then compared with the test rate-of-return to see if the scheme should be carried out. If a scheme was 'accepted' the details of the changes being made, the costs, benefits and rate-of-return were all written out to an economic data file. Then the link data for the new link (whether modified or not) was written out to the output network, with flow and condition data for 1985.

35. The process was then repeated for each link in the base network.

36. After the main program had been run a subsidiary sort program was used to group the schemes selected by different types (paving, dualling, resurfacing, etc), and sort them in order of rate-of-return. This program also produced cost and benefit totals for each type of scheme being carried out.

37 If required, the whole program could then be re-run for a second 5-year period, using the output 1985 network as an input, to calculate a further 5-year expenditure program and future network for 1990.

Use of the Program

38. In practice the program was run in two ways. First a 'blind run' would be carried out with few policy constraints and a low test rate-of-return merely to see what schemes were available. On the basis of the output a tentative policy could be identified, and further runs could then be carried out to study the costs and effects of this policy. By means of this trial-and-error approach it was normally possible to estimate what results any given budget could bring, choosing the test rate-of-return and suitable policy constraints until the required results were obtained.

39. The program was found to be very easy to use and modify: a one-day turn-round of testing a new policy was relatively easy to achieve. (This corresponded to approximately 2 hours of work on the computer.) The input data was largely fixed for each run, and the only changes needed were minor ones internal to the program.

40. The main findings of the initial work with the program were:-

a) The true costs of maintaining the existing network were considerably higher than expected. The program was able to predict the expected costs not only of routine maintenance, but also of resurfacing and repaving existing roads as their life expired. Earlier proposals had envisaged a capital expenditure over the 5-year period 1980-1985 of approximately $US1.5-2.0 Billion over 7000 km of route. However the program suggested that a more modest capital expenditure would provide a better return and release more funds for maintenance and repair projects.

b) In general in Argentina, surfacing roads with gravel as a stop-gap instead of paving them was a poor investment: if a road was worth 'improving' it was worth paving. Gravel roads appeared to entail high maintenance costs which made them more expensive in the long term than a paved road.

c) The provision of a third lane on two-lane roads was not normally worthwhile, given the speed-flow relationships obtaining in Argentina. The additional benefits of an extra lane gave only a poor return on the extra expenditure required to rebuild the road with a wider carriageway.

d) There were certain clear corridors in which systematic investment was desirable. It was apparent that the program was able to identify what parts of the network deserved 'preferential' treatment, and where links could (in contrast) be allowed to deteriorate, on economic grounds at least.

e) There were several routes where single links were at a poor standard relative to the rest of the route. Although upgrading of some such links was not justified by economic criteria, it might be favoured on grounds of policy.

41. In the light of these results, and analyses for the other modes, a broad pattern of expenditure - in terms of geography and the distribution among the various categories of upgrading - was drawn up for the National Plan. As to actual works, the Plan is a guiding rather than prescriptive document, and individual road maintenance programmes and improvements will of course be subject to detailed appraisal.

Future Development

42. The National Transport Plan is not a one-off study. A part of its purpose was to equip the Government with the means to keep its transport under review and adaptation. The main transport model will therefore be subject to repeated use and the input data to refinement and updating.

43. As part of the improvement of the model, it is hoped that the programs described here will be integrated more fully into the main model.

Acknowledgments

The authors wish to acknowledge their indebtedness to their colleagues in Consorcio Transplan - particularly Lic. Diaz Terrado, the project director, and Robin Carruthers, economic adviser; to the officers of the Direccion Nacional de Planeamiento de Transporte, particularly Ing. Pesce of the control group, and Ing. Kogan, the Director (who also kindly permitted the preparation of this article).

3

Toll road and road user charges applications in developing countries

S. K. HAMMERTON, PhD, and J. H. EBDEN, BSc, Halcrow Fox and Associates

Consideration of the merits of introducing toll roads and other forms of road user charges is becoming increasingly common in developing countries throughout the world. This paper discusses the practical advantages and disadvantages of introducing such revenue raising schemes, and the likelihood of their success in terms of offsetting road construction and maintenance costs. Reference is made to particular instances where such schemes have been or are being introduced.

1. INTRODUCTION

1.1 Studies of road user charges have formed an essential component in the planning of highway investments in both developed and developing countries. The evaluation of proposed highway projects, from inter-urban expressways to rural feeder roads, involves, either explicitly or implicitly, an application of the theory of road user charges. Numerous textbooks, papers and project reports have been written to describe the theories and elucidate their application.

1.2 It is becoming increasingly common for highway authorities in developing countries to review their policies with regard to road user charges and consider the introduction of more direct measures for the collection of revenues from road users. Specific attention has been given to the introduction of toll highways as an element in the development of a high capacity limited access inter-urban highway network.

1.3 In theory, revenue raised in this way provides a valuable source of funds for highway maintenance, upgrading and construction, while raising the awareness of road users to the costs of providing the highway from which they benefit. Furthermore, it is postulated that the development of toll highways minimises the necessity of raising very large loans and the penalisation of non-road users through the general taxation system.

1.4 The paper outlines the advantages and disadvantages of toll highway programmes, commenting in particular on the analytical difficulties involved in the analysis of toll applications, and the problems of determining financial returns, such that the funds so raised can make a real contribution in reducing the future burden to the taxpayer of network maintenance, upgrading and extension.

1.5 There are a range of taxes which constitute road user charges. The nature and level, in monetary terms, of each influences the perception of the vehicle user to the cost of travel. The various methods of charging for the use of roads was considered by Smeed and the principal elements are listed below:

INDIRECT — <u>Related to vehicle ownership</u>
- Annual licence
- Purchase taxes

<u>Related to vehicle usage</u>

i) Amount of usage
- Fuel tax
- Tyre and Parts taxes

ii) Amount and place of usage
- Differential fuel taxes

iii) Place and time of usage
- Parking tax
- Restricted licences

DIRECT — <u>Highway tolls</u>
- Open systems
- Closed systems

1.6 Toll charges are levied during the course of a trip and thus can be assumed to reflect the value to the user of making the trip by the specific route. The toll charged represents a composite valuation of the time and distance, or in the case of a river or water crossing, ferry costs, saved by using the toll highway.

1.7 The introduction of toll roads and their role in a highway network varies between developed and developing countries. It reflects the differences in attitudes to the overall balance of the various road user charges. In the United Kingdom and Northern Europe (Germany, the Netherlands, etc) there are only a few examples of tolled inter-urban highways, these being limited to large éstuarial and river crossings. Elsewhere in Europe (France, Italy, Spain, etc) a network of inter-urban toll roads has been developed.

2. PRINCIPLES OF TOLL COLLECTION SCHEMES

2.1 The main objective of a highway toll system is to provide a means of collecting revenue to repay the capital and operating costs associated with construction and maintenance. An effective system should:

i) provide a simple means of accounting for tolls collected;

ii) be easy to operate and convenient to use;

iii) be largely fraud proof to both users and operators;

iv) be economical in initial capital and operating costs.

2.2 The well established and most practical method of payment is 'stop and pay', in which the driver pays by cash, pre-purchased vouchers or credit cards at the time he makes the journey. Other methods of payment which do not require drivers to stop are being developed and tested experimentally in several countries, but these are not sufficiently robust yet for general use, particularly in developing countries.

2.3 Two systems of applying the 'stop and pay' method are possible:

i) the 'open' system in which a fixed toll for each type of vehicle is collected at suitable intervals along the highway. After paying the toll, the driver is free to leave the highway at the next intersection, or to continue to the next toll collection point.

ii) the 'closed' system, in which the driver takes a ticket at entry and pays on exit according to the distance travelled. After entering the system, the driver cannot leave it without paying a toll.

2.4 Open systems are common for estuarial crossings such as bridges and tunnels where there is no way of using the crossing without paying the toll, and a fixed toll for each type of vehicle is appropriate. It is less suitable for highways which have interchanges at frequent intervals because, unless toll collection points are located between each intersection, drivers will be able to leave and rejoin the highway if they wish to avoid paying the toll. In open systems in which the intersections, (and hence the toll collection points), are more widely separated, it is not necessarily important to ensure that 100% of journeys are intercepted, since it may not be worthwhile introducing barriers at locations where the revenue produced would not exceed the costs of collection.

2.5 Closed systems are in common use in Europe and America, and are suitable for long highways with frequent interchanges. Whatever distance is travelled, the driver stops twice only: once on entry and once on exit, the toll being related to the distance travelled. To operate effectively the system should be sealed so that no-one can use it without paying, and this means providing barriers at each entrance as well as each exit. If some drivers were able to enter without encountering a barrier, it would be impossible to ensure that the correct toll was being applied at the exit. The provision of toll barriers on every arm of each interchange involves considerable capital and recurrent expenditure which may exceed the revenue collected where volumes are low, particularly in locations where attractive and toll-free alternatives are available, or during the early years of the life of the highway before traffic volumes reach their design levels. One practical advantage of the closed system is that the operation has close control over access to the system, and can close parts of it in the event of accidents or other hazardous incidents.

2.6 In addition, hybrid systems can be designed: for instance, if in an open system the provision of a toll collection point between two important interchanges is impossible or undesirable, tolls can be collected on entry to the interchanges.

3. COSTS AND REVENUE

3.1 The provision and operation of a toll system will impose costs over and above the normal costs of constructing and operating a highway with unrestricted access. Some of the costs are listed in Table 1. The extent of these costs will obviously vary from country to country and application to application, but in a recent study in Malaysia the extra costs of the toll collection system amounted to approximately 7% of the total capital costs and 40% of the total maintenance and operating costs in the first six years of the life of the scheme.

TABLE 1: Costs associated with toll systems over and above normal highway costs

Capital Costs	Recurrent Costs
Toll booths and supervision building	Wages and Salaries
Land acquisition	Building and equipment maintenance
Highway works at toll plaza sites	Transport
Toll collection equipment	Office equipment and materials
	Depreciation charges

3.2 The revenue will depend upon the traffic levels and the tolls to be charged. The effect of charging tolls will be to reduce the amount of traffic: a very small toll will deter some drivers for whom the new road is only marginally more attractive than the old or, in financial terms, for whom the difference in the perceived costs of travel is less than the amount of the toll. As the rate of toll is increased, more and more drivers will find that the benefits of travel on the new road are outweighed by the toll, and they will divert to the old road. It is clearly important for the financial success of the scheme that the level of the tolls should be set at a level which will generate maximum or, near maximum, income.

3.3 There are many theories about the reason for drivers choosing between competing routes, and the problem of modelling traffic assignment must be resolved in all highway feasibility studies, whether or not they incorporate toll collection schemes. The various arguments are briefly summarised below.

3.4 The simplest assignment models find the route which minimises a particular travel parameter and then assign all or some of the traffic between a particular pair of origins and destinations to that route. The three commonly used travel parameters are time, distance, and cost. In inter-urban and regional traffic studies assignments based on minimum journey time have been found to give good correlation with observed travel behaviour. Distance based assignments are found to over-simplify peoples' perception of travel costs and are not often used: they are insensitive to different standards of highway design and capacity. The use of journey cost (or 'generalised cost') is an attempt to combine time and distance costs, and is particularly suitable for assignments between routes which incorporate tolls.

3.5 The 'cost' which is used in the assignment process is often called 'behavioural' or 'perceived' cost since it is the cost which is assumed to affect the decision making process of the trip maker. They are subjective, and different from economic costs which represent the real resources consumed in travel and transport and which are of principal concern to government rather than the trip maker. Generally accepted research findings have shown that people base their decisions on an imperfect understanding of the true costs involved and that they generally under-estimate the costs of running motor cycles and cars; for instance, depreciation and interest costs are rarely considered by the private motorist. An example of the ways in which perceived costs can be calculated is:

Category	Perceived Costs
Car and motorcycle	Fuel plus oil costs, and in vehicle-travel time
Buses and taxis	Fare plus in-vehicle travel time
Goods vehicles	Financial cost of travel

Research has also indicated that the valuation of travel time varies according to the type and standard of road. Trip makers tend to place a higher value on time savings for journeys that involve travel on a major highway as opposed to a minor road. In the case of a journey using a toll road, of course, the toll would also be included in the cost of the journey.

3.6 Once journey costs between the various pairs of origins and destinations by the alternative routes have been calculated, the journeys can be assigned to the various competing routes. Figure 1 shows a network of alternative routes for the Kuala Lumpur-Seremban Expressway, which incorporated an open toll system with two toll plazas. The 'all or nothing' assignment method assigns all the journeys between each origin and destination to one route, but this is generally considered to be unrealistic where alternative routes are close in their respective journey costs. The multi-route assignment method is designed to overcome the shortcomings of the all or nothing method, and assumes that drivers have an imperfect knowledge of the cost of using each link: a randomisation technique selects a link cost from a range of values distributed around a mean value. The method is designed principally for use in dense urban areas, and is not usually applied in simpler inter-urban situations, instead, the well tried diversion curve method is used. Diversion curves have been developed from experimental data, and have been based on time, distance, and cost. An example of a diversion curve based on cost is:

$$p_1 = \frac{1}{1 + T6}$$

where p_1 is the proportion of traffic using the new route,

T is the ratio C_1/C_2

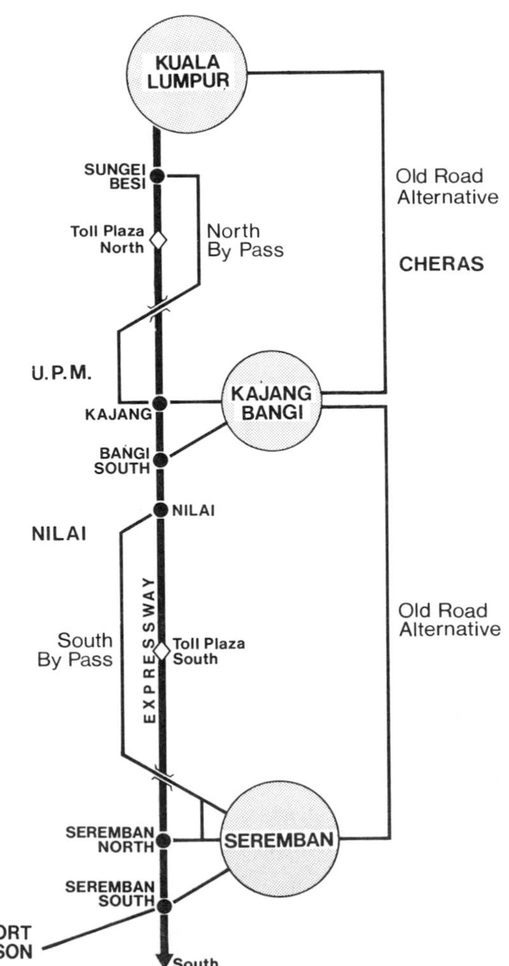

Fig. 1. Kuala Lumpur - Seremban expressway: alternative routes

C_1 is the cost of travel on the new route, and

C_2 the cost on the old route.

The formula produces an 'S' shaped curve, and a variety of modifications to the form of the curve can be made to suit local conditions in the light of local experience.

3.7 As the toll increases, the number of drivers electing to use the new highway will decrease. The revenue collected by the operation will be the product of the traffic in each class and the toll applied, and as the toll increases the revenue will increase to a maximum and then fall away. The toll which produces the maximum revenue is termed the optimum toll, and an example of the form of the curves is given in Figure 2.

3.8 Having developed a set of optimum tolls, there are practical matters to consider before establishing a scale of charges. The first is to calculate whether the income over time will exceed the capital and operating expenses and, to a degree, which will satisfy the financial

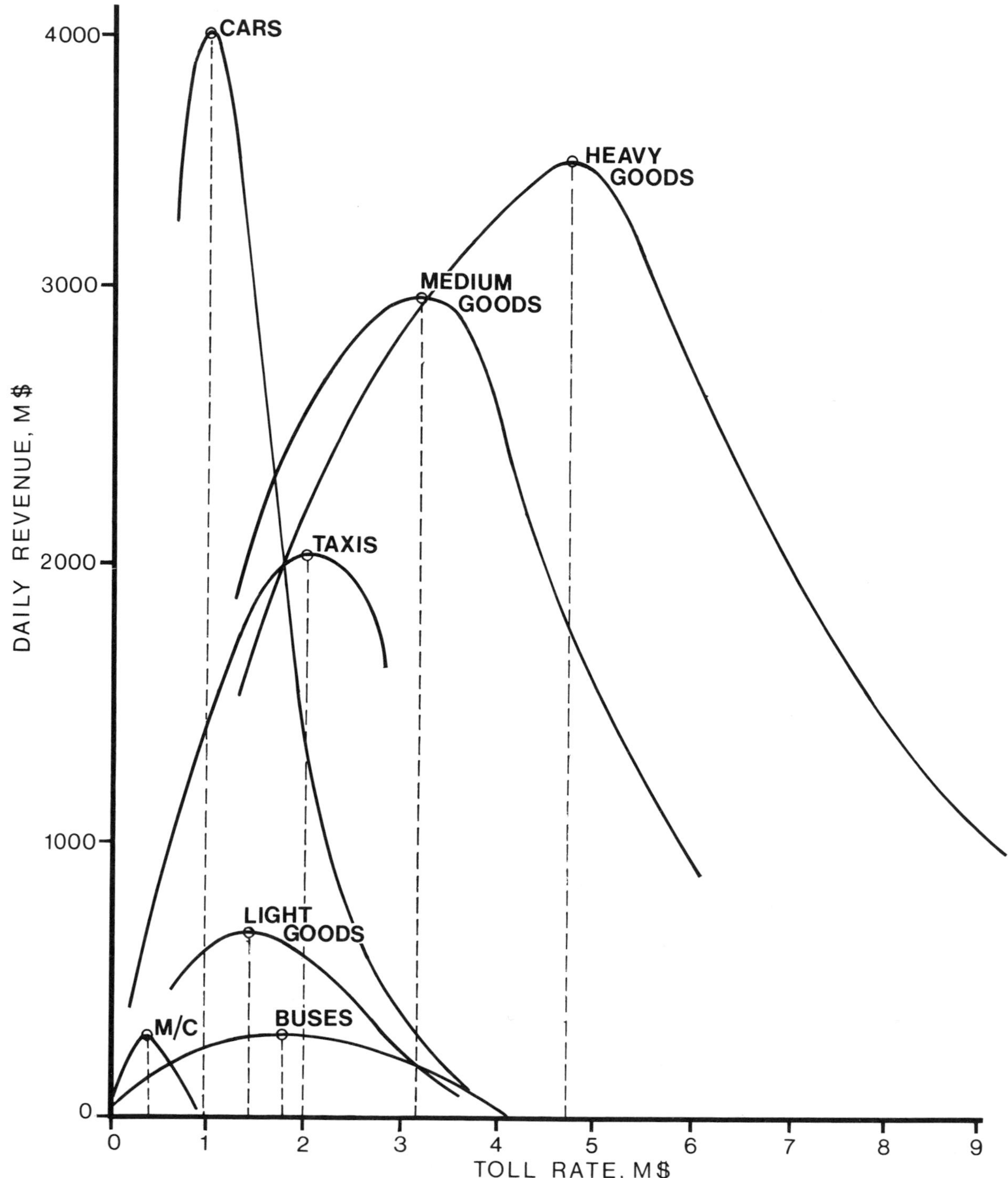

Fig. 2. Kuala Lumpur - Seremban expressway: revenue at various toll levels

criteria established by or imposed on the operation; if not, the system may need to be re-designed. It may be preferable to establish a scale of charges which would eliminate or minimise the need to give change at the toll booth, thereby simplifying the collection procedures. It may also, for policy reasons, be desirable to encourage particular types of traffic; for instance, buses could be exempt to encourage the use of public transport services. If the shape of the revenue/toll curves is fairly flat, it may also be worthwhile considering a toll set at, say one third lower than the optimum, which may reduce revenue by only 10%. This will make the new road more attractive; traffic will be higher (typically 40% higher), and therefore the economic, as opposed to financial, efficiency will be greatly increased although this is of more interest to government than to the operator. The use of lower charges are also of course likely to find political acceptance easier in countries which are un-used to tolled highways, and has the advantage of increasing the probability of financial targets being met thereby increasing the confidence which can be placed in financial planning.

3.9 Central to the results of the analysis are the assumptions made as to the rate of inflation during the evaluation period. Inflation will affect the financial return in a number of ways:

i) The actual costs of construction

ii) The magnitude of annual operating and maintenance costs

iii) The effect on vehicle operating costs, and hence the degree of diversion to the new highway

iv) The rate of interest at which the capital loan for the project can be secured.

3.10 Whilst this induces an element of uncertainty and hence risk to the investment decision, there are positive aspects. A high rate of inflation in the long term is beneficial to the financial viability of the project. As with all investments, once the capital has been borrowed and invested, the rate of inflation makes the invested capital appear smaller over time due to depreciation of the monetary unit by increasing price levels. Thus, provided that the actual tolls levied are increased at regular intervals the increase in revenue should more than cover any inflationary increases in operating costs and, at the same time, speed the reduction of the capital debt.

4. ADVANTAGES AND DISADVANTAGES OF TOLL HIGHWAYS

4.1 Arguments for and against toll highways have been investigated in detail by transport economists and others. In general it is postulated that there are very few cases where toll roads are justified in developing countries, the exceptions being the provision of urban toll roads for the main cities where there is some likelihood of congestion. The justification of inter-urban toll roads is more difficult. The main criticism is based on the premise that such roads result in the unnecessary duplication of facilities. Furthermore, if a case can be made for a toll-highway then an even better case can be made for a toll-free road.

4.2 Rejection of toll road policies has been based primarily on the assumption that demand for travel is elastic, that is that the introduction of tolls would result in a substantial reduction in forecast traffic levels and hence a reduction in the economic justification of the road. This may be so in cases where the extent of the total highway network is limited or where the level of economic activity is low. However, in the context of the more developed newly industrialising countries, this argument may no longer hold true. The past rate of economic growth, (typically in excess of 7% per annum), and increase in vehicle ownership are such to suggest that the elasticity of travel demand is decreasing.

4.3 Toll highway schemes have been proposed for a number of reasons and are viewed by their proponents as possessing a number of advantages; the combination of which varies from country to country:

i) They redress possible imbalances or distortions between the road user charge revenues raised through indirect means and the costs of providing new facilities

ii) They can lead to the establishment and self-funding of a highway maintenance unit (subject to a favourable toll revenue cash flow)

iii) They help to satisfy public objections to the overt concentration of investment in the more highly developed regions of a country as opposed to the investment, by Government, in areas of greater social and economic deprivation

iv) It allows, in certain cases, for the involvement of private sector funds in infrastructure investment (albeit balanced by the loss of such funds for investment elsewhere in the economy)

v) In the case of urban highways, they can be used as a policy tool in the control and restraint of traffic within and into congested city centres.

4.4 The validity of such arguments is dependent on the nature of the scheme in question and the reality of any alternative option. For example, the imposition of tolls might prove to be more acceptable and have fewer direct consequences than the overall re-alignment of indirect road user charges. The latter policy could invoke greater public reaction and have more effect, through general cost increases, inflation indices, etc. on consumers as a whole.

4.5 There are, nevertheless, several objections and disadvantages to the introduction of toll highways in addition to their principal effect of reducing the economic value of the investment:

i) They do not necessarily generate sufficient income to establish an increased fund for highway investment

ii) They can add to bureaucratic authority, and hence cost, further reducing their economic contribution. Evidence from the United States suggests that a substantial proportion of toll revenue was devoted to administration

iii) They require an element of over-design, and hence uneconomic provision, to cater for a final state where the toll may be removed. Such a case is converse to the theory that charges should be levied to reduce congestion and prevent wasteful investment in excess capacity

iv) In the case of a closed system, the cost of installing and operating toll stations at intersections where only limited flows are anticipated may result in a reduction in access to the highway in order to save costs, thus resulting in a reduced accessibility to the highway

v) They result in public disquiet if introduced retrospectively on roads which were originally provided 'free of charge' to users

iv) They may require the unnecessary duplication of investment and maintenance in competing toll-free roads.

On balance the choice of a toll policy cannot be made on the basis of a single criterion and demands that a balance be achieved between various conflicting economic and political criteria.

5. CONCLUSIONS

5.1 Financial anlayses indicate that returns are poor and accrue over a relatively long period, thus rendering any investment a high risk. In general financial returns are lower than can be obtained from investment in other sectors of the economy.

5.2 Following from the above the involvement of private sector capital is likely to be limited and in most events would require government guarantees on forecast traffic, and hence likely toll revenue returns.

5.3 The principal reasons for the poor financial performance are connected with the high rates of traffic growth forecast for the schemes. A high rate of growth demands the construction of highway capacity which may be relatively under-utilised in the early years of a scheme's life. If low traffic levels cannot generate sufficient revenue to cover operating and loan repayment and interest charges then the capital debt will increase. It is only in the later years of the scheme's life that a position of positive cash flow is reached.

5.4 One of the more problematic areas concerns the limited amount of local empirical research on travel behaviour and, hence, the likely reaction to the imposition of tolls. The completion and opening of toll roads in a number of countries over the next few years will present some reliable opportunities for the conduct of ex-post analysis of road user behaviour.

5.5 In general, the subject of road user charges is complex, involving not only a consideration of all road use, both urban and inter-urban, and that of other modes, particularly the railway system, but also involving political decisions about the level and distribution of the general taxation burden.

The object of this paper has been to show that the choice of whether to implement toll collection systems is not clear cut, and unavoidably involves development of a policy based on political rather than technical considerations. However they can, in certain situations, provide a source of income to finance particular schemes in a way which, superficially at least, is financed by the users rather than by the community as a whole.

6. REFERENCE MATERIAL

WALTERS, AA: The Economics of Road User Charges, IBRD, Washington DC. 1968

SMEED, RJ: Road Pricing: The Economic and Technical Possibilities. HMSO, London. 1964

OWEN, W and DEARING, CL: Toll Roads and the Problem of Highway Modernization. The Brooking Institute, Washington DC

Various Reports prepared by Halcrow Fox and Associates

MCINTOSH, PT and QUARMBY, DA: Generalised Costs and the Estimation of Movement Costs and Benefits in Transport Planning. MAU Note 179, Department of Transport, London. 1970

CLAFFY, PJ: Characteristics of Passenger Car Travel on Toll Roads and Free Roads for Highway Users Benefit Studies. US Highway Research Board Bulletin No 306.

Road user charges: some implications of charge policy

R. D. JARVIS, BSc, MS, MICE, MIMC, Atkins Planning

Over recent years, increased attention has been paid to ensuring a more integrated approach to the planning of transport investment at national level, recognising the interdependence of different modes and the importance of the economic context at the national level. Road user charges are significant in both these respects as they are an important factor in travel costs and they generate government income. The paper reviews the relationship between user charge policy, which establishes the basis for setting charges, and planning the development of the highway network. The links between user charge revenue and expenditure on roads, and between charges and traffic, are the main topics discussed. The paper emphasises the value of understanding the user charge system.

INTRODUCTION

1. The governments of developing countries are faced with the issues of how to develop their highway networks to meet objectives as diverse as improving accessibility in rural areas, easing inter-city movement, achieving greater social and political cohesion and coping with a rapidly growing number of private cars in cities. At the same time, governments are responsible for the management of the transport system, including the prices its users pay. It is the relationship between highway investment planning and user charges that forms the subject of this paper, which draws on experience gained on user charge studies in Mexico and Yugoslavia, and similar studies elsewhere. The paper does not attempt to cover all the implications of charge policy which, because of the general significance of transport costs, extend to many aspects of national economy.

2. The economic appraisal of investments in highways should be conducted as far as possible using real costs so that net benefits are assessed in real terms. Resources costs are measured using shadow border pricing, and all financial transfers and taxes are removed from the analysis. User charges, once identified, therefore have no direct bearing on economic evaluation but, as elements in the perceived cost of travel in the form of taxes and duties, on petrol for example, affect the volume of traffic. Vehicle ownership, modal split and traffic composition are all influenced by charge policy, and distortion away from efficient travel patterns will result from applying charges that do not reflect the economic costs of road use.

3. The economic level of charge for using road is not related to the cost of the initial highway investment but to the costs imposed by the user other than his or her own private costs, i.e. the short-run marginal costs (SRMC). This is the charge rate that results in the efficient use of the investment. In this case, SRMC consists of the variable costs of maintaining the road due to traffic and social costs such as congestion and pollution (ref. 1). It is the relationship between road damage and traffic (the so-called fourth power law) that gives rise to a pattern of economic charge rates in uncongested situations which are much greater for heavy lorries than for private cars. In practice, however, user charges are rarely determined by the SRMC criterion, and the effects of different policies on matters related to highway investment, including resources, traffic volumes and travel costs, are discussed below, following a review of the user charge system.

ELEMENTS OF THE SYSTEM

4. Road user charges generate income that generally accrues to the government, which disburses money for the maintenance, policing and signing of the road system and for the construction of new or improved roads. This is described diagramatically in Figure 1, which also indicates the link between traffic and the requirement for road expenditure. The external factors shown as affecting the charge policy decision are often highly significant, and may be over-riding in determing policy (such as the need to raise revenue for general taxation purposes). User charge policy defines the level and types of charge to be levied.

5. Aggregate user charges can include taxes which are levied on all goods and services in the economy, of which Value Added Tax is a good example. These should be regarded as 'general' charges and not considered as a matter of user charge policy. 'Specific' charges are those affecting vehicles or road users only. They can be grouped into two categories:

- fixed charges, i.e. those levied on a vehicle either only once or annually, such as a registration fee, special import duty or

Highway investment in developing countries. Thomas Telford Ltd, London, 1983, 21–24

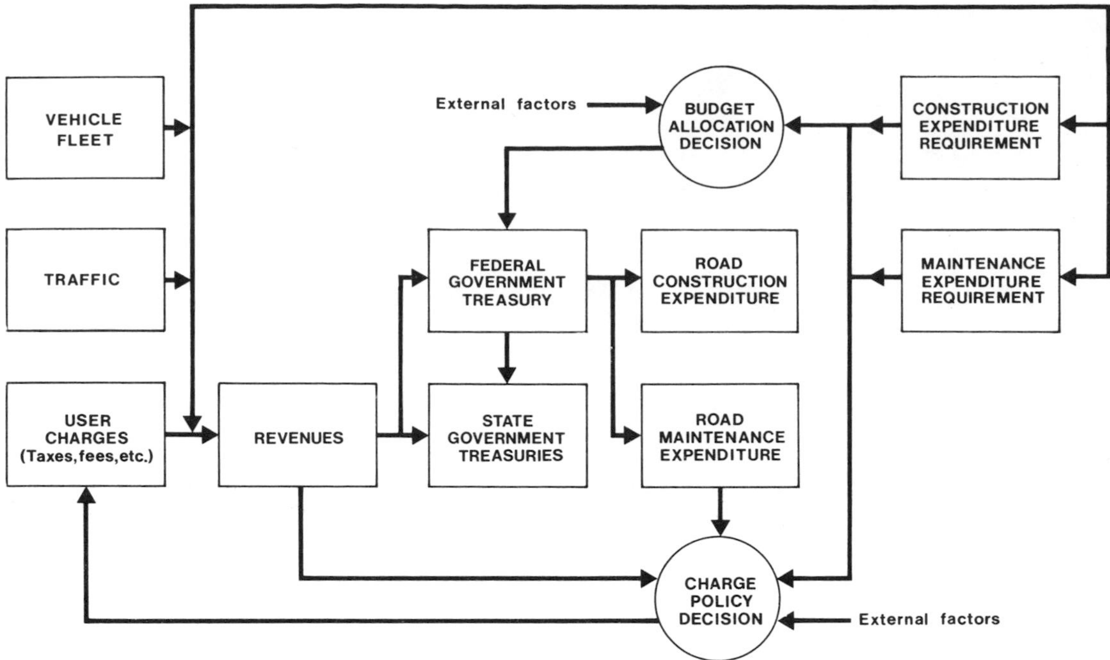

Fig. 1. The basic charge-revenue-expenditure model

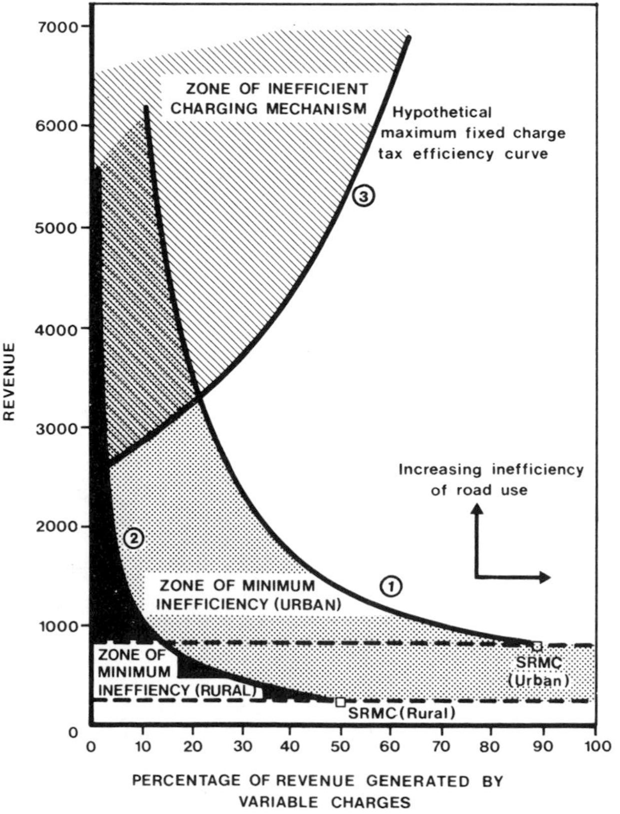

Fig. 2. Revenue/charge relationships

vehicle sales tax. Fixed charges can be sub-divided into those that are solely administrative, say to cover the issuing of a licence, and those that generate revenue. variable charges, i.e. those that are related to the amount of use of the vehicle, such as taxes on the sale of fuel and lubricants, and tolls.

Recognising a user charge is not always a straight-forward matter of obtaining tax rates, since in many developing countries the government exercises a degree of price control which must be taken into account. Similarly, the impact of user charges is closely linked to the form of regulation of the transport industry.

6. As far as highway expenditure is concerned, the development of a user charge policy should take into account the budgetary requirements for each type of spend (construction, reconstruction, maintenance), and category of highway, such as:

- national and provincial roads
- urban roads
- rural access roads.

The composition of the traffic using these different categories of road varies and, correspondingly, the pattern of charge generation by vehicle type is also linked to road type.

REVENUE AND EXPENDITURE

7. User charge policy has three components:

- amount of revenue to be raised
- charge distribution by vehicle type
- charge type,

of which the revenue level is the most important. The range of options for raising revenue is usually limited, and it is the quantity of income to be obtained that principally determines the user charge structure. Although the income to be generated is usually arrived at as the result of a combination of many factors influencing transport and financial policy, it is worth examining the objectives of a charge policy. These can be simplified to the following areas:

- utilisation of the transport system
- raising funds to pay for roads
- raising revenue for government use other than for roads.

Since in developing countries, where social costs are negligible outside urban areas, the economic charge rate is likely to be low, and expenditure on roads comparatively high, an objective of efficient use of the system will be incompatible with covering all highway expenditure from road user charges.

8. While the concept of a balanced budget for roads is an attractive one, there is little evidence of its having been adopted. Reliable statistics are difficult to obtain and are

Table 1. Total expenditure on roads as a percentage of road user taxation

Country	Year	$\frac{\text{Expenditure}}{\text{Revenue}} \times 100$ %
Ethiopia	1980	161
France	1978	48
Great Britain	1980	33
Hong Kong	1977	78
Italy	1979	55
Kenya	1976	1,475
Switzerland	1980	94
Taiwan	1979	136
Uruguay	1976	132
U.S.A.	1979	96
West Germany	1980	89
Zaire	1979	2,154

Source: International Road Federation, World Road Statistics 1976-80.

frequently not directly comparable. Table 1 gives an indication of the range of ratios between income from all road user taxes and total expenditure on roads for a selection of both developed and developing countries. Even so, the decision on the amount of revenue to be raised can have a bearing on the amount of money allocated to highway investment, and will certainly affect the amount of traffic. The question of how best to raise revenue above the economic level is essentially a political one, although economics can provide guiding principles. Since we usually have inadequate information on demand elasticities, the most satisfactory method would be to tax vehicles rather than their use. However, fuel tax is an easy tax to collect, while vehicle licences can be avoided. Figure 2 illustrates, in a format developed for Mexico, the reduction in efficient use of roads with increases in revenue and variable charges. Curves 1 and 2 indicate the efficient proportion of fixed and variable charges for any revenue level for the urban and rural SRMC cases. Curve 3 indicates the effect of assuming a hypothetical maximum for an effective fixed charge rate.

9. Developing countries often present a striking pattern of urban development, with one city developing very rapidly and becoming the focus of wealth and economic activity. The distribution of car ownership is characterised by a high concentration in the cities. Thus a high proportion of user charge revenue is generated in urban areas or even in the major urban centre. In Mexico, over 40% of the country's cars are registered in the Federal District of Mexico City, and a large proportion of the petrol consumption takes place there. It is rare for expenditure on roads to follow the same pattern so that,

viewed in terms of a highway budget, income from the city is spent on inter-urban and rural roads. Thus user charges can be an instrument of income redistribution and regional development. (Toll roads are a means of avoiding this issue by levying charges directly on the user.)

It will be readily appreciated that institutional factors can play a major part in the establishment of charge policy, particularly on revenue levels, since conflicting objectives are involved. For example, in Mexico the interested parties include the Ministries of Finance (responsible for the budget), Transport and Communications (responsible for transport policy and regulation) and Human Settlements and Public Works (responsible for building and maintaining roads), plus the Federal/States relationship. In the case of Yugoslavia, the problem of income received and the direction of expenditure is complicated by the constitutional independence of the individual Republics and Autonomous Provinces, who need to work together to achieve a coherent highway network.

TRAFFIC, VEHICLES AND COSTS

10. User charges influence the size and structure of the vehicle fleet and its utilisation. Consequently, they affect traffic volumes, and demand forecasts used for investment planning should take this into account. In any event, it is important to recognise the influences that charges have on traffic volumes. Two obvious instances of particular relevance in developing countries are vehicle import duties and fuel pricing. The first of these can be set at a level designed to restrict imports, either to encourage domestic industry (say in the assembly of vehicles) or simply to deter the use of foreign exchange for the purchase of private cars. Levies of this type can be very high, reaching 100% or more of the import value. The obvious effect is to constrain the level of vehicle ownership. A similar result is achieved by imposing a high initial vehicle sales tax or registration fee.

11. The fuel price situation is difficult to deal with in this period of uncertainty over oil prices. Nevertheless, this is a most important factor. Oil users in developing countries should not be shielded from the consequences of the increases in the price of oil any more than should those in the industrialised world. To do so, on other than a short term basis, is to risk distortions in the use of oil products and to store up problems for the future. Subsidies on petrol and diesel make travel cheaper, giving rise to more traffic, and potentially to greater road investment than would be economic. The costs to the country of oil products can only be perceived by reference to the shadow price. The same approach is essential in arriving at resource costs for economic evaluation purposes.

12. User charges can affect demand, and hence investment plans, through their effect on modal split and the structure of the vehicle fleet. This is particularly evident in countries where road and rail freight haulage is regulated. Tariff structures should, as far as possible, reflect real costs so that the most economic mix of transport is used. There may be other reasons (for example, to maintain employment levels) why the system is operated to favour one mode or another, and it is as well to have the facts clearly set out as to the costs of such a policy.

13. In Mexico in the mid-1970s there was an absence of diesel trucks, largely due to the differentiation in sales tax between petrol and diesel vehicles. This caused a significant weakness in the haulage industry, which found it difficult to adapt to the more efficient diesel units as petrol prices rose. Similar fleet distortions can be caused by charges related to vehicle size, and can create conditions under which operators resort to overloading.

SUMMARY

14. The assessment of highway projects requires, in the simplest terms, a knowledge of the construction and maintenance costs of the scheme, and of the traffic and economic benefits it will generate. Through user charges, governments influence the demand, in terms of both traffic volumes and composition. User charges may also be linked with the financial resources available for highway projects, not for economic reasons but for institutional or political reasons.

15. In view of these interactions, we should seek to minimise divergence away from the economic pricing levels, because distortions can lead to uneconomic highway investment decisions, but there can be no single solution to the problem of what charges to impose. The charge systems in each country are usually the result of the historical development of fiscal practice, and are difficult to adapt. (Within the European Economic Community, for example, there is a considerable range of systems, and convergence towards a common basis has so far proved impossible.) However, the past decade or so has seen a growing recognition of the advantages of developing a reliable data base and analysing the user charge system and its implications, so that the economic context of highway planning is more fully understood.

REFERENCES

1. WALTERS A.A. The economics of road user charges. I.B.R.D., 1968.

Discussion on Papers 1–4

Mr C.J. HOLLAND, Mr T.J. POWELL and Mr R.M. DUNMORE *(Introduction to Paper 2)*: Table 1 gives a sample of a possible expenditure programme for the 1980s, derived from the programme described in Paper 2. Although the program itself considers only economic factors, any number of policy inputs could be inserted, for example to include schemes which are not economically viable but are considered desirable on social grounds. The fast turn-round enables many different policies to be evaluated. The particular run represented in Table 1 contains a number of minor policy inputs.

Table 1 shows that relatively little of the modelled network merited an increase in capacity: 2100 km of paving of earth-and-gravel roads, and only 400 km of either widening or dualling of paved roads - out of a total network of 40,700 km. In contrast with earlier highway investment plans, the emphasis lies on maintenance and renovation, rather than on the provision of new or upgraded facilities. Of the total program of \$2956 million, only \$467 million, or 16%, is related to upgrading and improving road capacity. The remainder is assigned to maintenance, with a large element of \$2130 million, or over 70% of the total, being assigned to the reconstruction of life-expired paved roads.

For some activities, the program produced widely different expenditure plans for each five-year period. In the case of repaving, the higher expenditure in the first period results from there being a backlog of work on roads in very poor condition. Renovation of earth-and-gravel roads requires less expenditure in the second period because, for the most part, the same roads are being renovated a second time, so that less work is needed. Renovation costs per kilometre are consequently lower, and benefit/cost ratios higher. In practice, of course, the timing of these programmes would be adjusted to produce a more regular demand on maintenance resources.

Mr N.D. LEA *(N.D. Lea & Associates Ltd, Vancouver)*: The analysis models described in Paper 2 have similarities with models used by our firm for the past 10 years or more. From our experience, the rapid review of the total network - link-by-link, several times using variations in assumptions - is useful but it is frequently helpful to precede this by what we call switch-point analysis. This is essentially the intensive analysis of one link which has

Table 1. Outputs: Proposed capital and maintenance programme

Activity	Period	1000 km	Cost($ million)	Annual benefits/ annual cost	Cost ($1000/km)
Renovating earth-and-	81/85	9.9	63	1.08	6
gravel roads	86/90	9.9	14	3.40	1
Total		19.8	77	1.51	3
Paving earth-and-	81/85	1.7	162	1.33	96
gravel roads	86/90	0.4	53	1.05	132
Total		2.1	215	1.24	103
Repaving	81/85	14.3	2020	1.10	141
	86/90	0.8	109	1.15	129
Total		15.1	2130	1.11	141
Widening/dualling	81/85	0.2	104	3.36	669
	86/90	0.3	148	0.93	516
Total		0.4	252	1.92	570
Total capital	81/85		2350	1.20	
programme	86/90		324	1.47	
Total			2674	1.24	
Maintenance	81/85	40.7	141		3
programme	86/90	40.7	141		3
Total		81.4	282		3

been selected to be representative of certain conditions. This link is analysed for several conditions to determine the traffic volume at which the analysis shows a switch of the economic optimum from one class or standard of road to another.

The results of such switch-point analysis may suggest changes to the input assumptions or to the structure of the model as illustrated by the following.

(a) Surface treatment as an upgraded option is very important because under many conditions it has been found to be the economic optimum.
(b) Varying the length of the analysis period of salvage assumption can change the results significantly. It is therefore useful to do trial switch-point analysis for several assumptions concerning salvage value and length of analysis period.
(c) Disaggregate trucks into several types or classes according to their impact on the road surface.
(d) Divide occupants' time into paid time and unpaid time.
(e) Disaggregate public transport into publicly operated buses and privately operated buses.

It is very important to output such switch-point sensitivity analysis and compare the results with actual field experience. For example, is the finding in Paper 2 that gravel-surfaced roads are a poor investment borne out by experience in the field? My experience has been that most countries contain some regions and conditions where loose-surfaced and surface-treated roads, if well built and maintained, are an economically attractive solution.

Mr E. KONSTANTAS *(Doxiadis Associates Int., Athens)*: Paper 3 is concerned with toll roads in developing countries. Although they are not of a high standard, non-toll roads result in linear development and encourage economic activities alongside them. When development has reached its optimum stage and traffic is approaching capacity, a new high-standard highway (freeway) is constructed with access to specific places, invariably far away from the existing road and existing development.

The new toll road is found to have many disadvantages. The new alignment of the road has a social effect on the settlements served and on the development along the existing road. The new highway does not adequately serve all the economic activities along the old route due to the limited interchanges, nor does it serve areas of development, and people are discouraged from using the road because of the tolls. Therefore, although a great percentage of traffic using the old roads may have been expected to be diverted on to the new roads, this does not happen. The application of tolls seems more feasible for long trips. New roads with tolls are not attractive for short- or medium-length trips originated and terminated in locations where the economic activities exist. Furthermore, in most developing countries, the road user can not fully understand the benefits of using a new road; what he can understand is the obvious, out-of-pocket cost: the toll.

Mr J.S. YERRELL *(Transport and Road Research Laboratory)*: Paper 2 presents an interesting account of an important piece of work - the formulation of a National Transport Plan for Argentina. It described the planning of highway construction and maintenance but was part of a wider study which looked at the role of other modes as well.

Many developing countries are, in the light of increases in transport fuel costs, looking actively for policies to shift goods from road to rail. Future planning may well have to take into account the extent to which such policies are likely to be applied, and the extent to which they may succeed. Although Argentina may not be in this position, could the Authors outline how plans for other modes - especially rail - were able to affect the road network study described here? In other words, to what extent was rail investment (its scale, nature and so on) part of the criteria for planning highway investment, and how were any such interactions handled in the planning procedure?

Mr POWELL: With reference to Paper 1, the major factor which must influence the transport policy of landlocked countries is that they are ultimately dependent on the goodwill and stability of their neighbours. They must therefore plan their transport system so as to minimize the risk of delay or interruption to their traffic in the countries through which they gain access to the coast. One way of doing this is to develop where possible routes through more than one country and by more than one mode. The additional transport costs of sending a proportion of goods by a route which is not the cheapest may be well worth paying so as to ensure some continuity of access should they be denied access to the cheapest route for any reason.

Some countries are, however, dependent on only one access route. In this case they can only hope to improve their security of access by entering into long-term agreements with the access country on which they are dependent. In these circumstances it may be worth the landlocked country's agreeing to provide some or all of the specialized equipment required to transport its goods (e.g. by providing and paying for rolling stock on a neighbouring railway hypothecated solely for the use of the landlocked country or by paying directly for certain road construction and maintenance costs). It must also be prepared to accept that it will have to pay the full economic cost for the transport it requires. I suspect that one of the problems at the moment is that some landlocked countries are not paying sufficient to the access country to make it worth the latter's while to provide efficient through transit facilities.

Mr J.C. FIELD *(Independent Consultant, London)*: Although the problem of landlocked countries' access to the ports is well known (Paper 1), their problem when they are themselves transit countries can be enormous. In a specific case, Upper Volta - an extremely poor, landlocked country - provides transit for traffic from Lomé to Niamey. The roads used would not be priority links for the development of Upper Volta, yet, within a fairly restricted overall financial envelope on aid for construction and maintenance of roads, Upper Volta has to invest in these roads.

The problem hinges on the way costs and benefits are allocated, particularly when a transit country is itself too poor to construct and maintain a network essential for its own development.

Mr C.J.D. LANE *(World Bank, Washington)*: Would the Authors of Paper 1 outline what procedures UNCTAD has developed to help transited countries to meet maintenance costs on roads used by transit traffic which causes more cost to the transited country than it brings benefits?

Mr J.B. COX *(N.D. Lea & Associates, Jakarta)*: The Authors of Paper 3 state that financial returns from toll roads are low. One of the major reasons could be the high capital cost of these roads and I would query whether the freeway standards transferred from Western countries are appropriate. In general, most geometric and pavement design standards were produced in the 1950s and 1960s in the UK and USA when capital was available for freeways and interest rates were low. In fact, most of those rural road standards have already been reduced in developing countries because they have not been economic.

However, none of this economic analysis of construction standards seems to have been carried out for freeways/toll roads in developing countries and they appear to be severely overdesigned for the particular conditions in developing countries - limited capital, smaller vehicle dimensions, lower axle loadings and so on. One reason for this could be the psychological reason: standards are kept high because a toll is being charged. There are strong economic reasons, I believe, for reducing standards on toll roads in developing countries - even by just using ordinary two-lane and four-lane rural standards for both geometric and pavement design, e.g., no parking lane and so on.

Dr R.S. MILLARD *(World Bank, Washington)*: Mr Jarvis presents in Paper 4 a thorough and useful review of road taxation policies. One aspect which curiously has not appeared in any of the papers to the conference concerns vehicle and axle loading. It has proved quite impossible in Third World countries to control the weight of vehicles and vehicle-load spectra are generally much heavier than they are in Europe and North America.

It is now commonly agreed that road transport systems operate most effectively when heavy vehicles are permitted on road networks which have pavements strong enough to carry them. In Third World countries, the road pavements are far below this happy optimum condition; weak pavements are being rapidly destroyed by very heavy vehicles. Would Mr Jarvis comment on the prospects for using road-user taxation as an instrument to help in restraining the indiscriminate use of excessively loaded vehicles?

Dr J.B. METCALF *(Australian Road Research Board)*: In Paper 1 the Authors mention the difficulties of quantifying such matters as the quality of life. Could they suggest some measures? An approach such as that of Professor Hills and Professor Jones-Lee (Paper 23) might be a starting point. Would Mr Pusar and Mr Cabanius like to start off the matrix? After all, a box in a model labelled 'policy impact' is not enough.

In Paper 2 the Authors give a good example of the application of planning models, but I would have liked to have seen more detail. Can the Authors provide cost tables? Was there a breakpoint (say, average annual daily traffic) in reaching their conclusion that if a road was worth improving it was worth paving (sealing)? Table 1 shows the highest benefit/cost ratio for repaving earth-and-gravel roads. There were corridors requiring improvement - was this because of growth patterns and/or paving or for current inadequacy? Was there any pattern between the two?

Mr CABANIUS and Mr PUSAR *(Paper 1)*: Mr Powell raises the issue of route diversification. The last UNCTAD V in Manila in 1979 stressed the importance of providing each landlocked country with alternative routes wherever feasible in order to ensure against any difficulty that may arise on other transit routes. The additional cost due to simultaneous use of a number of transit corridors represents the price of the economic security and political independence of the landlocked countries. However, we have to recognize that such a principle raises practical difficulties. What capacity and quality should insurance routes possess - sufficient to carry all cargo or essential commodities? The insurance route may be uneconomic on conventional economic criteria. This may make it difficult to obtain capital for its development. To operate along the insurance route may be more expensive than the existing route. This may make it difficult to encourage direct traffic to the new route. How should the transit partner or the insurance route be compensated for the erratic use of its infrastructure?

Investment by landlocked countries in transit countries have been suggested. The principle is accepted when transit cargo is an important factor in the transport transit system. Otherwise, as it is in most cases, when the cargo of the landlocked country represents a peripheral economic issue for the transit country, the latter is not ready to countenance the proposal because it fears it may represent an interference in its state sovereignty. Even when increased transport charges against better services have been proposed, transit countries often have been reluctant in order not to give, in fact, transit cargo privileged treatment.

The point raised by Mr Field underlines the paradox of the prevailing situation, at least in the African continent, fragmented in many independent countries. Transportation infrastructure required for the economic survival of a country or less dramatically, for the expansion of its trade relations, may not be located in the country. The transit country, sometimes a landlocked country itself or listed among the least developed, will have to support the substantial cost of the improvement and maintenance at the expense of other national priorities while the benefits largely accrue to the neighbouring countries.

UNCTAD is presently working in formulating what could be a regional transport transit agreement, which when completed may provide appropriate and equitable guidelines for regional investment. It will define the rights and the obligation of both landlocked and transit countries in such financing arrangements. But obviously it will require simultaneously the creation of a new system of international financing. The creation of regional funds for improved access to landlocked countries, to funnel foreign financial aid for the construction and maintenance of priority transportation network appears to be a solution worthy of analysis and discussion. The effort undertaken by the Commission of the European Community with its EDF regional resources is a first important step in that direction.

Mr Lane asks what procedures have UNCTAD developed to help finance road maintenance on transit routes? The question is pertinent in a situation where countries pay less than the full economic cost for transport because of the internal pricing policy of the transit countries. The application of border tolls with weigh bridges have been suggested to several transit countries, helping at the same time to control overloaded transit vehicles. However, a more comprehensive system could also be suggested. This implies a system which records the type of vehicles used, classified by the number of axles and vehicles' payload as indicated in the custom declaration used for transit. Transit service charges then are calculated by relating them to the type of vehicle and its payload. They can be determined on road maintenance costs alone (and a relationship between axle loads and road damage will have to be assessed) or could also include a financing charge for the initial investment. Such a procedure will clarify what should be financially at least supported by either the transit country or the landlocked country.

Mr HOLLAND, Mr POWELL and Mr DUNMORE *(Paper 2)*: Mr Lea asks whether our finding that gravel-surfaced roads are a poor investment is borne out by experience in the field. Our model was in fact based on data provided by the respective provincial maintenance bodies. We have no reason to disbelieve them. The absence of suitable material for looser surfaced roads reduces the potential attraction of such roads and leads to the situation where in a number of cases there is no economic alternative between an earth- and a paved-road.

Mr Yerrell asks how did plans for other modes especially rail affect the road network study? The answer is that all modes were studied together and recommendations for common pricing policies for all modes were put forward as part of the National Transport Plan. Nevertheless, it was found that the interaction between the rail and road modes was relatively small. There was relatively little road traffic that could have been carried more economically by rail.

Dr Metcalf's request for further information could be supplied from the detailed study data. The conclusion that if a road was worth improving it was worth paving followed from the nature of the terrain. The gravel-road or the improved non-paved route was not normally economic at any traffic flow. Many of the routes which needed repavement cover major traffic routes which had been paved within the last 10-20 years and needed repaving. There were, however, a few paved routes where the traffic flow was too low to justify continued maintenance to paved route standard.

Mr JARVIS *(Paper 4)*: In response to Dr Millard's question, road-user charges typically involve so many policy issues that I would be reluctant to burden them with an enforcement role as well. In general, the problem with designing a user-charge system to combat overloading is that the cure may be worse than the disease. The implications of increasing transport costs (where the majority of goods are carried by road) or, in effect, reducing the carrying capacity of the vehicle fleet have to be thought through for each case. This is often further complicated by regulation of the road haulage industry.

Nevertheless, the use of differential charge rates can help by influencing the characteristics of the vehicle fleet. In particular, the use of a greater number of axles for a given vehicle weight can be encouraged. The rest of the charge system should reinforce the aim, so that, for example, there should be no excess sales tax applied to tyres.

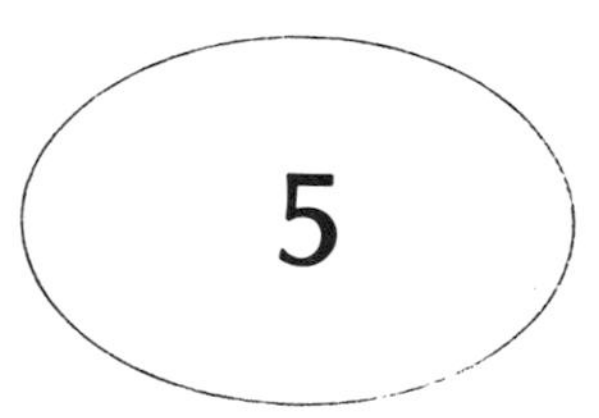

Leading issues in aiding road development

M. B. GRIEVESON, FICE, MIWES, and J. T. WINPENNY, MA, MPhil, Overseas Development Administration

As an aid agency active in a large number of developing countries, the ODA has evolved a certain "collective wisdom" on roads projects. In vouchsafing this, it might be convenient to organise the discussion roughly according to the sequence of decisions that are taken about a typical roads project. Thus, our paper will treat the following topics:-

I. The Decision to Aid;

II. Appraising the project;

III. Design Parameters;

IV. Construction Methods;

V. Maintenance; and

VI. Evaluating experience.

I. DECIDING WHETHER AND HOW TO AID

1. Aid is very different from charity in that it consists of an active collaboration between the donor and the recipient. The decision by the donor to become involved in roads normally results from a process involving initial approaches from the recipient, suggestions and hints by the donor agency, aid missions, policy discussions, consultancies and appraisal visits. There is invariably prolonged discussion at an official and professional level between the two parties about the best development of the road network, the standards to be employed, the way the work is to be organised, and the precise financial and technical involvement of the donor agency. These trite remarks are simply made to illustrate that the aid "decision" is neither simple nor quick.

2. The decision whether or not to aid roads rather than some other worthy projects is influenced by a few major considerations. An important one is the direction pointed by national transport policies. Many countries show a strong bias in favour of the roads sector, which discourages the other obvious modes like railways and water. It is not uncommon for developing country governments to be strongly influenced by the "highways lobby", since transport operators tend to be politically influential, and transport vehicles are a favourite form of investment by savers both small and large. The government can show its favours to the roads lobby not only by the pace of new road construction, but also over such policy matters as fuel pricing, licencing of commercial vehicles, axle load regulations, and the size of vehicle taxes.

3. This heavily influences the trend in modal split between roads and, say, the railways. The roads/railways split is largely coterminous with that between the private and public transport sectors, since road transport is typically in private hands while the railways are typically publicly owned. Developing countries have witnessed a strong trend in favour of road transport of both passengers and goods. This is partly because road systems have been able to respond more flexibly to changing geographical patterns of origin and destination of goods and people, partly due to the relative increase in the cost of constructing and operating railways since their heyday earlier this century. It is also due to factors like: the growing inefficiency of the railways, which have often been seen as a source of privileged employment for political favourites; the scarcity of skills to maintain railways systems; and policies on railway fares and finances. Although recent movements in oil prices have led to a revival of interest in railways, especially in Eastern and Southern Africa, in most countries there is still a trend towards carrying goods and people by road.

4. In developing countries the road network is still very small in relation to total needs, and financial and human resources necessary to build roads are very scarce. In these circumstances it makes sense to expand the system outwards from the more densely trafficked routes. This ensures that scarce resources by and large benefit the most travellers. It is not denied that roads can generate their own traffic, but it is futile to expect sizeable "developmental" effects in remote regions with few resources and population. One of the most persistent

myths about roads is that virtually any road can justify itself because of the traffic it generates. Evaluation studies (see below) do not support this general view; some roads have witnessed dramatic traffic build-up, but others have had disappointing results. Also, as road building in Scotland demonstrates, roads can depopulate as well as develop.

5. ODA is increasingly giving attention to the financial constraint when deciding on the form of its aid to the roads sector. Recently factors on the side of both the donor and the recipient have favoured the construction of new roads rather than the rehabilitation and maintenance of an existing network. The chickens have started to come home to roost in the shape of serious pressure on maintenance budgets and premature deterioration of the road network. Before agreeing to finance the building of new roads or an extension of the road network we need to be convinced that the maintenance budget is sufficient to support the addition, without taking away resources for maintaining existing roads. By the same token, we are increasingly exploring ways of supporting the local maintenance effort and we talk more about this in Section V. (Ref.1)

II. ISSUES IN APPRAISAL

6. The main problems that come to light revolve around lightly trafficked roads where existing user benefits are insufficient to justify the investment, and therefore where forecasting becomes especially important and the estimation of generated traffic is crucial. ODA is not the first to have discovered this difficulty (ref.2). The normal methodology for justifying roads relies on the quantification of cost savings to existing users of the route. Where traffic is very light, say less than a few dozen, this is not sufficient, and benefits from generated traffic have to be invoked. The exercise then becomes very tenuous and arbitrary. In these circumstances we would agree with the approach suggested by the World Bank to focus attention on the direct benefits to farmers or other producers in the area to be served. This requires some estimate, based perhaps on scrutiny of farm budgets, on the net gains to producers from improved transport, allowing for other constraints or changes in other variables which are likely to ensue.

7. The circumstances in which generated traffic is likely to be sizeable are illuminated by the growing body of evaluation studies of existing roads projects (see final section). In an interesting study of the "development" impact of road investment in the UK, it is found that such effects were very small, and mainly confined to relocating existing activity. It is suggested that in assessing the size of developmental benefits, attention is paid to the size of likely reductions in transport costs, the presence of preconditions for development, consistency with governmental regional policies, and an assessment of the specific benefits likely to the major commercial undertakings in the area. The results are of course obtained from the study of an already dense transport network in a mature industrial country, and scepticism about the small size of developmental effects should not be generalised to developing countries as well. Nevertheless, the check list of points to be considered remains valid. (Ref.3)

8. It has been customary in economic appraisals to assume adequate maintenance of the road investment. This assumption guarantees a certain stream of benefits in the future, associated with the preservation of the road in a satisfactory state. It also means that different types of roads, e.g. earth, gravel or bitumen, can be ranked according to a comparison of their expected streams of benefits compared to cost. The assumption of adequate, let alone "optimal" maintenance now seems to be extreme. If we were to assume inadequate, or even nil, maintenance the benefits from the investment would be very different, and might not warrant the initial outlays. Moreover, the ranking of different types of road might be different since the riding surface of an improved bitumen road, to take one example, might after a certain number of years be worse than that of an unimproved gravel road. Further thought needs to be given to the most realistic assumptions about maintenance when performing economic appraisals, and the consequences for the investment need to be spelled out.

9. In a sense capital funds for investing in a new road may be more "plentiful" to a developing country than the recurrent financial resources needed to maintain existing networks. (This is basically because foreign finance is more readily available for new investments than for supporting regular budgets.) This suggests that there might be an economic case for the designers to exploit the trade-off between capital costs and maintenance costs. It might be possible to spend more initially on a certain design of road which requires less subsequent maintenance. However, it is not clear how far this attractive theoretical possibility exists in practice, save in rather limited circumstances (e.g. concrete roads in certain cases). More research is needed.

III. DESIGN PARAMETERS

10. ODA has no predetermined and fixed set of design standards on which we insist before agreeing to finance a road scheme, but rather we try to retain a flexible approach and, provided they make general economic and technical sense, usually go along with the national standards which our aid recipients work to. Predicted traffic volume is the

starting point in determining pavement design which we have preferred to see based on TRRL Road Notes 29 and/or 31 using the Equivalent Standard Axle approach to the design of a flexible pavement with a 15 or 20 year design life. In the lower traffic density situations we have funded gravel or even unsurfaced roads looking towards a 'stage construction' development approach in which, having laid down the geometry, as traffic builds up over time the road pavement will be increased in strength through the "gravel" to the surface dressed stages and eventually into the "overlay" phase. Pavement and shoulder width are also looked at in relation to traffic volume, whilst curvature and gradient are viewed in relation to design speed although with lower traffic volumes tighter curves and steeper grades would probably be called for in more difficult terrain in the interests of economy. Generally speaking we would prefer to think in terms of a 15 year design life particularly in lower traffic situation because we believe it is extremely difficult to make reliable traffic projections on a longer time horizon in developing countries. Some of the design standards adopted on recently executed road projects are shown in Table 1.

11. The number of variables which affect optimum pavement design is very large and the inter-relationship complex making it difficult to predict how a pavement will behave over time. In the past ODA has generally favoured the flexible pavement with a stabilised gravel or crushed stone base and two coat surface dressing. With rising bitumen prices and varying maintenance situations we are now beginning to look at the concrete road as a possible alternative in some cases although we have not yet used this design option in other than the short feeder or industrial access road situation. In the "overlay" field recent research by TRRL on pavement strengthening schemes seems to indicate that the mechanism of failure in the tropics is very different from that in temperate climates. In the tropics the general tendency seems to be for the overlay to crack from the surface down rather than from the bottom up. The implication is that the appearance of surface cracking does not necessarily mean total failure of the overlay and that judicious maintenance could arrest a deteriorating situation.

12. Of continuing and indeed increasing concern is the effect of the exponential

Table 1. Some typical road standards

Project	Pavement Width m	Shoulder Width m	Embankment Slopes h/v	Cut Slopes h/v	Maximum Gradient	Minimum Radius of Curvature	Design Speed
Rural Access Roads - Kenya (gravel)	4.00	0.25	1/1	1/3	desirable 8% absolute 12%	desirable 30m absolute 15m	30 km/h
Lakeshore Road - Malawi	5.5	1.5					
Songea-Makambako Road - Tanzania	6.0	1.2 cut 1.5 fill	3/2	1/5 rock 1/2 consolidated 1/1 unconsolidated	10%	90m	50 km/h (Wino-Njombe) 100 km/h
Thuchi-Nkubu Road - Kenya	6.5	1.0		1.5/1	8-10%	125m	60 km/h
Mumias Sugar Roads - Kenya			low / high	low / high			
- Major(bit)	6.5	1.0	4/1 / 1.5/1	4/1 / 1.5/1	6.75%	400m	80 km/h
- Feeder (gravel)	5.5	1.25	4/1 / 1.5/1	1/1 / 1/1	n.a.	n.a.	n.a.
- Access (gravel)	4.0	1.0	2/1 / 1.5/1	1/1 / 1/1	n.a.	n.a.	n.a.
Dharan-Dhankuta Road - Nepal	5.5	Usually 0.5 (Formation 6.5)	2/1	Variable	10%	25m	30 km/h

relationship between increase in axle load and pavement damage. The tendency to allow ever increasing axle loads and laxity in enforcing axle load legislation can only place a growing burden on developing countries capital programmes with a need to build stronger pavements, and on their recurrent budget in demanding more road maintenance expenditure. It is true that relatively small increases in pavement thickness can lead to significant increases in carrying capacity as the relationship between the two factors is at a power lying between 6 and 10 but this assumes adequate maintenance - a factor which is not always present. Without adequate maintenance the relationship is very difficult and unquantifiable.

IV. CONSTRUCTION METHODS

13. In recent years we have funded projects to be implemented by a variety of means. These have included:-

a. direct labour by the aid recipient's own organisation, sometimes with a small expatriate input as Technical Cooperation support;

b. direct labour with an expatriate management team;

c. direct labour as in b. but in conjunction with a system of labour only sub-contracts;

d. the usual admeasurement type of contract;

e. target cost contract.

14. In many countries direct labour is only suitable for the lower categories of road because there is a local shortage of construction skills and/or equipment to carry out the more sophisticated forms of construction, and their management organisations may be very overloaded with few experienced project managers available. This type of situation can, however, be improved upon by Technical Cooperation support and the Rural Access Roads Programme in Kenya is a very good example. Beginning in 1974 ODA funded a number of small units to build rural access roads by labour intensive methods. Other bilateral and multilateral donors joined the programme in the ensuing years and more than forty units have been operating in the field in 1981. Expatriate technical assistance was provided by donors for a number of years at both the central organisational and at the field level but this is now declining as more Kenyan engineers move into the programme. Over 3500 km of road have been constructed or improved in the rural areas.

15. Where higher standards of road are required we have funded equipment (in some instances including operating costs) and provided larger and more formal expatriate management teams to assist local direct labour organisations execute road construction projects. In some cases the expatriate team has been advisory and in others executive. Various organisations have been engaged to provide the teams, including Property Services Agency of the Department of the Environment (PSA), the Crown Agents, Consultants and the Royal Engineers. The management team approach offers considerable scope for local training provided counterparts of the right skills and calibre can be obtained and retained on the project for sufficient time to benefit. From the donor point of view it is better to engage an organisation to provide the team than try to recruit a series of individuals. Although the engagement of an organisation may appear at first sight more expensive, it reduces the administrative burden on the aid agency, provides better back-up and leads to more effective staff continuity. In a scheme in Nepal with a management team from PSA supported in design and quality control functions by a consultant, a system of labour only small contracts has also been used in conjunction with a direct labour operation. This has proved quite successful although there were some social problems in the early days until the contractors were induced to pay adequate attention to the welfare of their labour force.

16. The usual admeasure type of contract continues to be used with large expatriate contracting firms bidding against designs, bills of quantity and specifications produced by consultants who are also engaged to supervise the work. An additional complication which we have encountered in these operations in the recent past is in countries with serious foreign exchange shortages. It has been necessary in some cases to negotiate a system whereby the contractor entered into agreement with a major oil company for the direct importation of fuel and the oil company was paid "off shore". Only in this way could continuity of fuel supplies to the projects be ensured and acceptable completion dates be achieved. On a major road project in Tanzania we have tried another innovation - the target cost contract. This was used to enable the work to start at an early date without going through a long drawn out and detailed redesign process (the original design was not thought to be the most cost effective) and also in the interest of keeping contractors' bids to realistic levels in a situation where the risks and imponderables were unusually large. The contractors' bid a target cost made up of an actual cost and a percentage management fee; all plant and materials are supplied to the contractor who runs an open book system subject to continuous on-site audit. If the works are completed at greater than the target cost the contractors' fee reduces by a sliding scale to a lower percentage. If works are completed within the target, in addition to his fee the contractor receives half the saving. There are clauses within the contract which allow for the target cost to be adjusted with inflation and significant changes in the scope of the work. A contract for the first phase, the 150 km northern section of the Songea-Makambako Road was let in 1979. It has been sufficiently

successful for a second contract with the same contractor for Phase II, a further 150 km at the southern end of the route, to be negotiated. The whole system calls for a radical change in attitude at site level and in the boardroom to make this type of operation work but it does offer considerable scope for future prospects.

V. MAINTENANCE

17. No one can deny that road maintenance is generally a neglected area in many developing countries yet one in which it has been very difficult to get local governments or/and donors seriously interested. The latter have tended to the view that they cannot support recurrent budgets directly and other than provide equipment and sometimes technical assistance at management level have not done much to help improve the situation. Yet substantial investments in new roads are at serious risk because of the inadequacies of the maintenance organisations which have to look after them. ODA has recently taken a policy decision that in road projects in future it will build in a maintenance element if this seems to be necessary. Such a maintenance input will not necessarily be confined to the new road being built - it might be more appropriate in some circumstances to think in terms of a parallel district or regional maintenance scheme, first of all to ensure that the new road is properly maintained but also to see that this is not at the expense of the maintenance of other roads built first because they were of greater economic importance. Such maintenance support would in each project be on a decreasing scale over time.

18. There are three main objectives in any road maintenance programme; to prolong the useful life of the road by reducing the rate of deterioration, to keep vehicle operating costs at an acceptable level and to provide appropriate standards of service and safety. These are dealt with under five different forms of road maintenance activity. First of all there are the routine maintenance tasks - drain and culvert clearing, grass cutting etc which must be carried out on a continuing basis. Then there is recurrent maintenance - a series of tasks carried out at regular intervals of a frequency dependent upon topography and traffic volumes. This includes grading of earth and gravel roads and pot-holing and edge repairs on surfaced sections. Thirdly we have periodic tasks which need to be carried out after some years - regravelling, resealing and surface dressing. Fourthly there is special maintenance where sections of road are strengthened or upgraded and minor improvements are made to the geometry. In some circumstances these activities may be more properly considered as new capital investment rather than as a recurrent expenditure item. Finally there is emergency maintenance, clearing land slips, repairing flood damage, repairing collapsed bridges, etc. In putting together a road maintenance project to be aid funded the most appropriate combination of these activities to be locally and off-shore funded needs to be worked out.

19. Virtually all road maintenance programmes are dependent upon the will to grasp the problem and the application of appropriate management and organisational skills. Research we have funded into why road maintenance works are so poorly carried out, in addition to financial issues highlights a number of frequently recurring management faults.

a. Over optimistic planning with too little allowance for plant down time, the poor organisation of materials supply, and bad labour utilisation;

b. failure to set priorities (this is typified in attempts to carry out maintenance tasks on sections which have deteriorated beyond a point at which they can be effectively maintained rather than concentrating on areas where maintenance can still be effective)

c. trying to expand the responsibilities of central road maintenance organisations to take over works neglected by other authorities without an appropriate increase in resources;

d. excessive concentration on the easier to organise and more obvious tasks (e.g. grass cutting) at the expense of more important items;

e. poor supervision of maintenance operations;

f. shortage of managerial, supervisory and technical skills which centralised training schemes could not cope with as effectively as properly organised on-the-job training;

g. over centralised management more involved with paperwork than supervising effective maintenance programmes.

All of these are issues which need to be dealt with in the Technical Cooperation element of any road maintenance management package.

20. In considering road maintenance projects however there is always the need to think about getting roads to a standard or condition at which it would be desirable that they be maintained. The absence of this condition should not however pre-empt road maintenance schemes. In many parts of the world highways have deteriorated to below a desirable standards. This deterioration must be arrested if some form of communication link is to be preserved whilst betterment and improvement programmes get underway.

VI. EVALUATING EXPERIENCE

21. The ODA has an Evaluation Unit whose purpose is discovering how existing projects have been performing in comparison with original expectations. A number of studies have already been done of projects in the roads sector (ref 4). Since it is impossible to summarize our experience with all roads projects, we will

conclude by mentioning a few of the points of interest from some of the reports.

22. In Sierra Leone it was found that the construction of a new road in the interior led to increased polarisation of economic activity in the major towns served, at the expense of more scattered locations along the old route. The full potential of the road on agricultural marketing was not realised because of the absence of a complementary feeder roads programme. The movement of population to locate along the new road alignment was restricted by land rights. Land was held in communal tribal fashion and it was difficult for people to transfer to the new alignment. The majority of trips (over 70%) were made for apparent "social" reasons. This was also the finding of a study of feeder roads in Swaziland, though these "social" motives included such important activities as visits to schools and clinics, as well as relatives and friends.

23. The Swaziland study also showed that in well populated areas with agricultural potential the construction of feeder roads helped to increase the consumption of modern farm inputs such as high yielding seeds and fertilizer. It greatly facilitated the spread of the exchange economy. However, less well populated areas lacking such potential not surprisingly showed little of these effects.

24. One point of interest in the study of a road in Belize was that a condition of UK aid was the imposition of a toll whose receipts were earmarked for maintenance of the road. This did not apparently affect the growth of traffic, which exceeded expectations, but the toll was not very rigorously or fully collected. If it had been, it would have covered most of the annual maintenance costs, provided the fund had not been raided for other governmental purposes.

25. A "base-line" study in Tanzania was done in order to establish the initial or "without" case against which the effects of a major road could be assessed in future. This study uncovered a number of problems likely to hamper traffic growth. The shortage of oil products which resulted partly from the war in Uganda and partly from foreign exchange difficulties had severely restricted vehicular movements. Both the response of producers and the distribution of consumer goods along the road were likely to be affected by the state controlled marketing system and the prices offered for crop procurement. Crop collection in particular was haphazard. The fertilizer subsidy in use was having a big effect on the use and transport of fertilizer. Villagisation policies, and controls on geographical movements of the population, were affecting people's freedom to move in response to economic opportunities.

VII. CONCLUSIONS

26. There is near-unanimity about the bleak prospects faced by most developing countries over the next few years. This is especially serious for the net oil importers and those without buoyant sources of export revenue. Prospects in the major developed countries are also likely to remain bleak for the foreseeable future. In these circumstances, the remarks in this paper about the importance of cost-effective design, staged improvements, and proper maintenance of existing roads as an alternative to new construction, should be underlined. (Ref. 5). To make a virtue out of necessity, however, a climate in which both aid donors and host governments are looking for better value for money from what they do in the roads sector may produce lasting benefits.

REFERENCES

1. "A Guide to Transport Planning within the Roads Sector for Developing Countries" by Bovill, Heggie, and Hine, HMSO for Ministry of Overseas Development, 1978.
2. "The Economic Analysis of Rural Road Projects", World Bank Staff Working Paper No. 241, August 1976.
3. "The Effect of Road Investment on Economic Development in the UK" by Mark Parkinson, Government Economic Service Working Paper No. 43, August 1981.
4. "The Evaluation Activities in the Overseas Development Administration", available from the ODA's Evaluations Unit.
5. The recent advance in policy on road maintenance within the ODA is encapsulated in a new Policy Guidance Note, "Aid for Road Maintenance".

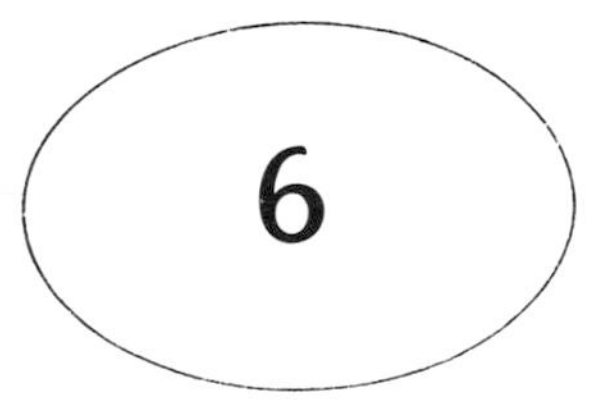

Methods of highway project evaluation in the national transportation plans of the Southern Cone countries of Latin America

I. THOMSON, B(Econ), United Nations Economic Commission for Latin America

All but one of the countries of the Southern Cone of Latin America have undertaken at least one nationwide study in recent years whose objectives included the development of an investment program in highways. This paper outlines very briefly the general modeling approaches used in these studies and the methods used for highway project evaluation. It concludes by assessing the adequacy of the evaluation criteria in the prevailing regional environment.

INTRODUCTION

1. The Southern Cone of Latin America, for the purposes of this paper, comprises the seven republics of Argentina, Bolivia, Brazil, Chile, Paraguay, Peru and Uruguay. The Ministers of Public Works and Transport of these nations regularly participate in annual meetings designed to facilitate transportation between the countries concerned. In 1979 this Meeting of Ministers requested the United Nations Economic Commission for Latin America (CEPAL) to carry out an "Estudio Integrado de Transporte", or Integrated Transport Study, which study facilitated the assembly of the information upon which this brief paper is based. No more than an idea of the subject matter can be provided in a paper of this length. More information is provided in ref. 1, whilst the more interested reader is directed to refs. 2 and 6.

2. All but one of the countries involved have prepared national transportation plans within the past ten years using analytic spatially disaggregated techniques, the one exception to date being the Republic of Chile. These national transportation plans generally had several goals but in every case these goals included the evaluation of highway projects and the formulation of investment plans for the highway sector. This paper is concerned with the principles used in traffic estimation and, more particularly, project evaluation, concentrating on facilities of national rather than local significance, as was generally the emphasis of the national transportation plans themselves.

3. The "Plan Nacional de Transporte" of the Republic of Argentina derives from a Government decree dating from 1977 which authorized the preparation of such a Plan, the financing for which was partially covered by a World Bank loan. The First Phase of the Plan has recently been completed, being carried out by the "Dirección Nacional de Planeamiento de Transporte", a dependency of the Secretary of State of Transport and Public Works. The studies were undertaken by Argentinian professionals, the training of some of whom formed part of the overall Plan package, and several specialized teams of foreign consultants. The First Phase of the "Estudio Integral del Transporte" of the Republic of Bolivia has also recently been completed. This was carried out by a firm of consultants based in the U.S.A. in conjunction with a group of Bolivian professionals. Financial assistance was provided by the United Nations Development Program (UNDP) whilst the World Bank served as executing agency, an arrangement common in Latin America. This study was the second of its kind in Bolivia, the first being undertaken in the late nineteen-sixties.

4. In the Federative Republic of Brazil the equivalent of the Argentinian and Bolivian studies mentioned in paragraph 3 is the "Plano Operacional de Transportes", the National Transportation Plan of Brazil being concerned with overall policy guidelines rather than the spatially disaggregated analysis of specific transport problems. The first two phases of the "Plano Operacional de Transportes" have been completed by GEIPOT, the planning division of the Transport Ministry. (The third, and final, phase would update the first two phases.) In the late nineteen-sixties a study with somewhat similar objectives was carried out under the direction of foreign consultants under the title Brazil Transport Survey. The "Estudio Integral del Transporte" of the Republic of Paraguay was undertaken between 1971 and 1973 by French and local consultants, involving the UNDP and the World Bank as a source of financing and executing agency respectively. Subsequently two technical assistance projects, carried out jointly by Argentinian and Paraguayan teams, have carried out more limited studies to update the findings of the "Estudio Integral del Transporte".

5. The Republic of Peru has not pursued the idea of undertaking an integrated transpor-

tation study as such, but has developed two long term transport plans, one an updating of the first, which are somewhat similar in nature. These were developed by the "Oficina Sectorial de Planificación" of the Ministry of Transport and Communications using a model earlier developed with the aid of a visiting foreign specialist. (Insufficient information is currently available about the techniques used in the long term transport plans of Peru and thus they will not be dealt with in this paper.) An "Estudio Integrado de Transporte" was carried out in the Republic of Uruguay in 1976 and 1977 by Italian consultants, involving also UNDP and the World Bank. Subsequently a quite distinct national transport plan was developed in Uruguay which concentrated on general policy guidelines and suggests a medium term investment program. This did not involve spatially disaggregated analysis and is not discussed in this paper.

GENERAL PLANNING METHODOLOGIES

6. In all but the two technical assistance projects undertaken in Paraguay, the studies for the national transportation plans in the Southern Cone countries adopted what is basically a traditional modeling structure. This comprises, as a first stage, the analysis of the production (or importing), consumption (or exporting) and marketing of the major products served by the national transportation system. In most cases person movements were also studied, usually in less detail, as befits a region in which freight movement often dominates interurban transportation by surface modes. The output from the analyses of major products was typically a set of origin-destination matrices by mode for the base year and a set of listings of productions and consumptions by zone for future analysis years. Sometimes all-mode matrices were also prepared for the future analysis years. In two cases these were disaggregated by mode, albeit in a provisional manner subject to subsequent modification. The most detailed product studies yet undertaken in the region were probably those carried out in Brazil where 80 different products were examined in 20 sectorial studies, the origin-destination matrices being specified at the level of the 445 zone plan used by GEIPOT. Files were maintained of the routes taken between each origin and destination pair.

7. The subsequent stages of the analyses varied considerably between countries. In Argentina it was initially intended to employ a modeling procedure which would have been the most advanced yet used at the nationwide level in the Southern Cone, involving such sophistications as a two-level logit model for modal split and the use of the logit principle in trip assignment to allocate trips between competing routes. The model was later somewhat simplified for the purposes of the first phase of the Plan. Six main major product types and passenger flows in the major corridors were analyzed. For grain crops a doubly constrained gravity model was used for distribution, the cost input was a composite of the separate modal costs, whilst simpler methods were used for other products. Modal split was simulated by a single stage logit formulation whilst assignment used all-or-nothing principle. The simplified model actually used partially recognized congestion, a factor which was not considered in the simulation models used in other Southern Cone countries. Congestion costs were however recognized in project evaluations in Brasil (and Argentina), as is described later in this paper.

8. In Bolivia a traditional form of transport model was used, the successive stages of which were relatively independent of each other. However, the model includes certain interesting component features. Distribution was simulated either by a gravity model, by a model based on the principle of transport cost minimization, or by a model which uses the principle of proportional distribution, for each one of the 39 product groups analyzed, the model considered most appropriate being chosen in each case. Modal split was dealt with by a simple two-dimensional model, the calibration of which presented considerable problems, as was the case also in virtually all the other studies. Assignment was by the all-or-nothing principle. The Bolivian study included a component to split truck flows by type of truck and another to estimate empty truck flows.

9. The main objective of the Brazilian "Plano Operacional de Transportes" was the identification of those parts of the transportation network upon which congestion would develop, were no avoiding action taken, thereby threatening the efficient transportation of important commodities, and, subsequently, the evaluation of alternative forms of avoiding action. The projected matrices were firstly assigned to the modes and routes used in the base year (modified as necessary in the light of new and improved facilities.) To simulate that assignment pattern which might result once railroads become more competitive, high volume flows were then rerouted according to a cost minimization principle. Links having insufficient capacity were identified and, for highway links, economic evaluations were carried out. Passenger traffic was not explictly modeled in the Brazilian study, but was allowed for in the determination of capacity requirements.

10. In the paraguayan "Estudio Integral del Transporte" future cargo flow matrices were derived from those constructed for the base year by growth factor methods. The modal split for the horizon year was assumed to be the same as at the base year except when the competitive balance between the different modes was expected to change significantly, in which cases a simple diversion curve

method derived from European experience was used. Assignment was based on the all-or-nothing principle. The two subsequent technical assistance projects estimated flows independently for each link in the network, thus needing no formal distribution, modal split, nor assignment model. One study based future traffic growth on past trends whilst the other determined growth on the basis of trends in fuel consumption and toll revenues and indicators of expected future regional growth.

11. The modeling component of the "Estudio Integrado de Transporte" in Uruguay was conceptually quite distinct from other in the Southern Cone countries since the preferred network was generated with the aid of an optimizing model. Future freight matrices were first of all estimated by a linear program whose objective function sought to minimize transport costs. Then an attempt was made to adjust (or calibrate) the transport costs on the various links in the network so that the base year modal split and route choice pattern could be adequately simulated were a model which directs traffic to the lowest cost mode and route combination to be applied. Although none of the sets of adjustments considered was deemed wholly satisfactory, the preferred one was input to the optimization model, along with (i) proposed projects for each link, (ii) the trip matrices, and (iii) a budget restriction. The optimization algorithm involves the successive consideration of sets of feasible projects, i.e. sets which satisfy the budget constraint, for each of which an assignment is made. The optimized set of projects is that feasible set which results in minimum transportation cost. The model was run several times with different proposed projects per link and different budget restrictions. The set of projects produced by one of the runs was selected and each of the projects was evaluated individually, as described later.

METHODS OF HIGHWAY PROJECT EVALUATION

12. In general, the studies for the national transportation plans of the Southern Cone countries used economic efficiency as the only criterion in the economic evaluation of highway projects. Simple procedures were generally employed, lacking in sophistication but usually reliable enough to permit an initial sifting of projects for inclusion in an investment program. Only in one case, that of Argentina, was a serious attempt made to denominate the economic evaluations in terms of shadow prices, but tax and subsidy elements were generally extracted from the costs used in the evaluations. The set of projects selected according to economic efficiency criteria was sometimes merged with another group of projects considered to be justified on non-economic criteria, such as to promote regional development or for strategic reasons, and with those projects which had already been authorized. In no case, however, were criteria other than economic efficiency internalized in the project evaluations.

13. The first phase of the Argentinian "Plan Nacional de Transporte" sought to estimate a global sum for investment in the main interurban highway system of the country by determining investment needs on each section of highway in a preliminary manner (which relied very little upon the results from the transport model). The procedure used commenced with the determination, independently of particular sections of highway, of the traffic level which would warrant the replacement of an existing road state by an improved state given: (i) growth rates in traffic by vehicle type, (ii) breakdown of existing traffic flow by vehicle type, (iii) vehicle operating costs on the existing and improved highway, (iv) assumed amounts of diverted and generated traffic were the highway to be improved, (v) the capital cost of the proposed improvement, (vi) estimated maintenance costs of the highway in the two conditions, and (vii) where necessary, arbitrary adjustments to allow for possible changes in accident costs and the avoidance of closures of unpaved roads in wet weather. Thus, given the existing state of each link and the existing traffic level, it was possible to determine its optimum state during the planning period. Benefit:cost analyses of the projects so determined for each link were undertaken and used to formulate the suggested investment program. This program was completed by projects deemed to be worthwhile on non-economic grounds and by other projects considered to be worthwhile economically but which were not able to be analysed by the standardized evaluation methodology.

14. The Bolivian "Estudio Integral del Transporte" included a relatively intricate procedure for highway project evaluation. First of all, in order to avoid the costly running of the transport model separately for different combinations of projects and to minimize the necessity of making manual reassignments of traffic for the purposes of project evaluation, for each candidate project a first year rate of return was estimated on the basis of the existing traffic level and an assumed growth rate. Only those projects thereby shown to have a reasonable chance of being demonstrated to be economically viable were included in the test network for which the computerized assignment was made. The economic evaluation of projects proceeded in different ways depending on the nature of the project. Major schemes involving construction or reconstruction of major highways were assessed by the consumer's surplus method, utilizing traffic flow estimates produced by the transport model. Projects involving the opening up of relatively inaccessible areas, which cover a large part of Bolivian territory, were considered as part of development

projects and evaluated by the producers' surplus methodology. For the analysis of the existing main road network extensive use was made of the World Bank Highway Design and Maintenance Standards model (HDM model - see ref. 7). This model simulates highway condition through time given: (i) the existing pavement condition, (ii) present traffic levels and expected growth rates, (iii) maintenance policy, and (iv) environmental conditions. It also calculates the corresponding total transportation cost given input vehicle operating and road maintenance cost functions. It was used to investigate pavement design, road improvement policies and maintenance policies. An attempt was made to evaluate all projects in a comparable manner so that they could be jointly considered by a computerized version of another procedure developed by the World Bank which selects the optimum set of economically feasible projects subject to an overall budget constraint. (See ref. 8.) However, for technical reasons, not all types of project could be evaluated on exactly the same basis and the estimated returns tended to be higher for projects for certain modes than for others. Thus, recognizing the political impossibility of accepting the results from the procedure to optimize budget allocations in such circumstances, inter-modal budget allocation was determined on the basis of past trends and judgement. The optimizing procedure was used for certain intra-modal allocations.

15. The objectives of the "Plano Operacional de Transportes" in Brazil included the identification of those highways whose capacity would not be sufficient to deal efficiently with predicted traffic flows. For each such highway link a set of possible capacity expansion projects were hypothesized (once it had been decided that operational improvements would not resolve the supply problems) and each was evaluated by estimating transportation costs in the with and without project cases. Congestion costs were assessed, when appropriate, in the without project cases. Projects were ranked according to benefit:cost ratio and assigned to the recommended investment program according to their optimum starting years. (Although not forming a part of the "Plano Operacional de Transportes", it should be mentioned that Brazil is undertaking a massive research effort on the interrelationships between costs of highway construction, maintenance, and utilization, in order to update and calibrate for Brazilian conditions the HDM model. See ref. 11.)

16. In the "Estudio Integral del Transporte" carried out in Paraguay a procedure similar to that applied in Argentina was used, but somewhat more detailed in that it considered the possibility of staged construction. (In certain other respects the Paraguayan model was less refined than that used in Argentina.) The transport model produced six traffic estimates for each link (for different assumptions about the state of the transportation network and generated traffic). Projects justifiable with the minimum of the six levels of traffic predicted were first of all included in the investment program. The program was completed with projects on links which would be justified were traffic levels to be greater than the minimum estimated. A simplified version of the same general approach was used in the two technical assistance projects. The first of these included in its recommended investment program projects required for general development or strategic purposes, although these were not evaluated. The second technical assistance project considered a wider variety of highway projects than the first, namely projects involving pavement strengthening, feeder roads, and widening of existing highways. These extra categories were included in the recommended investment program but were not generally evaluated by economic criteria. Non economic considerations were recognized in a general manner in the formulation of the investment program and an attempt was made to give the program a regional balance acceptable from the political and social standpoints.

17. The network optimization model used for the "Estudio Integrado de Transporte" in Uruguay was applied five times with different sets of candidate projects and different budget restrictions, all of which yielded optimized project sets with benefit:cost ratios of less than unity. The set which would result in the least unfavorable ratio (but which, however, would be expected to include some worthwhile projects) was selected for detailed analysis. For each project contained in it a simple economic evaluation was performed, considering cost savings to existing traffic (and growth in existing traffic) adjusted in a coarse manner to recognize diverted traffic. The proposed investment program was prepared considering both optimum starting years and internal rates of return.

CONCLUSIONS ON THE EVALUATION CRITERIA USED

18. The basic criterion of economic efficiency used in project evaluations for the formulation of the highway investment component of the national transportation plans of the Southern Cone countries is without much doubt the best single criterion to use. But there exists a need both to expand the concept of what is "economic" and to recognize non-economic factors as well.

19. It has been noted that only in one case was a serious attempt made to denominate the project evaluations in terms of shadow price adjusted costs. In some Southern Cone countries liberal economic policies imply that were shadow prices used in project evaluations rather than market prices or market prices net of taxes and subsidies there would be little impact on the conclusions

from these evaluations. On the other hand, the region also contains protected economies subject to a considerable degree of government intervention in market mechanisms where the use of shadow prices could make significant differences to project evaluations. As a further point of interest it is noteworthy that considerable inconsistency has existed within the region as to what the extent the economic efficiency criterion embraces the value of personal travel time.

20. The studies often recognized that non-economic considerations are important in transportation planning. Generally, however, such recognition was merely registered and had no influence on evaluation methodologies, which continued to give attention solely to economic aspects. On some occasions non-economic factors were allowed to affect the recommended investment program, but in a conceptually unsatisfactory manner and only after the evaluations had been carried out solely from the economic efficiency standpoint. Considerable economic research has been directed to the matter of distributional and regional factors in project evaluation (see refs. 9 and 10) but very little research has been conducted into the matter of the consistent incorporation into project evaluations of strategic considerations or sentiments of nationalism or patriotism, all of which are particularly important in Latin America. In one Southern Cone country it has been suggested that the transportation economist should evaluate projects purely in economic or socio-economic terms and then request that those ministries other than transport which might wish to see the project implemented, such as the Ministry of Defense or the Ministry for the Interior, state how much they would be willing to contribute to the capital costs of the project. The economist could then revise his benefit:cost calculations considering just that part of the project cost not covered by these other ministries. This suggestion has decided attractions. However, rather than resolving the problem of how to incorporate strategic and other factors in project evaluations, it merely transfers the problem away from the transportation economist to somebody else.

21. It might be claimed that the methods used for highway planning for the national trasportation plans of the region have sometimes been somewhat coarse and capable of yielding only an approximate idea of the most suitable highway investment program. Planning at the national level rarely permits greater levels of precision. However, there does seem to be a need to include national transportation planning studies, or as subsequent follow-up exercises, the type of analysis conducted in Bolivia using the HDM model. One problem of the application of the HDM in the latter country is that it was being used in an environment very different from that prevailing in Kenya where it was largely developed. The ongoing research in Brazil to update the model and calibrate it for Brazilian conditions will provide that country with a very useful highway planning tool. However, the Brazilian experience may not be immediately transferable to other countries in the region (where vehicle weights and dimensions, and environmental conditions are often different). Most of these other countries lack the resources needed by a research effort on the Brazilian scale.

22. Systematic transportation planning at the national level in the Southern Cone has typically been concentrated on large scale analyses at irregular intervals. Sometimes there has been an inadequate level of continuity of effort in the interim periods. This has presented a number of problems. One of these is that highway projects are authorized in the interim period without being guided by any overall planning framework, possibly entailing that when the large scale planning exercise takes place the analysts find that a large proportion of the resources likely to be available for highway investment have already been committed. The large scale analyses usually have to meet tight deadlines and this can affect the quality of the results obtained. The CEPAL study for the governments of the Southern Cone countries has examined transport planning methodologies and the general transport planning environment in the region, making a number of suggestions as to how some of the problems encountered in the past might be addressed in the years to come. These are discussed in ref. 1.

REFERENCES

1. COMISION ECONOMICA PARA AMERICA LATINA. La planificación del transporte en los países del Cono Sur: Un análisis de las metodologias aplicadas, E/CEPAL/R.287, CEPAL, Santiago de Chile, 1982.
2. E/CEPAL/R.287/Addendum 1. Las metodologías aplicadas en Argentina. 1982.
3. E/CEPAL/R.287/Addendum 2. Las metodologías aplicadas en Bolivia. 1982.
4. E/CEPAL/R.287/Addendum 3. Las metodologías aplicadas en Brasil. 1981.
5. E/CEPAL/R.287/Addendum 4. Las metodologías aplicadas en Paraguay. 1981.
6. E/CEPAL/R.287/Addendum 5. Las metodologías aplicadas en Uruguay. 1981.
7. HARRAL C.G., WATANATADA T. and FOSSBERG P. The Highway Design and Maintenance Model (HDM): Model structure, empirical foundations and applications. World Bank, Washington 1979. (Paper presented to the Second International Conference on Low Volume Roads, Iowa State University, August 1979.)
8. WATANATADA T. and HARRAL C.G. Determination of economically balanced highway expenditure programs under budget constraints: a practical approach. World Bank paper B-20, April 1980. (Presented at the World Conference on Transport Research, London, April 1980.)

9. MARGLIN, S.A. Public Investment Criteria. Unwin University Books, London, 1967.

10. SQUIRE L. and VAN DER TAK H. Economic Analysis of Projects. World Bank/Johns Hopkins University Press, 1975.

11. REPUBLICA FEDERATIVA DO BRASIL/PROGRAMA DAS NAÇOES UNIDAS PARA O DESENVOLVIMENTO. Pesquisa do inter-relacionamento entre custos de construção, conservação e utilização de rodovias. GEIPOT, Brasília. (Relatorio preliminar - conceitos e metodologia - maio 1976, and subsequent volumes.)

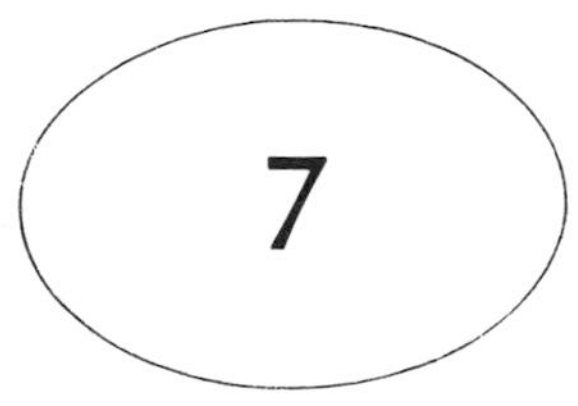

Method of estimating construction costs on large rural projects over varying terrain in developing countries

W. K. CROSS, MBE, TD, FICE, FIHE, FASCE, Rendel Palmer & Tritton

Construction costs were estimated for each of nine route corridors varying between 540 and 800 kms in length in Papua New Guinea. The cost estimates were used as a basis for making economic comparisons between the routes to enable one of them to be selected for more detailed study. The routes lay over practically uninhabited mountainous terrain and coastal swamps. A method of terrain evaluation in terms of engineering difficulty related to construction costs is described.

INTRODUCTION

In Papua New Guinea there had been considerable speculation for many years as to the viability of constructing a trans-island road which would connect Port Moresby, the capital lying on the south coast, with Lae, the country's principal port and industrial centre on the north coast. No such road link exists and the central mountain range effectively cuts this developing country in two.

2. Nine possible broad corridors for such a link had been established. The Government of Papua New Guinea therefore asked that each corridor be studied to obtain estimates of construction costs and of the potential economic and social benefits which would accrue from construction of a road within the corridor. The routes varied in length between 540 and 800 kms and the study period allowed was only three months.

3. It was therefore necessary to evaluate and compare within that period road construction costs for some 6000 kms of mountainous , densely forested and inaccessible terrain. This paper describes the methods adopted to undertake this task. No description is given of the evaluation of economic or social benefits or of the succeeding phase of the study, which comprised a more detailed study of the preferred route.

CHARACTERISTICS OF THE STUDY AREA

Topography

4. Topographically the study area is dominated by the Owen Stanley Range, which runs the length of the mainland of eastern Papua New Guinea. This range forms the main watershed between the Solomon Sea and the Coral Sea off the northern and southern coasts respectively. Several peaks rise to over 3000m. Other mountain ranges occur on either side of the Owen Stanley Range. Ranges of hills occur along the south-west coast and are interspersed with low lying planes associated with the main rivers and the coast. The area is almost entirely covered by dense forests.

5. Some coastal areas are permanent or seasonal swamps with seasonal fluctuations of water depth varying between 1 and 2m. Wide rivers with braided channels and shifting meanders characterise the main drainage pattern. Other coastal areas form a savannah landscape with high smooth ridges rising abruptly from concave foot slopes and erosional plains. Moderately dense drainage reflects high runoff from impermeable surfaces with extensive sheet flow on gentle slopes.

6. Further inland foothill zones of increasing altitude and higher rainfall are encountered. Relief is much more pronounced and slopes tend to be uniformly steeper than the coastal hill areas. Land forms range from plains to low hills. A closer network of streams occurs, of which the larger ones are perennial.

7. Characteristic land forms of the upland areas are ridges of greater altitude ranging from hilly to mountainous, with deep valley re-entrants and extensive ridge and plateau remnants. The terrain is intersected by a dense network of perennial streams.

Climate

8. The lowland areas of less than 300m in altitude have fairly even temperatures, usually ranging between a mean maximum of 28-34°C and a mean minimum of 20-25°C. In the highlands at about 2400m the temperature ranges between a mean maximum of 20-29°C and a mean minimum of 10-18°C.

9. Rainfall everywhere is over 2400mm per year with the exception of limited areas principally around the coastal strips east and west of Port Moresby where the rainfall is lower. Some parts of each of the route corridors receive more than 4000mm per year.

Engineering Geology

10. The bedrock comprises volcanic and sedimentary rocks, with unconsolidated alluvial sediments in the plains and valleys and varying depths of colluvial, talluvial and residual soils related to the respective parent bedrock

in the hills and mountains. Rivers in ravine and mountain tracts more or less everywhere contain considerable quantities of gravel and boulders as erosion rates are high and supply of rock debris large.

11. The engineering characteristics of the study area are directly affected by the terrain, climate and geology. In hilly areas of high rainfall greater depths of residual soil may be anticipated, especially in flatter areas with low rates of erosion. In mountainous areas of less rainfall more rock excavation is likely to be encountered. Scarcity of coarse construction materials in flat coastal areas would probably require suitable materials to be brought in from more distant borrow areas.

12. Slope problems are generally caused by a variety of failure mechanisms which may become progressive, sometimes leading to deep seated rotational failures. Such slips occurring in weathered coarse grained sedimentary rocks are often considerably larger than in the fine grained rocks, where they tend to be more numerous but smaller.

13. The wider meandering rivers of the low lying plains require particular care in the selection of stable crossing sites. Where swamp areas are unavoidable, careful attention has to be given to suitable embankment and crossing designs.

Access Within the Study Area

14. No road link exists across the central mountain range and road access was not available for a considerable proportion of each of the routes to be studied. The rural road development which exists in the study area has been based mainly upon foot access trails originally located by missionaries and villagers. They have been constructed over many years by hand labour, mainly as inter-village links or to water points. The engineering standards of location, design and construction for these roads are necessarily low. Inadequate roadbases and drainage frequently results in erosion and stability problems.

15. The remoteness of some parts of each of the route corridors from airstrips and rural access roads, and the short period allowed for the study, required a methodology which enables estimates of construction costs to be made on a realistic basis for large lengths of route without the benefit of a full walk-over or drive-over survey. The study team were well aware that the study could not be entirely "desk bound" as this could lead to large errors in estimating construction costs and entirely the wrong conclusion being reached as to the route most suitable for development.

STUDY METHODOLOGY

Introduction

16. The following data and mapping was made available to the study team:

- 1:100,000 scale topographical maps
- 1:250,000 scale geological survey maps
- complete aerial photography coverage.

In addition 1:500,000 scale terrain maps of Papua New Guinea produced in 1971 by the Department of Defence of Australia were made available approximately half way through the study period.

17. The study team comprised a Project Manager and a Highway Engineer full time, with considerable assistance from an Engineering Geologist and a Cost Estimator. The economic and social benefit analyses were undertaken by additional staff.

18. The study team had considerable experience of road design in Papua New Guinea and thus started the Project with an appreciation of the terrain and problems to be encountered. This was supplemented by a brief fly-over survey, using a helicopter, of the corridors to establish the variations in terrain which would be encountered.

19. Desk Study. It was first necessary to define the type of road considered most appropriate to the initial requirements of a trans-island road linking Port Moresby with Lae. It was considered that this should provide an all-weather link for two wheeled drive vehicles. A preliminary economic analysis indicated that the road need only be sufficient for low volumes of traffic, travelling at slow speeds in Hilly/ Mountainous terrain.

20. The design standards adopted were those of the lowest class of rural road constructed in Papua New Guinea, with certain exceptions. These exceptions included steeper maximum gradients over short lengths in very severe terrain. It was also envisaged that the whole road width, including shoulders, would be sealed where gradients were over 6% in order to reduce maintenance costs to a reasonable level.

21. The standards adopted were refined later in the Study. It should be noted that the standards chosen did not affect the way in which the Study was undertaken.

22. The nine broad route corridors were refined by the examination of the Papua New Guinea 1:100,000 scale topographical maps. From this examination certain "tie points" were identified through which the route might suitably be located. Such tie points included low passes over mountain ranges, feasible river crossings and similar criteria. The difference in altitude of such tie points in relation to an assumed average gradient within the adopted standards gave a route length which would have to be developed between them. This length of route was then located on the 1:100,000 scale topographical maps, making the best possible use of the terrain. Here the word "terrain" takes into account such factors as steepness of side slopes, swamp areas and narrow valleys which might constrain freedom of final location. Aerial photographs were used to confirm the adequacy of bridge crossings used as tie points.

23. The refined routes were then transferred to the 1:250,000 scale geological survey maps of Papua New Guinea. This scale proved adequate to

establish with sufficient accuracy the division of each route into geological units. For each of the geological units traversed by one or more of the major routes the major engineering geological criteria were then estimated. These criteria included possible depths of water table, ease of excavation of parent rocks and residual soil, bearing capacity of parent rock and residual soil, and stable cut angles in parent rock and residual soil.

24. By an inspection of the topographical and geological maps, 15 tentative terrain units were established within each of which it was considered that the criteria affecting road construction costs could be taken to be constant. These criteria were assumed to be those listed in para. 23 above plus the topographic relief, estimated number of drainage structures per kilometre and the potential availability of fill and aggregate materials.

25. The terrain units are described in Table 1 in terms of topography. The units were given a reference letter indicating terrain type, followed by a number indicating the relative difficulty in terms of engineering construction. The reference letter S referred to steep terrain, P to plateaux, B to basins, C to coastal areas and A to alluvial flood plains and terraces.

26. This phase of the Study was ultimately simplified by the 1:500,000 scale terrain maps of Papua New Guinea becoming available. This map is delineated by 25 terrain units, each of which is characterised by distinctive topographic features. The method described in this Paper would be equally effective in areas where no such terrain mapping existed. In this case the division into terrain units could have been made purely on the basis of topographical and geological (or geomorphological) mapping, but a more detailed field reconnaissance or more field staff would have been required.

27. Field Reconnaissance. To this stage the work had generally been a desk study. The next stage was to undertake a field reconnaissance in as much detail as possible in the time allowed. Where roads or tracks existed along or adjacent to the routes, these were utilised. However helicopters provided the most convenient form of transport over much of the length of the routes. The helicopter was landed wherever convenient to inspect potential bridge crossings and other major tie points.

28. The field reconnaissance was used to confirm and refine the original terrain unit division of the routes and to calibrate the units against the topography. In particular, the criteria listed in paras. 23 and 24 were substantially revised as a result of the field work. The number of major culverts was obtained by counting stream crossings in each kilometre over many

Table 1

Terrain Group		Terrain Unit Topographic Descriptions
Mountainous Severe	(S7)	V-shaped valleys and severe rugged ridges of relief greater than 300m, with deeply dissected slopes at angles greater than 30°, or greater than 40° if not dissected.
	(S6)	V-shaped valleys and rugged steep ridges of relief between 100m and 300m, with deeply dissected slopes at 30° - 40°.
Mountainous Moderate	(S5)	V-shaped valleys and sharp steep ridges of relief between 100m and 300m, and moderately dissected slopes less than 30°.
	(S4)	Convex slope and concave valleys of relief between 100m and 300m.
Hilly	(S3)	Upland ridge and valleys of relief varying between 30m and 100m
	(S2)	V-shaped ridge and valleys of relief varying between 30m and 100m in the foothill areas.
Undulating	(P2)	Low plateaux strongly dissected in foothill areas.
	(S1)	Gently convex slopes and broad valleys of relief varying between 30m and 100m.
	(P1)	Low plateaux gently dissected in foothill areas.
	(B2)	Poorly drained upland flood plain and basin in intermontane valley areas.
Coastal Wet	(C3)	Perennial swamps in littoral plain areas.
	(A1)	Flood plains subject to inundation in fluvial plain areas.
Coastal Dry	(A2)	Higher flood plains and alluvial terraces in fluvial plain areas.
	(A3)	Plains and higher alluvial terraces in fluvial plain areas.
	(B1)	Upland plains, valleys and basins of relief less than 10m in intermontane valley areas.

Table 2; Terrain Unit Engineering Descriptions

TERRAIN UNIT	A1	A2	A3	B1	B2	C3
Relief and Water Table	$<$ 10m H.W.T.	$<$ 10m L.W.T.	$<$ 10m L.W.T.	$<$ 10m V.L.W.T.	10–30m L.W.T.	$<$ 10m H.W.T.
Estimated Minor Culverts/km	8 Small	8 Small	6 Small	5 Small	5 Small	12 Small
Estimated Major Culverts/km	1/5	1/4	1/3	1/3	1/2	1/2
Estimated Excavation of Residual Soil	Easy by Conventional Wet	Easy by Conventional Dry	Easy by Conventional Dry	Easy to Moderate by Conventional Dry	Easy to Moderate by Conventional Dry and Some Wet	Easy by Conventional Wet
Estimated Excavation of Underlying Bedrock	N/A	N/A	N/A	N/A	N/A	N/A
Estimated Safe Bearing Capacity of Residual Soil	80 kPa dry 50 kPa wet	100 kPa dry 50 kPa wet	100 kPa dry 50 kPa wet	160 kPa dry 80 kPa wet	160 kPa dry 80 kPa wet	25 kPa wet
Estimated Safe Bearing Capacity of Bedrock	N/A	N/A	N/A	N/A	N/A	N/A
Estimated Natural Slope Failures of Residual Soil	Minor Under-Cutting by Streams	Minor Under-Cutting by Streams	Minor Under-Cutting by Streams	Occasional Stream Undercutting with Falls	Occasional Stream Undercutting with Falls	River Bank Slopes Adjusted
Estimated Natural Slope Failures of Bedrock	N/A	N/A	N/A	N/A	N/A	N/A
Estimated Cut Slope Angle of Residual Soil	$1\frac{1}{2}$h : 1v	3/4h : 1v Dry	3/4h : 1v Dry	$\frac{1}{2}$h : 1v	$\frac{1}{2}$h : 1v	2h : 1v
Estimated Cut Slope Angle of Bedrock	N/A	N/A	N/A	N/A	N/A	N/A
Estimated Fill Borrow Potential	Only Selected Material from Roadside	Random Material from Roadside	Random Material from Roadside	Random Material from Roadside	Random Material from Roadside	Only Selected Material from Roadside
Estimated Rock and Aggregate Borrow	River Gravel or From Nearest Tertiary Outcrop	River Gravel or from Nearest Tertiary Outcrop	River Gravel or from Nearest Tertiary Outcrop	Select Material from Near Road	Select Material from Near Road	River Gravel or from Nearest Tertiary Outcrop

Table 2 (Cont). Terrain Unit Engineering Descriptions

TERRAIN UNIT	P1	P2	S1	S2	S3	S4	S5 S6 S7
Relief and Water Table	10-30m L.W.T.	30-100m V.L.W.T.	10-30m L.W.T.	30-100m V.L.W.T.	30-100m V.L.W.T.	100-300m V.L.W.T.	> 300m V.L.W.T.
Estimated Minor Culverts/km	5 Small	5 Medium	5 Small	5 Medium	5 Medium	4 Medium	5 Medium
Estimated Major Culverts/km	1/4	1/4	1/3	1/4	1/2	1/2	1/2
Estimated Excavation of Residual Soil	Easy to Moderate by Conventional	Easy to Moderate by Conventional	Easy by Conventional Dry	Easy by Conventional Dry	Easy by Conventional Dry	Easy by Conventional Dry	Easy by Conventional Dry
Estimated Excavation of Underlying Bedrock	Moderately Difficult. May Require Ripping	Moderately Difficult. May Require Ripping	Moderate to Very Difficult Ripping	Moderate to Very Difficult Ripping	Moderate to Very Difficult Ripping. Some Blasting.	Moderate to Very Difficult Ripping. Some Blasting.	Moderate to Very Difficult Ripping
Estimated Safe Bearing Capacity of Residual Soil	100 kPa dry 50 kPa wet	100 kPa dry 50 kPa wet	80 kPa dry 50 kPa wet	80 kPa dry 50 kPa wet	80 kPa dry 50 kPa wet	80 kPa dry 50 kPa wet	80 kPa dry 25 kPa wet
Estimated Safe Bearing Capacity of Bedrock	> 200 kPa	> 200 kPa	> 200 kPa	> 200 kPa	> 200 kPa	> 200 kPa	> 200 kPa
Estimated Natural Slope Failures of Residual Soil	Minor Under-cutting by Streams	Undercutting By Streams with Falls and Slips	Rare Stream Undercutting with Falls	Stream Under-cutting with Falls and Slips	Common Stream Undercutting with Falls, Slumps and Slips	Common Stream Undercutting with Falls, Slumps and Slips	Common by Undercutting with Falls Slumps and Slips
Estimated Natural Slope Failures of Bedrock	Not Exposed Naturally	Not Exposed Naturally	Rare Stream Undercutting with Falls	Rare Stream Undercutting with Falls	Occasional Streams Undercutting with Slips and Falls	Occasional Stream Undercutting with Slips and Falls	Occasional Stream Under-cutting with Slips and Falls
Estimated Cut Slope Angle of Residual Soil	3/4h : 1v	3/4h : 1v	3/4h : 1v	3/4h : 1v	3/4h : 1v	1/2h : 1v	1/2h : 1v + benching
Estimated Cut Slope Angle of Bedrock	1/2h : 1v	1/2h : 1v	1/2 : 1v	1/2h : 1v	1/2h : 1v	1/2h : 1v	1/2h : 1v + add. Works
Estimated Fill Borrow Potential	Random Material from Roadside	Random Material from Roadside	Random Material from Roadside	Random Material from Roadside	Random Material from Roadside	Random Material from Roadside	Random Material from Roadside
Estimated Rock and Aggregate Borrow	From Deep Cuttings and Local Quarries	From Deep Cuttings and Local Quarries	From Deep Cutting and Local Quarries	From Deep Cutting and Local Quarries	From Deep Cuttings Local Quarries and Rivers	From Deep Cuttings Local Quarries and Rivers	From Deep Cutt-ings, Local Quarries,Rivers

kilometres. The locations where large bridges would be required were individually recorded and assessed. Existing cuttings and major slips were inspected to ascertain depths of residual or colluvial soils and stable cut angles in soils and bedrock. The availability of suitable sources of aggregate was assessed, particular care being taken to establish potential sources in the coastal and alluvial areas.

29. Despite the relatively small Study team, the use of a helicopter enabled all nine routes to be completely traversed. Table 2 presents a summary of the major engineering geological characteristics of each of the 15 terrain units after refinement as a result of the field reconnaissance. This Table was used as a basis for the estimation of construction costs.

30. During the field reconnaissance "route factors" were derived. Route factors were obtained by comparing several map measured distances with corresponding distances travelled by vehicle along roads and tracks over the same terrain. It was found that distances travelled over flat and undulating terrain were 1.1 times longer than the map distances. In hilly terrain the actual distances were 1.2 times longer, and in all mountainous terrain they were 1.3 times longer than the corresponding distances measured from the maps.

DERIVATION OF CONSTRUCTION COSTS

31. A preliminary assessment of road construction costs in each of the terrain units made it possible to allocate each of the 15 terrain units found occurring in the nine route corridors to one of six Group categories in which construction costs were taken to be constant. These Terrain Groups are shown in Table 1. Within each of these groups the items which had the greatest effect on construction costs were examined.

32. In view of the difficulty in evaluating the effect of so many variables upon construction costs, it was decided to update costs for each of the six main elements of road construction for roads recently completed or under construction in each of the six Terrain Groups.

33. The updating included any provisions resulting from under-estimating, which could subsequently be recovered as claims. The results of interviews with several contractors concerning their difficulties in working in each of the Terrain Groups enabled weightings to be made to the cost estimates.

34. Factors affecting costs for the six main elements of road construction in each of the Terrain Groups were considered as follows:-

Clearing: Access and working difficulties increase abruptly from the relatively level flat coastal areas to mountainous terrain of increasing side slope. The problem of disposal, whether by controlled burning or selective extraction of commercially important timber, is also increased with remoteness of operations.

Earthworks: In dry coastal areas most of the embankment material would normally be obtained from shallow longitudinal borrow areas designed to be self draining. In marshy coastal areas there would be less likelihood of such material being suitable for building embankments. This may require particular borrow material to be hauled some distance. Special foundation designs would almost certainly be required where embankments were built across swampy areas. As the terrain became progressively more hilly/mountainous, excavation should be expected to be in harder materials. In certain areas of steep side slope, some restriction might have to be placed upon indiscriminate side tipping of material to spoil. Operating costs would increase as the work became more distant from base camp and plant yards.

Rock: Excavation in rock would be broadly expected to increase in proportion to the ruggedness of the terrain and increasing relief. Similar remarks apply concerning any restrictions on tipping to spoil, and on increasing costs as the work became further away from base camp and maintenance facilities.

Pavement: Here the term pavement includes roadbase and any sub-base courses as well a as surfacing layer. Whilst some sands and fine gravels were noted in certain of the larger river beds in coastal areas, generally the availability of suitable pavement materials was likely to be insufficient. These would have to be hauled from borrow areas some kilometres away. In marshy areas, due to the probability of lower strengths in wet embankment materials, the likelihood of increased thickness of base and sub-base courses would have to be considered. Hence the costs for pavement construction would be increased. As hilly/mountainous terrain is encountered, fines from river sources are likely to be less available and selective rock crushing may have to be used. Again the cost of operating and distributing from quarry/crusher sites may be expected to increase with increasing distance from base resources.

Culverts: More culverts are required per kilometre in wet coastal areas than in other Terrain Groups. This is for the purpose of hydrostatic equalisation in addition to catering for drainage flows. As the terrain becomes more dissected, culverts may be more directly related to drainage channels. In some areas the density of such dissection may determine the number of culvert sites more than rainfall runoff. Major culverts or medium bridges would be required in hilly/mountainous terrain to maintain the alignment at re-entrants in such terrain of high relief.

Bridges over wide rivers: These rivers occurred mainly in the coastal areas. Whilst site conditions would determine design detail for particular locations, the simple multi-span Bailey bridges on piled trestles were seen to have proved adequate for a number of years under similar conditions. A rate per linear metre including contingency and engineering was adopted for these bridges and the costs were shown separately in the estimates for each route.

Factors affecting percentage additions to construction costs

35. Contractors establishment percentage. The contractors establishment costs are those incurred in setting up base depots, offices, plant yards, and in obtaining and transporting vehicles and equipment necessary to do the work. The costs are directly affected by remoteness from supplies and difficulty of access to construction sites. A contractor also needs fairly flat well drained areas to establish plant yards and stock piles as well as availability of water supplies. Allowance was also made in the establishment costs for soils laboratories, as well as offices and services supplied to the supervising engineer. Such establishment costs were estimated as a percentage of construction costs, which for the trans-island road routes varied from 10% when operating in well drained undulating terrain to 25% in severe mountainous regions.

36. Engineering Percentage. The engineering percentage is for services customarily supplied by a Consultant. It usually includes the cost for survey work, sampling and testing required for site investigation, detailed designs, specifications and preparation of contract documents, as well as assistance to the Client with issuing invitations to tender and with the evaluation of tender replies. This percentage also includes supervision of the contractor during the construction stages, as well as further site investigation and materials testing as the work proceeds. The rising percentages as the terrain becomes progressively more difficult reflect the increasing engineering services anticipated.

37. Contingency percentages. Contingency percentages are those customarily added to cater for any item of expenditure which could not reasonably be foreseen. In a project of this nature, at the preliminary engineering feasibility stage, in which construction costs have been estimated for several alternative routes in difficult terrain, a minimum contingency of 15% rising to 20% in the mountainous areas was considered appropriate.

Tabulation of Construction Costs per kilometre by Terrain Group.

38. With the information now available, it was possible to construct the following Table 3. The costs are given in local currency (Kinas) at 1980 prices. The conversion rate is approximately K1.3 to £ Sterling

39. Determination of route distances in each Terrain Group. The length of route located in each Terrain Group was measured directly from the 1:500,000 scale terrain maps. This measured length was increased by the "route factor" for each terrain unit. The derivation of route factors is described in para. 30.

APPLICATION OF THE METHOD

40. It may be appreciated that a table such as Table 3, used in conjunction with a map from which route distances through the various terrain groups may be measured, enables a rapid first estimate of construction costs to be made. The table may be updated and refined with use over a period of time.

41. The engineering aspects of the earthworks and materials are primarily presented in Table 2. These were largely compiled from rapid field observations, para.(29), almost completely unsupported by any investigation data and therefore must be regarded as preliminary.

42. A plan of a typical part of a route in steep terrain is shown in Fig.1. It was obtained by plotting developed route distances between tie points, as explained in para.(22), and after applying the appropriate route factors.

Table 3. Construction Costs per Kilometre by Terrain Groups

Description	Coastal Dry A2 A3 B1	Coastal Wet C3 A1	Undulating S1 P1 P2 B2	Hilly S2 S3	Mountainous Moderate S4 S5	Mountainous Severe S6 S7
Clearing	1000	2500	10000	15000	15000	15000
Earthworks	8000	35000	30000	50000	90000	155000
Rock	1000	1000	5000	10000	20000	26000
Pavement	25000	37500	30000	45000	55000	65000
Culverts	13000	33000	20000	20000	30000	42000
Bridging	32000	36000	12500	25000	25000	30000
Sub-Total	80000	145000	107500	165000	235000	333000
Establishment	15%	20%	10%	12½%	20%	25%
Engineering	10%	12½%	12½%	18%	20%	20%
Contingency	15%	15%	15%	20%	20%	20%
Total:	116380	225112	152985	262845	406080	599400
Adopted:	116500	225000*	153000	263000	406000	600000

*Additional K60,000 (including establishment, engineering and contingency) to be added for perennial swamp (C3)

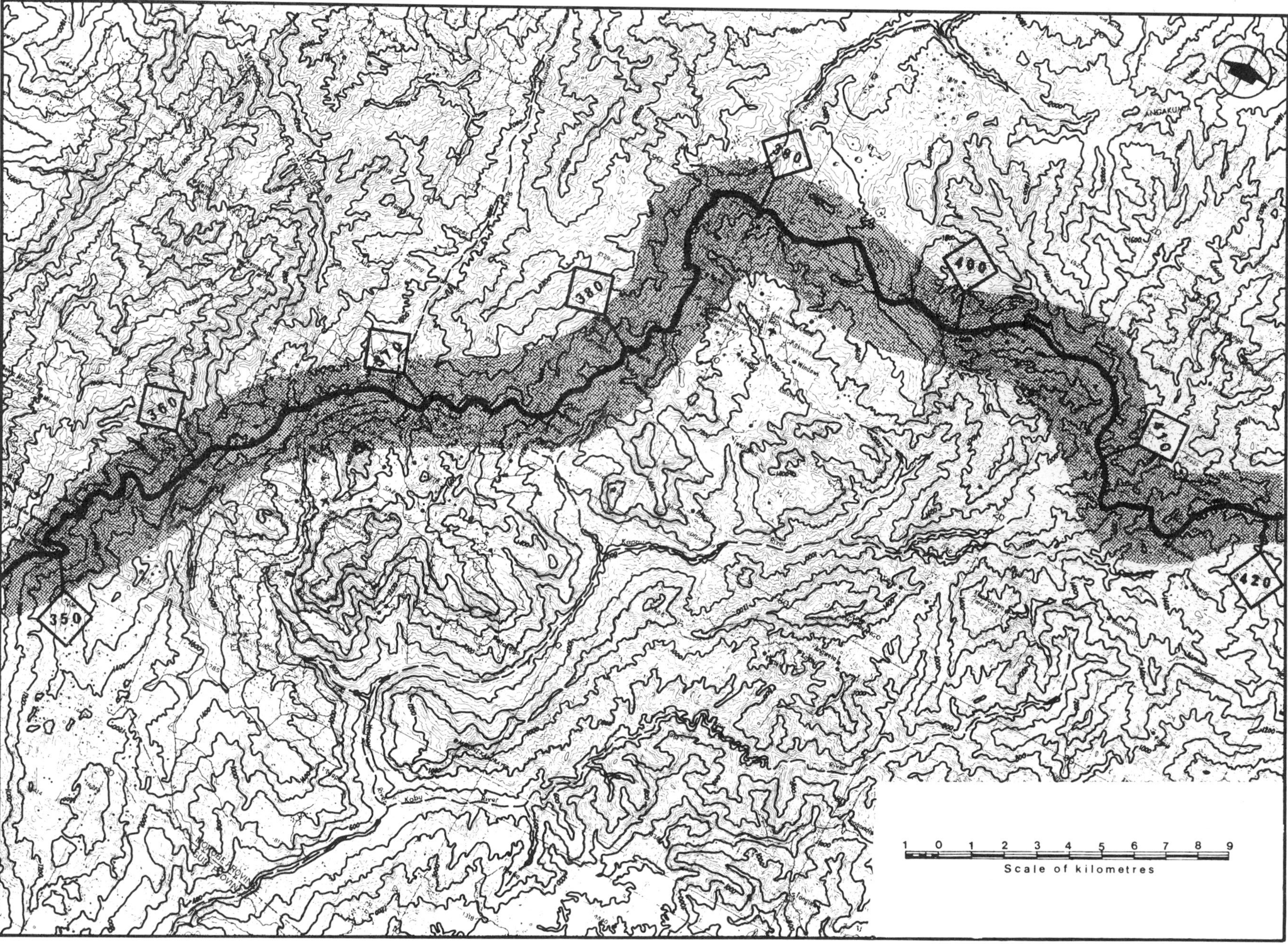

Fig. 1. Route corridor, km 348 to km 420

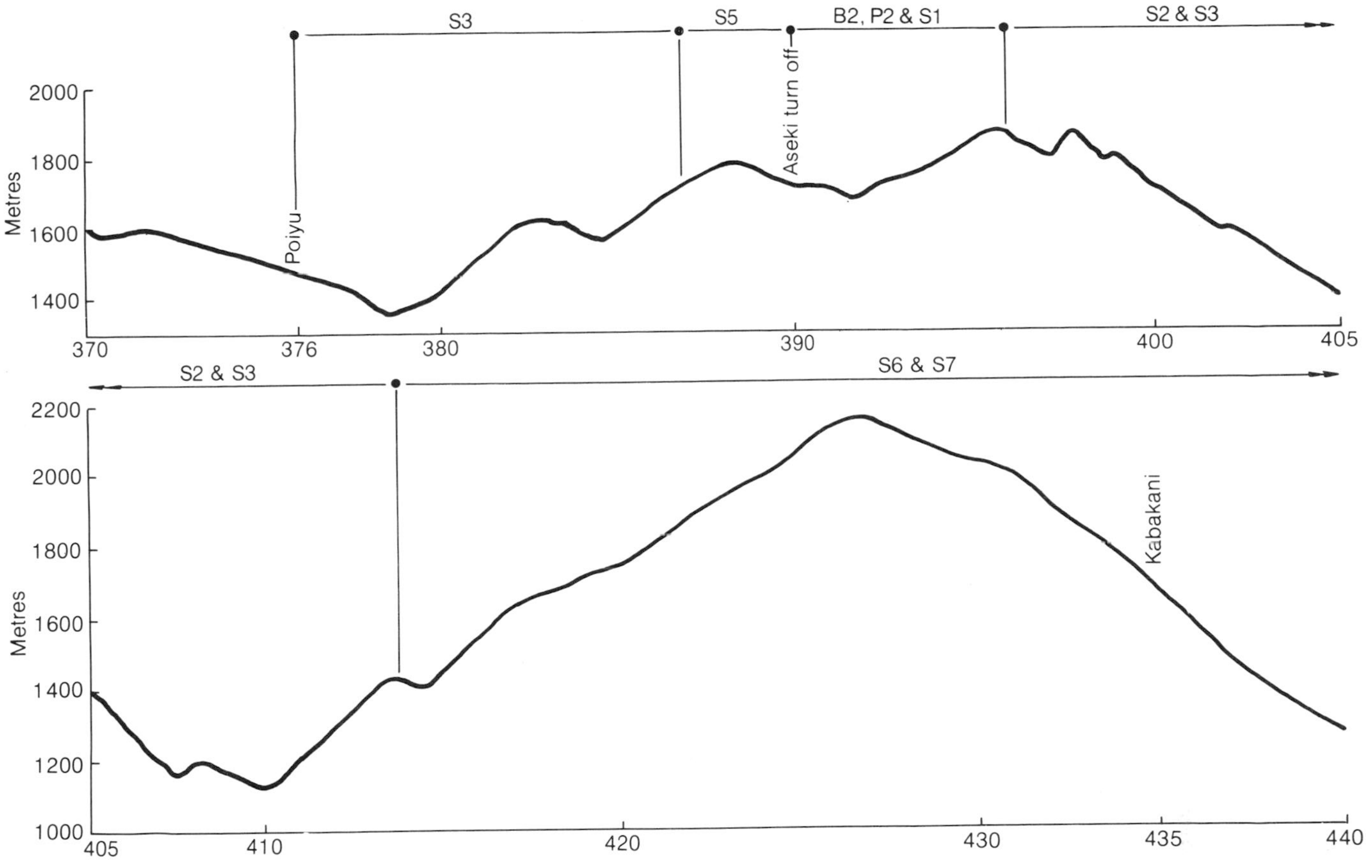

Fig. 2. Long profile of route, km 370 to km 440

43. A long profile covering part of this route is shown in Fig.2. This was obtained by conventional plotting from Fig.1. The Terrain Groups through which the route was located have been plotted on the profile after making the field reconnaissances, para.(27), and referring to the terrain map. Thus if required the construction costs for any part of a particular route may be estimated with a knowledge of the route length in the relevant Terrain Group and the cost per kilometre of road construction in that particular group.

44. The method may be further refined by supplementing the information obtained in the table of terrain unit engineering descriptions (Table 2) by information obtained from a more detailed helicopter fly-over reconnaissance with a greater number of landings in critical areas where possibl

EXAMPLE OF SUPPLEMENTARY ENGINEERING INFORMATION

45. The following paragraphs give an example of the additional engineering data which was derived from the field reconnaissance. The example given relates to the area covered by Figures 1 and 2.

46. The route section shown in Fig. 1 is generally one of moderately steep slopes and high rainfall. Therefore colluvial and residual soils may reach many metres thickness and borrow and slope stability may become major problems, particularly on the steeper slopes at higher altitudes. In addition, gravel bearing rivers may be long haul distances from the alignment, and limestone terrain with its own peculiar advantages and disadvantages occurs widely.

47. Slope failures on cut slopes in thick colluvial and residual soils will be very frequent unless slack slopes are used. However these will often not emerge from cut on steep mountain sides. Therefore structural support may be necessary or frequent benching or selection of a route avoiding those parent rocks which tend to have particularly low shear strength and a thick cover of colluvial and residual soils. There will probably not be enough rock generally available for extensive gabion works to be used and simple earth retaining methods should be considered in the design phase.

48. Particular care must be taken to prevent erosion by water, which is particularly common in the local soils and weathered rock. Ditch lining will undoubtedly save on maintenance and extensive remedial works attempting to keep the road open.

49. Pavement and concrete aggregates will be in short supply locally, especially at high altitudes. Special bedrock quarries may have to be opened if long hauls are to be avoided. Fills from any colluvial or residual soil will require careful control of lifts, and compaction at optimum moisture content for good performance characteristics. Granular soils and rock fill will be scarce except near large rivers and in talluvial areas.

EXAMPLE OF COST CALCULATION.

50. The estimated construction costs of new roads built to the adopted standards for a

specimen section of one of the trans-island routes is calculated as follows:

Section Chainage	Terrain Units	Length (kms)	Cost/km (Kina)	Construction Cost (Kina)
Km 376	S3	10.9	263,000	2,866,700
to	S5	3.0	406,000	1,218,000
Km 440	B2 P2 S1	5.9	153,000	902,700
	S2 S3	17.8	263,000	4,681,400
	S6 S7	26.4	600,000	15,840,000
				25,508,800

The average cost of constructing a new road to this standard over this length of the route is therefore Kina 398,575 per km. This will not necessarily bear much relation to costs in other countries, where factors affecting construction costs would probably be quite different.

DISCUSSION OF THE METHOD

51. Other applications. Clearly the method of cost calculation is capable of extension to other applications, such as road upgrading or maintenance. In the case of the trans-island routes it was found possible to incorporate certain lengths of existing tracks, after upgrading, into the proposed trans-island routes. This achieved cost savings which were taken into account when compared with new and possibly shorter routes. It was also of considerable interest to compare maintenance costs for the routes. In certain Terrain Groups these were significantly higher than in others, and in some cases might have reduced savings made by locating a shorter route through such Terrain Groups.

52. This method also proved to be very adaptable to a sensitivity analysis. A range of construction costs in each Terrain Group was established about the estimates given as Table 3. The effect of these variations of the cost in each Terrain Group was then tested against the ranking of total costs for each of the nine routes to prove that the most economical route had been established.

CONCLUSIONS

53. The method of cost estimation described is quick, flexible and capable of refinement according to the number of land forms into which it is possible to separate the terrain.

54. The establishment of the method relies heavily upon the use of an experienced Engineering Geologist to make assessments of the engineering properties of soils and rocks based on largely visual inspections.

55. Once established for a particular area, the method enables a rapid preliminary estimation of costs for new roads to be made in terrain where detailed reconnaissance on the ground would be exceptionally difficult and time consuming.

ACKNOWLEDGEMENTS
The study was carried out by Rendel & Partners (a subsidiary of Rendel Palmer & Tritton, London) for the Department of Transport and Civil Aviation, Papua New Guinea.

The Author wishes to thank the members of the Study Team, and in particular Dr. P.G. Fookes (Consultant Engineering Geologist), for their part in the original study. Thanks also to Mr. D.V. Hattrell and R.M. Stephens (both of Rendel Palmer & Tritton) and Dr. P.G. Fookes for their assistance in editing this paper.

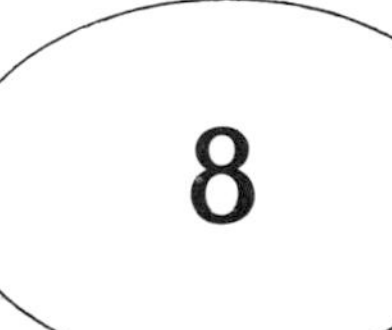

A new approach to setting highway design standards

N. A. PARKER, BE, ME, PhD, PE, AMASCE, MIET, Professor of Civil Engineering, University of Dar es Salaam, Tanzania

A procedure is outlined for determining appropriate highway design standards for final alignment, during the route location stage. An iterative two-step process combining linear programming and shortest path techniques is used to search out a near-infinity of general locations with different combinations of horizontal and vertical alignments and surface type, between the termini of a proposed route. The objective is to choose the location(s) with the greatest likelihood of containing the final alignment that minimises the total cost of construction, maintenance and vehicle operation. Inputs to the linear programming step are a regularised digitized broad-band terrain model between the route termini, various maximum gradients, and control elevations and gradients at the route termini. The output from this step is a model of likely cuts and fills which, together with the terrain model, constitute a basic model for the estimation of the total cost components of any horizontal alignment. The estimation of these costs is done in the shortest path step in which all the horizontal alignments are evaluated and the best one chosen as a function of the expected traffic over the design life of the road, and of the surface type. The consequence of this approach is that each highway project can be viewed as a unique design problem with no preconception of the design standards as read from a design chart or table, and the design standards that are output from the route location process can reflect the physical and economic environment of the road project itself. An important feature of the process is that recommended gradients are piecewise along the proposed general alignment.

INTRODUCTION

1. Highway design standards are conventionally predetermined for a given project from prior analysis and experience. Not surprisingly then, these design standards differ from one country to the other, ostensibly because the conditions of environment, technology and human behaviour differ from one country to the other. Admittedly, design standards might better be interpreted as design guidelines or recommendations, but their very existence tends to lend them a stamp of legal specification and defensibility, especially when transferred from a developed to a developing country which has not yet evolved its own standards, based analytically upon its own conditions.

2. The non-existence of a rationally based set of standards in developing countries offers, however, a unique opportunity for the development of new approaches to highway design. By capitalising on recent advances in the comprehensive analysis of construction, maintenance and vehicle operating costs (ref. 1,2,3,4,5), and on location modeling (ref.6,7), new methodologies could be developed for specifying both location and design guidelines unique to a highway project.

3. The objective of this paper is to outline, for discussion, a new approach to setting rural highway design standards as a by-product of the route location exercise, on a project by project basis. While the methodology is not yet fully developed and a great deal remains to be done with respect to field testing of the various submodels, the author is convinced nevertheless that there is now sufficient research experience and results to permit the eventual simultaneous selection of general horizontal and vertical alignments as a function of surface type, in such a way as to maximise the likehood that final alignments will minimise the total costs of construction, maintenance and vehicle operation.

ROUTE LOCATION DESIGN

4. The route location engineer bears a tremendous responsibility for defining the general location for a highway link between two points. While there is still the overiding tendency to overemphasise the minimisation of construction costs -- understandably so in developing countries, especially, where in most cases these costs are externally financed -- vehicle operating and maintenance costs over the economic life of the facility may well relegate construction costs to insignificance where traffic is expected to be heavy. These life cycle costs -- especially vehicle operation costs -- are particularly burdensome for non-oil producing, non-industrialised countries struggling with escalating balance of payments problems. It is no longer sufficient, therefore, for the location engineer to approach the problem primarily from consideration of grade-controlled and non-grade controlled alignments, but also from the point of view of total life cycle costs. Thus, the recommended location must specify the optimum combination of general horizontal alignments, profiles and surface types, which will most likely minimise total life cycle costs for the traffic mix and intensity expected. Since the conditions of environment, technology, economy and human

Highway investment in developing countries. Thomas Telford Ltd, London, 1983, 51–55

(driver) behaviour differ from project to project, it follows that with this approach the optimum design combinations would differ accordingly.

Elemental cost estimates

5. The Road Transport Investment Model, RTIM (ref.4), considers the following cost components in determining total road transport costs: for construction costs, site clearance, drainage, earthworks and pavements; for road maintenance costs, grading and regravelling for unpaved roads, patching, surface dressing and overlaying for paved roads, and drainage and overheads for all roads; and for user costs, including vehicle operation, time, fuel and oil, replacement parts, tyres, depreciation and interest. Some or all of these components may be included in the route location stage in an explicitly generalised form.

6. A basic cost model. A number of the above-mentioned components are functions of gradient, e.g. earthwork, grading and regravelling of unpaved roads, and fuel consumption. Some rational basis for the estimation of the likely profiles of all possible alignments between two route location end points, prior to the actual choice of the general alignment, is therefore essential. A model under development by the author (ref.8) seems to provide one such basis.

7. The model consists of a smoothed "construction surface", computed through the terrain between the end points, over which a specified design gradient is not exceeded in any direction, and which is constrained at the end points by their design elevations and gradients. The difference in elevation of the construction surface from the terrain at any point on the surface is a measure of the likely cut or fill at that point. Thus this model provides a basis for estimating earthwork costs, as well as piecewise gradients along any alignment.

8. The surface function is assumed to be a polynomial which minimises the total sum of the differences in elevation between the construction surface and the natural terrain. This optimisation is done subject to the constraints mentioned above, the inputs being a digitized terrain model between the end points. The basic model concept is illustrated in Fig. 1.

Route selection

9. Typically route selection models employ dynamic programming (ref. 7, 9), or shortest path (ref. 10, 11, 12) techniques, to choose one or more best paths from a synthetic network of links and nodes (see Fig. 2) constructed between the end points. The shortest path approach is preferred by the author because it permits the inclusion of cut-back solutions.

10. An N-best path model. A shortest path algorithm for selecting the best and next best routes from a grid network, such as in Fig.2, is being developed by the author (ref.13). Inputs to this first version of the model are the basic cost model in digitized form, traffic estimates, including vehicle mix, over the life of the road, and unit cost estimates for construction, maintenance and vehicle operation. Construction cost estimates include all the RTIM (ref.4) components, plus land compensation costs and bridge costs. Maintenance costs also include all the RTIM components, while vehicle operating costs only include the fuel and oil components. RTIM relationships are used for estimating the fuel and oil costs. All cost components are on option by the user.

11. The model is a modified version -- for a grid network -- of the Road Research Laboratory (RRL) algorithm (ref. 14), in which the cost of each link is computed by the model itself at the time at which it is considered. The next best path embellishment is based mainly on the concept of path deviations (ref.15), in which new paths are constructed from a portion of the current next best path, starting at the destination, a deviation link to a non-best path adjoining node, and the shortest path from the adjoining node back to the origin. A typical output from the model would include a coordinate trace of the best general path or paths accompanied by a graphic representation, summary of the construction, maintenance and fuel oil cost estimates, plus total costs, and a linkwise recommendation on gradients (see Fig.3).

Optimum route determination

12. The typical output in Fig.3 is for a given surface type with a given maximum gradient over the construction surface. The route location engineer is, however, interested first, in seeking out the optimum maximum gradient for a given surface type, and second, in finding the optimum combination of surface type and maximum gradient. The user must therefore iterate between basic cost model computations and best path model selections.

13. A route location procedure. A computer-aided procedure for providing the basis for selection of the optimum route is diagrammed in Fig. 4. Basically it prescribes that for each surface type, the best path, or paths, be determined for a range of gradients spanning the gradient that produces the minimum total costs. The basic cost models being gradient dependent, rather than surface type dependent, might be computed for the expected range of gradients, prior to deployment of the best path model. In any event, the result of the exercise is expected to be a summary of costs of the best locations by surface type, and therefore a rational basis for decision-making.

AN APPRAISAL OF THE METHODOLOGY

14. It is instructive to review some basic concepts of highway engineering that impinge on route location design. For example, horizontal and vertical alignments are not independent of each other, and should therefore be chosen simultaneously. Construction costs tend to decrease with increasing gradient in a given terrain, ceteris paribus, while vehicle operating costs tend to increase. With maintenance costs being generally weakly dependent on gradient,

7	1940	2200	2320	2150	2020	2100	2180	2240	2160 D
6	1900	2170	2160	2130	1940	1960	2020	2140	2000
5	1860	2160	2000	2040	1900	1950	1980	1990	1960
4	1880	2080	2020	1980	1880	1970	1980	1890	2000
3	1920	2060	2020	1960	1880	1920	1920	1860	1980
2	1860	2000	2060	2000	2000	1860	1970	1860	1980
1	1800 O	2070	2190	1990	1900	1840	1860	1840	1980

a. Terrain Surface

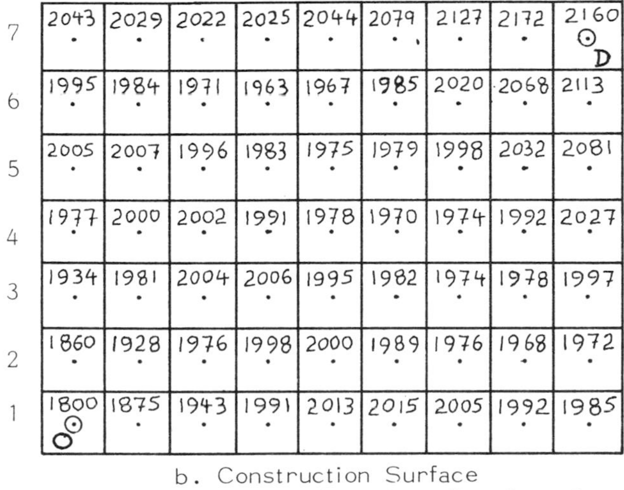

7	2043	2029	2022	2025	2044	2079	2127	2172	2160 D
6	1995	1984	1971	1963	1967	1985	2020	2068	2113
5	2005	2007	1996	1983	1975	1979	1998	2032	2081
4	1977	2000	2002	1991	1978	1970	1974	1992	2027
3	1934	1981	2004	2006	1995	1982	1974	1978	1997
2	1860	1928	1976	1998	2000	1989	1976	1968	1972
1	1800 O	1875	1943	1991	2013	2015	2005	1992	1985
	1	2	3	4	5	6	7	8	9

b. Construction Surface

Fig. 1. Illustration of the basic model concept

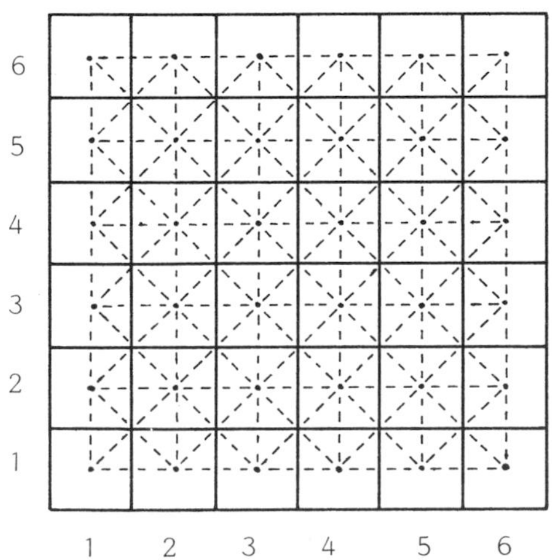

Fig. 2. Illustration of a route location network

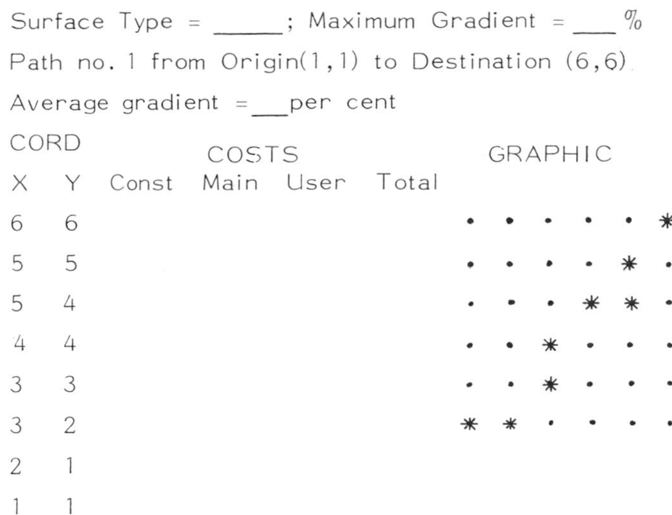

Fig. 3. Format for N-best path model output

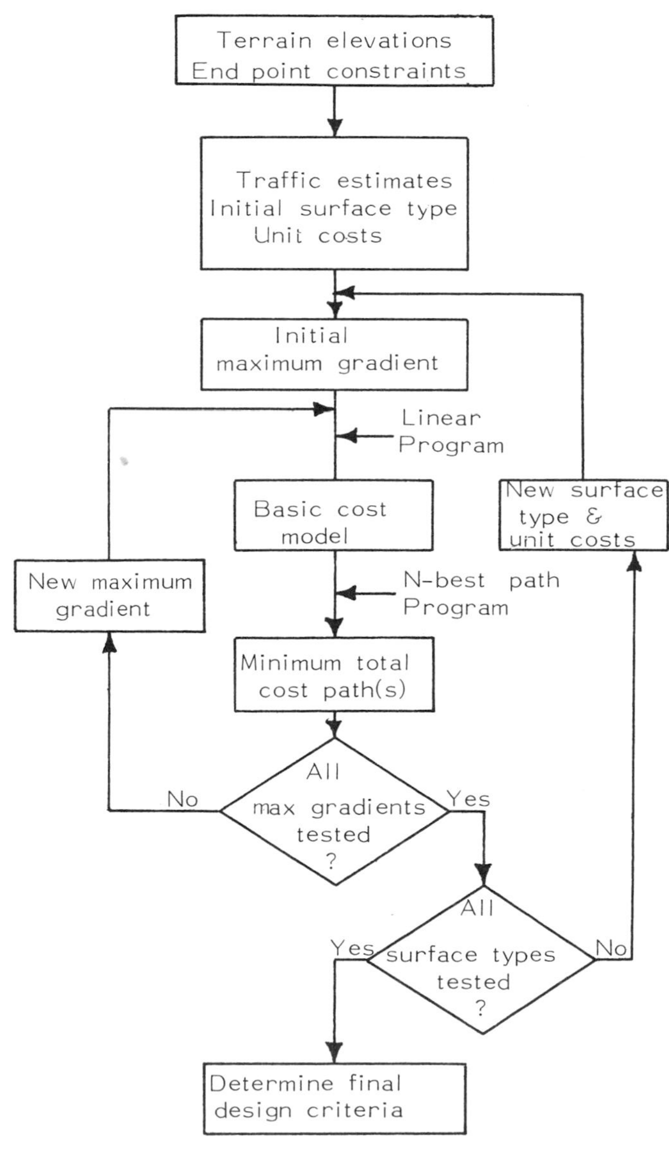

Fig. 4. Proposed computer-aided route location design framework

except for earth and gravel roads, one would expect to find a gradient that minimises total unit costs. The net effect on the route location is difficult to predict, with each case presenting a unique compromise between total length of route, horizontal curvature and combination of gradients.

15. The type of surface also affects fuel consumption, as well as oil consumption, tyre wear and spare parts replacement. However, the advantage of higher realisable speeds on high type surfaces are somewhat offset by increased fuel consumption. In developing countries with inadequate resources to police speed limits effectively, and a high percentage of foreign exchange going for the purchase of fuel, the net effect on route location may be to introduce additional horizontal curvature, to bring the effective design speed down to the level of a fuel efficient speed limit. This would further lengthen the route and offset the advantage of reduced fuel consumption. Again the net effect of all these interactions is difficult to predict, but it would seem obvious that the optimum location between the same two points in the same terrain would change for different surface types.

16. The methodology makes no apriori determination of surface type, gradient and horizontal curvature for a given project. While it is appreciated that existing design standards, regardless of origin do appear to seek an appropriate compromise between these three design parameters on the basis of unit total costs (ref.16), it is also clear that the generality of these standards, and indeed the very existence of them, mitigates against the setting of design criteria uniquely appropriate to each highway project. By the methodology described herein, these three critical design parameters can be determined at the route location stage as a result of a comprehensive trade-off analysis based on total life cycle costs.

Computational limitations

17. The infinity of alternative locations between the end points is approximated by a large finite subset of links and nodes, as shown previously in Fig. 2. The density of this network is limited by the computing facilities available to carry out the linear programming steps, but a 6 x 6 grid is the recommended minimum size for any meaningful analysis. However, because the output from the linear programme is in actuality a polynomial surface function, the construction surface elevations of the infinity of points could be estimated. Thus, the profile of any horizontal alignment could be approximated, and subsequent evaluations of construction, maintenance and vehicle operation costs made. Ideally, the distance between grid points should approximate the desired corridor width used in the detailed design of the alignment.

18. Most modern computer systems make linear programming packages available to the users, in which case it would be necessary to write a data preparation programme to fit the requirements of the package. The programme developed by the author (ref.8) is self-contained and system independent for ready deployment.

Model validation

19. The validity of the methodology in this paper rests heavily on the validity of the basic cost model. The underlying assumption is that evaluation of total costs is an extension of the evaluation of earthwork costs, which, in turn, require an ability to define feasible profiles for all possible alignments, in order to rationalise the earthwork estimation process.

20. One objection that can be raised is the fact that the basic cost model does not discriminate between the directions of passage of alignments through a point, in estimating the likely cuts or fills. While this is an understandable objection from the perspective of final design, it is the author's opinion that it would suffice for route location design as long as it could be shown that the resultant location, chosen solely on the basis of earthwork costs, is consistent with engineering judgement. However, while the limited experience does seem to suggest that the model is valid in the terms described above (ref.6), an extensive comprehensive model validation programme is currently underway.

CONCLUSION

21. The outlines of a methodology for the simultaneous selection of alignment, gradient and surface type, based on minimization of total life-cycle costs, have been presented. It is advanced that this approach is both feasible and appropriate for all countries, but especially for developing countries which are most sensitive to the financial and economic consequences of providing and maintaining their road-based transport infrastructure. With vehicle operating costs and maintenance costs relegating construction costs to comparative insignificance, even at relatively low traffic volumes -- with a predominance of commercial vehicles and a tendency to overload -- developing countries could no longer afford to ignore the explicit interrelationships of location and design, and must approach each highway project as a unique problem, with unique solutions, right from the route location stage.

REFERENCES

1. ABAYNAYAKA S.N., HIDE M., MOROSIUK G. and ROBINSON R. Tables for estimating vehicle operating costs on rural roads in developing countries. Department of the Environment, TRRL Report LR 723, Transport and Road Research Laboratory, Crowthorne, 1976.

2. HIDE H., ABAYNAYAKA S.W., SAYER I. and WYATT R.J. The Kenya road transport cost study: research on vehicle operating costs. Department of the Environment, TRRL Report LR 672, Transport and Road Research Laboratory, Crowthorne, 1975.

3. OSTLER P., BAILEY A.C. and STOTT J.P. Estimating highway construction costs: program COSMOS. Department of the Environment, TRRL Report LR 863, Transport and Road Research Laboratory, Crowthorne, 1979.

4. ROBINSON R., HIDE H., HODGES J.W., ROLT J. and ABAYNAYAKA S.W. A road transport investment model for developing countries. Department of the Environment, TRRL Report LR 674, Transport and Road Research Laboratory, Crowthorne, 1975.

5. WORLD BANK. Highway design and maintenance standards model -- HDM. International Bank for Reconstruction and Development, Washington D.C., 1979

6. PARKER N.A. Rural highway route corridor selection. Transport Planning and Technology, 1977, 3, 247-256.

7. STOTT J.P., CALOGERO V., HICKMAN D. and ROUMELIOTIS P. Broad band route selection-program BRUTUS: test on a section of motorway. Department of the Environment, TRRL Report (in preparation), Transport and Road Research Laboratory, Crowthorne.

8. PARKER N.A. A linear programming basic cost model for rural road location. Department of Civil Engineering, Research Project Report (in preparation), University of Dar es Salaam, Dar es Salaam.

9. OBRIEN W.T. and BENNETT D.W. A dynamic programming approach to optimal route location. Paper presented to a Symposium of Cost Models and Optimisation in Road Location, Design and Construction, PTRC, London, 1969.

10. GLADDING D.F. Automatic selection of horizontal alignments for highway location. Unpublished M.Sc. thesis, Massachusetts Institute of Technology, Boston, 1964.

11. TURNER A.K.F. Computer-assisted procedures to generate and evaluate regional highway alternatives. Joint Highway Research Project No. 31, Purdue University, Lafayette, 1968.

12. PARKER N.A. A systems analysis of route location. Unpublished Ph.D. thesis, Cornell University, Ithaca, 1971.

13. PARKER N.A. and KASEKO M.S. An N-best path algorithm for rural road location design. Department of Civil Engineering, Research Project Report (in preparation), University of Dar es Salaam, Dar es Salaam.

14. WHITING P.D. and HILLIER J.A. A method for finding a shortest route through a road network. Operational Research Quarterly, 1960, 11(1/2).

15. HOFFMAN W. and PAVLEY R. A method for the solution of the Nth best path problem. Journal of the Association for Computing Machinery, 1959, 6(4), 506-514.

16. PARKER N.A. Towards appropriate design criteria for roads in developing countries. Proceedings of the IX·th IRF World Meeting, 1981, Roads into the Future· GS, 51-69.

Discussion on Papers 5–8

Mr T.J. POWELL *(Introducing Paper 6 on behalf of Mr I. Thomson)*: It is significant that six out of seven countries in the Southern Cone of South America have carried out national transport studies - some of them more than once. There are various reasons for this, including an element of competitive cultural emulation, but one reason is the need to plan for further development of the highway sector. Many countries have built up a substantial road construction capability over the past 20-30 years to cater for the growing efficiency and reliability of the motor vehicle, which is tending to replace the older rail- and water-based transport as the prime mode in many countries. However, they are now finding that the more obvious improvements have been completed and there is a danger of diminishing returns from further investment.

One of the main advantages of national transport plans is the fact that they are comprehensive, covering all the country and all the competing transport modes. Often their most useful contribution is systematic collection of information on the state of the current infrastructure and also on current origin and destination movements. However, I would caution against making these studies unduly complex. Unnecessarily complex and sophisticated mathematical modelling processes are often unhelpful to a true understanding of the transport problems in a particular country.

Every country should carry out a national transport plan about once every 10 years, and it is significant to note than many countries in South America are now engaged on their second such plan.

Professor N.A. PARKER *(Introduction to Paper 8)*: References 8 and 13 of Paper 8 are intended to be first edition manuals detailing the logic, use and interpretation of results of the computer program packages for derivation of basic cost models (ref. 8) and selection of routes (ref. 13). The current situation is that substantial progress has been made on reference 13, while no progress has been made on reference 8, due to the lack of sufficient computer capacity, caused by a delay in the purchase and delivery of the expected system. The program logic on which the preparation of reference 8 is based had been designed and tested at the Transport and Road Research Laboratory (TRRL), and it is at the TRRL that further development would be pursued.

Regarding reference 13, the program logic has already been well documented. We have incorporated the facility to evaluate and select route location designs as a function of construction, maintenance and fuel costs, as well as traffic volume and mix, discount rates, and design life, in any desired combination. The program is written in Fortran IV, and is currently being run on a MINC 11 minicomputer in the Faculty of Engineering at the University of Dar es Salaam. The results obtained so far appear to be consistent with expected location sensitivities, and sample print-outs are available.

Dr J.B. METCALF *(Australian Road Research Board)*: The topic of national and provincial roads brings together the problems of criteria to be applied to social and political issues such as development of remote areas and provision for national security and the problems of geometric design standards. We cannot leave issues out of consideration because they are difficult to quantify; we cannot have an investment model with a box labelled 'policy issues' without examining and at least attempting to quantify them. I am attracted by the approach of Professor Hills and Professor Jones-Lee (Paper 23) to the use of accident costs, with different applications of costs being linked to different and appropriate costing bases. Can we not work towards a matrix of investment criteria linked to the scale and objectives of investment and in turn linked to data and methodology needs?

Turning to the question of geometric standards, current practice in Australia is turning away from fixed standards in favour of an investigation of the effects of different geometric standards on traffic performance. This is aided by the use of a simulated program (TRARR) which estimates the hard-time consequences of various road alignments. With a small amount of local data, TRARR can be used for most types of roads and traffic (ref. 1).

Mr J.J. GANDY *(Scott Wilson Kirkpatrick & Partners)*: Messrs Grieveson and Winpenny (Paper 5) refer in paragraph 5 to factors on the side of both donor and recipient which have favoured the construction of new roads rather than the rehabilitation and maintenance of the existing network. Although these factors are not listed, one must assume that failure by governments to provide adequate maintenance funds and the damaging effect of high axle loads on inadequate pavements would be important in this respect. More than one Director of Roads of a developing country has said that there is no point in wasting maintenance funds on a heavily trafficked road which, in any case, needs a new pavement. It is better to let it deteriorate to the point where a new road can be built on a better alignment with a pavement adequate for the traffic. External finance is more readily obtainable for new work and scarce maintenance funds can be better employed elsewhere.

In this situation it is unlikely that the main finance input will be 'adequate' in the sense used in paragraph 8 of the Paper and 'optimal' maintenance is rarely, if ever, found. The Authors suggest that the benefits from the investment would be very different if inadequate or nil maintenance is assumed. This is indeed the case and in recent years at Scott Wilson Kirkpatrick and Partners we have carefully assessed the existing maintenance record when performing economic appraisals.

With regard to the trade-off between extra-construction cost against maintenance costs, it would appear to be the case that, in general, the benefits accruing from better maintenance are much greater than those arising from more expensive construction. However, there are situations in which the reverse might be the case. Our recent work in Libya has been characterized by a determined policy of decreasing the maintenance effort through improved construction. This is government policy and is justified by the enormous distance between inhabited places, the general nature of the terrain and the climate. The cost of suppertime maintenance teams under those conditions is extremely high.

As far as 'stage construction' is concerned (paragraph 10), whilst we would agree generally, we would point out that it is not always easy to predict the life of a road at any construction stage when the traffic is heavy and the axle loads high. We know of one such road where the number of standard axles exceeds 10 million per annum and is likely to increase. Under such circumstances, if the various stages are not implemented at the right time, the road will fail even if the maintenance is first-class.

Mr I.G. HEGGIE *(Freelance Consultant, Oxford)*: In Paper 5 Messrs Grieveson and Winpenny make three observations: there is a temptation to favour the construction of new roads rather than the rehabilitation and maintenance of the existing network; road evaluations tend to assume optimal maintenance rather than likely maintenance, and consequently overestimate the benefits from rehabilitation and improvement; the evaluation of developmental roads is particularly difficult, although the method proposed by the World Bank (ref. 2 of Paper 5) now offers a satisfactory way of dealing with induced agricultural production.

The substance of their observations is that the balance of expenditure between constructing new roads, rehabilitating and improving existing ones, and maintaining the existing network does not correspond to the priorities implied by now normal methods of economic analysis. These methods invariably suggest more spending on maintenance and less on new construction and improvement. But, and this is the real issue, are the local officials in these countries shortsighted and foolish or are our methods of evaluation? I think there is probably a bit of both, but I am increasingly convinced that our methods of evaluation are biased.

The Authors of Paper 5 observe that if we use likely, rather than optimal, maintenance, we reduce the priority of rehabilitating and improving existing roads. Other adjustments further emphasize such reductions. For example, road improvements almost invariably lead to redundancy among drivers and mechanics and a shadow image is usually appropriate. This will again reduce priorities. So there are good reasons to suppose that the benefits of rehabilitation and improvement are often seriously overestimated.

But what about new roads? Messrs Grieveson and Winpenny think that we now have a satisfactory method of evaluation: I disagree. We can indeed now measure net value added in a consistent way, but what about transport cost-savings? Without roads, goods are carried on heads, backs, animals and canoes. What is the saving when these are carried by vehicle? Resource costs naturally go up, and even with shadow prices, it is extremely difficult to put a realistic value on the change in transport costs. And then there is the value of access *per se*: access to schools, hospitals and other social facilities. They must have some value and yet we exclude them. Paper 15 by Messrs Hine and Riverson, has a telling conclusion in paragraph 14: their analysis suggests that the advantages of access far outweigh the quality of the road surface. I agree and my conclusion is that our methods of analysis still underestimate the benefits of developmental roads.

My overall conclusion is thus that local administrators - who in my experience are a pretty canny lot - are at least partly responding to the serious desires of their constituents. People prefer new roads to improved maintenance and new roads to rehabilitation and improvement. It is our methods of evaluation that perhaps need revision.

Mr R.L. MITCHELL *(University of Zimbabwe)*: I support the philosophy of Professor Parker (Paper 8) - save in one significant aspect. He suggests that in developing countries there are inadequate resources to police axle loads. Without design loads, the engineer can design nothing - why should roads be different? In Zimbabwe, as in South Africa, we have learned that we cannot afford <u>not</u> to police axle loads - and reference 4 of Paper 12 has proved the economic necessity to restrain axle loadings, at least on low-count rural roads. Surely it is the political, not economic, restraint which has forced Professor Parker to an otherwise indefensible position?

Similarly, he states that we should reduce design speeds as we cannot enforce speed control. Steep grades and small curvature increase resistance and fuel consumption, not to mention accidents. High design speeds, with enforced lower legal speeds, are the correct solution. I made a ten-year study in Zimbabwe which showed that 100 km/h was the economic speed as a compromise between energy and journey time.

In road matters, law enforcement should be subservient to the engineer, not vice versa.

Mr C. IRWIN-CHILDS *(Taylor Woodrow International)*: I applaud the advice given by Mr Grieveson (Paper 5) that attention should be given to how a new road is to be constructed at the same time as the feasibility of the road is assessed. Construction costs are a primary factor in assessing project viability. Design, programme and construction techniques are significant cost factors comparable with materials, manpower and equipment resource costs. A much more satisfactory conclusion is likely to be achieved by integrating all relevant factors before formulating a course of action.

Mr Grieveson has advised increasing flexibility in contract procedures, which is encouraging in view of his reported experience using direct labour work forces. I suggest that consideration be given to using a management contractor to maximize local participation and bring appropriate technical assistance and construction resources to bear. The advantages of associating with a management contractor include possibilities of participation at the feasibility stage, integration of design and construction, reducing delay due to unforeseen circumstances, labour training and supervision of maintenance.

It it commendable that the EEC Commission has ensured that aid is only given to organizations who have made financial provision for maintenance. Maintenance requires proper finance, resources and motivation. Sympathetic management and careful direction of the maintenance team is required to compensate for the loss of prestige and job satisfaction which the men involved in the work generally suffer as a reflection of the inconvenience which they cause to the public, use of public funds and discontinuous nature of the work.

All materials are subject to deterioration in certain conditions and once disruption has occurred it will generally spread rapidly and cause user-inconvenience as well as increasing maintenance cost. Effective maintenance is a rigorous procedure dependent on continuous inspection and is a completely different type of activity to construction although it uses similar resources.

One should expect increasing involvement in maintenance and repair activities as road systems expand throughout the world and existing systems deteriorate. Increasing awareness of costs and physical characteristics of materials have established a trend to more economic designs which implies increasing maintenance and repair activities.

Mr J.R. McLEAN *(Australian Road Research Board)*: Professor Parker (Paper 8) has proposed a procedure whereby appropriate design standards can be specified as an output from the route-location exercise rather than the conventional adoption of rigid standards. Australian road designers have been aware of problems, both engineering and economic, that can arise through the routine application of arbitrary standards, and in 1975 the National Association of Australian State Road Authorities (NAASRA) established a working group to review the technical and economic bases of geometric standards. This review was supported by research carried out at the Australian Road Research Board (ARRB). As Australia already has an established road network, most non-urban road design is associated with reconstruction of route segments. Within this context, the working group considered that the geometric design selected for individual projects should be consistent with the requirements of the regional network and budgetary constraints. That is, the design standards appropriate in one region are not necessarily appropriate in another, and a national design policy should allow scope for inter-regional variations. This work concluded with the publication of the Interim guide to the geometric design of rural roads (ref. 2) which emphasizes design guidelines and criteria rather than rigid standards.

A number of specific items which arose from the NAASRA review and associated ARRB research are as follows.

(a) The conventional design speed approach to specifying alignment design standards carries implicit assumptions regarding driver behaviour which have not been substantiated by empirical research (ref. 3). An alternative design procedure has been developed which ensures compatibility between alignment standard and observed driver-speed behaviour (ref. 4). Emphasis is placed on consistency and driver expectancy rather than absolute minimum standards, and this should result in safer operations, particularly for low-standard alignments.
(b) For roads in hilly terrain, the optimum balance between horizontal and vertical alignment standards is strongly dependent on traffic mix. In general, trucks gain most from higher standard vertical alignment whereas cars gain most from higher standard horizontal alignment (ref. 5).
(c) Intermittent overtaking lanes provide a cost-effective method for improving traffic flow on two-lane roads (ref. 1). In difficult terrain they are likely to be more effective than realignment to improve sight distances.

Finally, I endorse the item raised briefly in paragraph 20 of Paper 8 regarding the need for engineering judgement, particularly for final design. Design decisions must often be based on factors other than vehicle operating costs and earthworks (e.g., construction methods, drainage requirements, access), and these cannot readily be incorporated in design optimization programmes. Australian experience suggests, while route optimization and design packages can be of considerable value for the evaluation and refinement of design alternatives, they should not be regarded as a replacement for design expertise.

Mr MITCHELL: Messrs Grieveson and Winpenny (Paper 5) have produced an excellent summary of the road problems of developing countries. They refer to the maintenance chicken coming home to roost: a colleague swears he saw an ostrich nesting in a pot-hole in Zambia! 'Some chicken, some neck!' In Rhodesia, the tradition was to fire any supervisor who failed to properly repair a pot-hole within 48 hours - now some remain for as many months. This is partly due to the lack of experienced maintenance staff, whose salaries do not keep pace with inflation, and whose experience is never rewarded, and partly due to delayed reseals, necessitated by the shortage of foreign currency for bitumen. Such currency, in Zimbabwe, is currently available on international funding for new construction, but not for maintenance, which is, as it should be, traditionally funded from local revenue. Shortage of skill and experience (which cannot be replaced entirely by training courses) and of foreign currency will surely remain in developing countries at least for this century.

The Authors postulate that, as overdesign is cheap, it might be a good insurance against poor maintenance. Design method MPD1 (ref 6) gives the cheapest designs of any known method, and was proved in Rhodesia over 30 years - but on conditions of enforced axle-load control and of perfect maintenance. However, although I have doubts about the validity of MPD1 in the absence of maintenance, all designs of bituminous pavements will fail under such conditions. If reseals are delayed, water will cause base failure irrespective of thickness.

As the Authors hint, the only relatively maintenance-free pavements are concrete which call for less skill than flexible pavements. In Zimbabwe (assuming good maintenance on flexible pavements) it has been shown (ref. 7) that they only compete with flexible on financial grounds where premix surfaces are the alternative (traffic exceeding some 5 million E80 axles). However, in economic as opposed to financial terms, where the shadow price must be considered, it is likely that they could be economic for lower traffic when future reseal costs are included in the analysis.

In supporting the Authors' thoughts on the future of rail, particularly when locally produced electric power is available and oil has to be imported, I would like to see minimum design speeds of 100-120 km/h, not 80 as postulated, for grades and curvature considerably increase energy demands; as railway engineers well know, high design speeds do not necessarily demand raising legal speeds.

Mr P.W.D.H. ROBERTS *(Overseas Unit, TRRL)*: It is not merely a convenient exaggeration to refer to a total lack of maintenance. Nor is it restricted to new road projects. It is the reality for the majority of existing roads in developing countries. Under these circumstances there are very strong reasons for not investing in the construction of new roads. Without adequate maintenance a new road will have a much-reduced life and therefore the anticipated benefits will not be realized in full. A new road will increasingly stretch meagre maintenance capacity with consequently more rapid deterioration of existing roads and higher user costs on these roads. Many of the resources required for construction could bring much higher returns by being employed in maintaining existing infrastructure. (Indeed, with the benefit of hindsight, it can be seen that, in several countries, established maintenance organizations were largely undermined by the great concentration on new construction over the last two decades.)

The willingness of the Overseas Development Administration to support maintenance is very welcome, but a word of warning is in order. Experience in a number of countries indicates that whilst there are potentially very considerable benefits from such investment, only a part of these benefits can be achieved in the short term. Indeed it is necessary to recognize that only a few of the present roads can be effectively supported until indigenous maintenance capacity has been expanded by the necessarily slow process of training and institution development. It is here that attention should be concentrated.

In spite of an increasing awareness of the need to take a more complete view of projects, it is still common for a technical solution to be considered separately from the problems of implementation and subsequent operation. The same dangers lie in the development of evaluation processes for rural transport investment, be they based on economic, social or a hybrid of criteria. It is really only the communities themselves that can identify their needs and priorities (which will probably change from time to time). Their commitment must be established from the start to ensure that the long-term support for administration and maintenance is generated among the local users and beneficiaries. 'Institution building' reveals an undesirable approach in so far as it implies the creation of additional organizations to cope with newly identified problems. It is extremely important to become aware of the existing institutions (many of which, operating at the local level, may not be readily apparent to the outsider) and their capacities.

Mr GRIEVESON and Mr WINPENNY *(Paper 5)*: Dr Metcalf's account of Australian procedures endorses our plea for a flexible approach to design standards not constrained by predetermined and fixed ideas. Such an approach would also cover Mr Gandy's point that some situations may benefit from higher construction standards to offset future maintenance. In deciding such issues, however, we would expect the economic analysis for an appraisal to cover the total transportation costs of alternatives, including future maintenance, and, where concrete pavements are being considered as an alternative to flexible pavements as suggested by Mr Mitchell, to also cover for periodic resealing of the latter. Mr Gandy's point on the need to implement the various steps in stage construction at the right time is very valid.

Far from believing that local officials in developing countries are 'short-sighted and foolish' put forward as one possibility by Mr Heggie, we think they are operating under constraints and pressures which inevitably force them into a situation where they must give priority to new construction over maintenance. Where such constraints and pressures are induced by foreign financiers they are to be regretted. We also find it difficult to accept as a generalization Mr Heggie's view that 'people prefer new roads to improved maintenance and new roads to rehabilitation and improvement'. It may well be true for those who have no road at all but it would not necessarily apply to those who already have to travel long distances over a deteriorating existing system. The statement that 'road improvements almost invariably lead to redundancy amongst drivers and mechanics' we find difficult to accept.

Mr Irwin-Childs makes a plea for the use of management contractors and this is indeed a further attractive possibility but not one in which we in ODA yet have any direct experience in the roads sector. We have, however, used it in process engineering and in gas well drilling. We go along with Mr Roberts' point that training to increase local maintenance capacity can be a slow process and on institution building we believe that first thoughts should always be in the direction of strengthening existing institutions rather than in creating new and additional ones.

Mr THOMSON *(Paper 6)*: While transportation planning studies in Latin America may have been stimulated by 'an element of competitive cultural emulation' (introduction to Paper 6, Mr Powell), I feel that a more important influence in the Southern Cone, regarding planning at the national level at least, has been the promotion of such studies by the World Bank, which has participated as executing agency in the majority of them. The Bank has also been able to influence their content and in recent years it has given an impetus to the inclusion of maintenance and general institutional issues in transport planning, after many years of overconcentration on investment-project analysis. In one case in the mid-1970s, a national transportation plan was initiated (under the normal regime with the World Bank as executing agency, the United Nations Development Program as the source of financing, and with extra-regional consultants charged with the technical analyses) only to be terminated at a comparatively early stage due to differences of opinion between the consultants and the host nation as to the future spatial development of the country. This is one indication that matters of more fundamental import take precedence over competitive cultural emulation.

It is true that transportation facilities exist on all important arteries but, in the case of the Southern Cone, it is less certain that the more obvious improvements have been completed.' With the growth in overall demand, existing facilities sometimes tend to become inadequate and it is not always obvious what should be done to expand the supply side. This is one of the prime justifications for large-scale national transportation studies. Moreover, it is not rare to find that transportation facilities in the sub-region tend to be undermaintained and/or improperly utilized, thereby creating a need to evaluate highway reconstruction or railway rehabilitation, and the best way to go about such schemes is not always self-evident.

Mr Powell's warning against making transportation studies unduly complex should be heeded by all concerned. However, transportation model structure can sometimes be quite complex while the model itself can be quite easy to use. Moreover, complex model structures do not necessarily imply the need to collect more data nor extra calibration problems. Were some of the components of the overall transport models to be integrated, it might be argued that there would be an increase in structural complexity, but the result might be a model that is easier to use. Also, models with a sounder behavioural base, which are sometimes quite complex, do provide a better understanding of transportation problems (although there are severe limits as to the extent that behaviourally sound models should be used since they often present very great difficulties from the standpoint of data collection and processing).

The only remark made by Mr Powell with which I am not in complete agreement is that in which he says that countries should carry out a national transport plan about once every ten years. I think it might be better if transport planning were conducted on a continual basis (although a particular effort might be made at ten-year intervals, particularly if the country concerned produces a national development plan every ten years). There would be advantages were such a continual planning conducted by national professionals and technicians, with extra-regional consultants being contracted only for training purposes and for specific tasks for which their expertise is particularly desirable. This *modus operandi* has recently been gaining favour in the Southern Cone; for instance, it was the stance adopted by the first phase of the Argentinian National Transport Plan.

Professor PARKER *(Paper 8)*: Mr Mitchell has interpreted my comments regarding axle loads as an abandonment of S design loads. In many developing countries, design axle loads are set at unrealistically low levels, even to the point of being incompatible with the existing truck technology, much of which is either imported or imposed as aid. These design loads must be raised and axle loadings above the limits be levied at a level that reflects the economic impact on road maintenance. In most developing countries enforcement of axle loads will continue to be ineffective in the foreseeable future, and it makes both political and economic sense to plan for heavy axle loads.

Mr Mitchell also seemed to miss the point about design speeds. While I do advocate in Paper 8 that design speeds should be reduced to desired speeds, I also proposed a methodology for choosing alignments and grades that would realize a desired speed S compatible with a minimization of fuel costs for the expected vehicle volume and mix. It may well be that for Tanzania, for example, the desired or optimum speed would be 100 km/h as Mr Mitchell suggests has been found to be the economic speed for Zimbabwe.

Regarding the comments and issues raised by Mr McLean, I can only state that I am in total agreement with him. I trust, however, that the route optimization methodology currently under development would in fact enhance the development of design expertise.

REFERENCES

1. HOBAN C.J. Overtaking lanes and stage duplication on two-lane rural highways. Australian Road Research Board, 1981, AIR 359-3.
2. NATIONAL ASSOCIATION OF AUSTRALIAN STATE ROAD AUTHORITIES. Interim guide to the geometric design of rural roads. NAASRA, Sydney, 1980.
3. McCLEAN J.R. Review of the design speed concept. Australian Road Research Board, 1978, 8(1), 3-16.
4. McCLEAN J.R. An alternative to the design speed concept for low speed alignment design. Proc. 2nd int. conf. on low volume roads, 1979, Transport Research Record 702, 55-63.
5. McCLEAN J.R. Cost implications of design speed standards. Australian Road Research Board, 1980, Vol. 72, Aug., 333-354.
7. MITCHELL R.L. A cost analysis of the use of concrete as a substitute for flexible pavements in Rhodesia. Ministry of Roads and Road Traffic, Government of Rhodesia, Harare, Zimbabwe, 1974, Lab. Record 13/74.

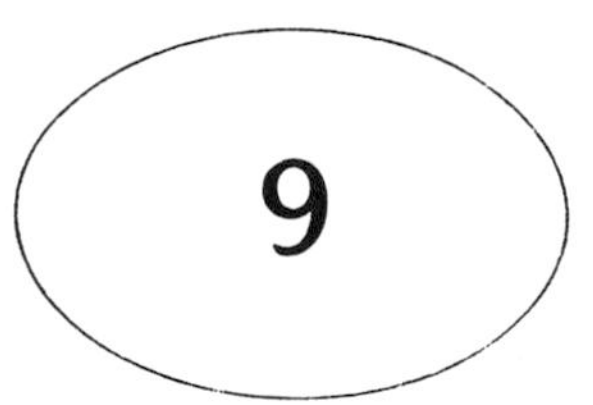

Institutional aspects of rural road projects

O. J. RAHKONEN, C. COOK and S. CARAPETIS, World Bank

The growth of institutional capacity is a key aspect of the development process and critical in the design of project identification, planning and implementation procedures. In the past, projects have often partly failed because they have not been adequately tailored to a particular institutional and cultural context. Rural roads projects are particularly "institution-intensive" both because they depend heavily on local human resources for their implementation and because their effectiveness often depends on simultaneous efforts by other agencies involved in the rural development process. Seven major themes emerge in discussions of the institutional aspects of rural roads projects: (1) Centralization vs. decentralization; (2) New vs. existing institutions; (3) Public vs. private sector; (4) Political commitment; (5) Beneficiary participation; (6) Inter-agency coordination; (7) Human resource development. Current knowledge indicates that it is possible to establish only the most important principles and common factors influencing these institutional issues. Therefore, project designs should be tailored to specific local circumstances, with the goal of making the most effective use of resources to serve development objectives.

INTRODUCTION

1. The World Bank has long recognized that the growth of institutional capacity is a key element in the development process.* Over the past ten years, the Bank's transport sector staff has shown increasing interest in institutional issues as they relate to the design and evaluation of transport sector projects. An early Bank evaluation of completed highway investments pointed out the importance of institutional factors in project success (ref. 1). These factors included both internal aspects of the implementing institution and external factors, involving coordination between the implementing institution and other agencies whose action or lack of action affected project success.

2. Nevertheless, in the early years of Bank operations, primary attention was given to technical, financial and economic aspects of project preparation rather than to the institutional and human resources needed to achieve project objectives. Changes in the composition of Bank lending, particularly since the early 1970s, have increased the emphasis on domestic resource utilization, and lessons learned from past projects now point to institutional development as a key factor in project success. Thus institutional appraisal, and programs to improve institutional capacity, have become critical factors in the selection and timing of future investments.

3. This paper presents some of the central themes which arise in discussions of the institutional aspects of rural roads projects, based on a review of Bank experience. It is not intended to provide a comprehensive overview of Bank lending for rural roads projects, nor is the focus limited to Bank criteria for project design and evaluation. Rather, institutional issues related to rural roads are viewed as matters which need to be considered on a national basis, taking into account the general goals of the national development process and the appropriate allocation of human, fiscal, and organizational resources. We conclude that the goal of institution-building components in rural road programs should be to develop self-sustaining institutions which continue to function because they succeed in identifying and meeting community needs efficiently and economically.

INSTITUTIONAL DEVELOPMENT: AN OVERVIEW

4. Institutional development has been defined as the process of creating or strengthening the capability of institutions to make effective use of resources in the pursuit of

* The authors are indebted to much earlier work done by Bank staff in this area. In respect to rural roads this applies in particular to Mr. Curt Carnemark, who laid the groundwork for this paper through his earlier work, assisted by Ms. Christine Kimes. Mr. Christopher Willoughby has been keenly interested in this topic throughout and has provided many valuable comments, as have numerous Bank staff members who have been consulted on specific issues and projects. Regarding institutional development in general, the work done by Messrs. Arturo Israel, Francis Lethem, Heli Perrett and William Smith has been particularly useful. Mr. Terry Walbert provided research assistance in reviewing file data on selected sample projects.

Highway investment in developing countries. Thomas Telford Ltd, London, 1983, 63–70

development objectives (ref. 2). By "institutions" we mean those formal and informal social arrangements through which decisions are made and actions are taken to assure the provision of facilities, goods or services for a community. Institutions may be found in the public or private sector; at the national, regional, or local level; and they may range from complex organizations with specific operational tasks and objectives to loosely organized community groups and tribal associations. The particular characteristics of these institutions constrain their effectiveness in relation to specific types of tasks and specific situational conditions.

5. A review of the institutional development literature suggests that the following factors are generally associated with success in developing institutional capacity (ref. 3):

(a) Host country commitment to institutional development objectives;

(b) Host country participation in the design of institutional project components;

(c) Intensive involvement by interested technical staff;

(d) Effective management of personnel systems, including adequate incentives for improved performance.

These factors generally relate to improving the internal effectiveness of a particular institution, and indeed this has been the focus of most institutional development components in Bank lending for projects in the highway sector up to the present time.

6. Recent Bank work has also focused attention on the requirements of interagency coordination, political support and beneficiary participation with respect to rural development projects (ref. 4). This work stresses the need for strong patterns of control and coordination at the local level, together with a balanced distribution of power and responsibility throughout the delivery system (ref. 5). This analytic approach, which focuses on the identification of significant actors ("stakeholders") in the rural development process, the analysis of resources available to and constraints on the action of each actor, and the development of exchange networks to maximize the effective use of resources under existing constraints, offers a potentially useful perspective in considering institutional aspects of rural roads projects.

7. Rural roads projects may be seen as being particularly "institution-intensive," for two reasons. One is that rural roads projects rely heavily on the use of human resources at all stages of the planning and implementation process. This implies a need for strong vertical linkages between such projects and the central sources of political support, on the one hand, and between the projects and the beneficiary communities, on the other. A second reason is that rural roads projects are often conceived and implemented as components in multisectoral rural development projects. This implies a need for horizontal linkages to permit coordinated action between the different agencies involved in such projects.

8. A number of identifiable issues or themes emerge in discussions of the institutional aspects of rural roads projects. These include: (1) degree of decentralization; (2) new vs. existing institutions; (3) public vs. private sector institutions; (4) political commitment; (5) beneficiary participation; (6) interagency coordination; and (7) human resource development. Each of these issues has different implications at the three stages of project implementation: planning, construction, and operation and maintenance. The development impact of rural roads projects can be enhanced to the extent that such institutional issues are identified and resolved at the planning stage, rather than being resolved on an _ad hoc_ basis after they have been identified as obstacles to project implementation. Nevertheless, it must also be recognized that circumstances change, often drastically, over the lifetime of a project. Institutional aspects of project design must therefore retain enough flexibility to be able to respond in creative ways to a changing political and socio-economic environment.

Centralization vs. decentralization

9. Traditionally, transport sector planning has focused on the development of national and international networks designed to serve major commodity flows. Such systems have been planned and designed by central government agencies with a small, highly competent technical staff, using sophisticated analytic techniques applied to detailed data on a relatively small number of investment alternatives. Where rural road programs have been planned and administered by these same agencies, it has proved to be much more difficult for them to carry out their task. Comparable levels of data collection and analysis for individual projects are simply not feasible in the light of the relatively limited total investment resources available for rural roads and comparatively small costs per kilometer of improvement. Thus, planners must often rely on inadequate data in making decisions. In addition, centralized agencies frequently fail to develop institutional capacity at the lower levels of the organization. The lack of strong patterns of local control and coordination makes such programs relatively unresponsive to local needs and priorities.

10. A second model is the planning of rural road improvements in the context of specific regional development projects. Such projects are sometimes carried out through sectoral

agencies such as agriculture, forestry, or irrigation, or by ministries with broader responsibilities such as planning, rural development or interior. This approach to planning is more decentralized, sometimes more flexible in response to variations in local conditions, and often more appropriate in terms of design standards, unit costs, and use of local resources. On the other hand, it may result in a wide variety of improvement types with duplication of services to some areas and total neglect of others. Most such projects have assumed that the central government agency will take over responsibility for maintenance of the improved roads. This is often difficult to accomplish in practice, due to differences in design standards, conflicts over agency "turf," and resource constraints.

11. Those countries that have developed national rural road programs have arrived over time at a set of institutional arrangements that jointly maximize the benefits of both models. Rural road needs are locally identified, and the evaluation of economic priorities takes into account local resource availability and development prospects, including the need for complementary investments in other sectors. On the other hand, proposals for different parts of the country are evaluated using common technical and economic criteria to assure an effective and equitable distribution of resources and the integration of planned improvements with ongoing maintenance programs. This decentralized model for planning comes closest to the objective of building self-sustaining institutions through rural road programs. However, the degree of decentralization which is appropriate in a particular country depends upon the availability of qualified staff and the degree of local and regional autonomy allowed by the national approach to development planning.

New vs. existing institutions

12. An important issue in using rural road projects to build up a nation's institutional capacity is whether to try to strengthen existing institutions so that they become capable of taking on new responsibilities, or to create new institutions for this purpose. There are advantages and disadvantages to both approaches. Existing institutions carry with them pre-existing definitions of institutional objectives, norms of professional and bureaucratic behaviour, and established exchange relationships with other agencies, clients, and funding sources. New institutions can be more easily tailored to meet the needs of rural road projects, but must negotiate new relationships with other agencies involved in the rural development process, with the beneficiary communities, and with the planning and budgeting authorities. Social change is more difficult to implement through existing institutions than through new ones, but its effects will be more lasting if the new institutional objectives and activities are fully integrated into those of the existing institutional structure.

13. Planning for rural roads projects requires a different perspective from that taken in traditional highway projects. It requires inputs from other agencies and from the communities to be served. It requires the development and use of methods for rapid screening of many proposals rather than the detailed study of a few well-defined alternatives. To the extent that new information, new analytic methods, and new external linkages are required, a new institutional setting may make it easier to implement change in the planning process. On the other hand, rural roads planning needs to be coordinated with planning for major infrastructure improvements, for road maintenance, and for the development of alternative transport modes. To the extent that this information is commonly used for planning in the existing road agency, the process will be facilitated by expanding agency capacity to include consideration of rural roads requirements.

14. At the implementation stage, a choice must be made between assigning responsibility to the existing road-building agencies, creating new project units with a special focus on rural roads, or developing the capacity of contractors or communities to implement rural road improvements. Traditional programs developed through the highway sector have tended to expand the implementation capacity of existing institutions; to some extent this is also true of irrigation and forestry projects. Agricultural and rural development projects in the seventies showed a strong tendency to create special project units, often with expatriate technical support, to execute road improvements needed for project purposes. Such units proved to be politically vulnerable because of their identification with specific project activities and their dependence on outside support for success. Despite their often excellent technical performance, they generally failed to show any lasting effect in terms of the development of institutional capacity to implement rural roads projects.

15. The use of contractors implies the development of an existing social institution rather than the creation of a new one. It may, however, be necessary to redefine the traditional role of contractors in order to allow them to participate in the execution of rural road programs. This means that contractors must develop new technical, financial, and managerial skills as well as relationships with a new set of employees and clients. Building up the capacity of the domestic contracting industry is sometimes a more difficult task than expanding the role of existing public institutions, but it is likely to prove more rewarding in terms of contribution to development capacity.

16. Proposals to implement rural road improvements through construction by community groups may either build upon existing institutional structures at the community level or attempt to create new ones. The success of such proposals depends on community acceptance of the rural road-building task as a relevant and appropriate

use of community resources. Traditional patterns of leadership and support, norms of social responsibility, and existing social control mechanisms may either support or come into conflict with the institutional structure proposed for the rural road program. Such programs therefore require intensive involvement by the beneficiary communities in planning as well as in implementation. They also frequently require the provision of managerial and technical assistance to communities by the road-building agency.

17. Few existing institutions have successfully addressed the problems of rural road maintenance (ref. 6). Most rural roads are unpaved, and the tasks required are largely labor-intensive. Maintenance activities are generally accorded a lower priority than new construction both by agency staff and by beneficiary communities. Poor management is one of the main obstacles to more effective use of manpower and equipment for maintenance purposes. Rather than expand the scope and budget for the road-building agency to cover this new responsibility, some countries are experimenting with institutional alternatives such as contract maintenance, the "lengthman" system, and developing community capabilities (ref. 7). However, the supervision and management of maintenance activities still has to be carried out by the central or provincial government.

18. An institutional arrangement which has been found effective in some cases is to create a separate branch within the traditional road-building agency for the maintenance and improvement of rural roads. This approach provides a channel for allocating funds exclusively to rural road improvements and permits the development of new methods for planning, construction and maintenance. It can help to create an esprit de corps among the professionals assigned to the program, particularly if it receives strong support and creative leadership from the top. Such an organization enjoys the legitimacy and the channels of communication of the larger agency, while at the same time it becomes a focal point for introducing institutional change. In the long run, the activities of such organizations should be focused on helping to develop community capabilities in the areas of road construction and maintenance.

19. In general, our experience indicates that it is better to try to modify existing institutions to meet new needs rather than to create new institutions with attendant stresses and strains on the social system. Existing institutions should not be limited to the traditional road-building agency, however. Other sectoral agencies whose resources and management systems are compatible with rural road work (such as irrigation or forestry agencies) may be appropriate focal points for action. Regional development planning authorities, if they are well established with a secure mandate to coordinate sectoral programs, may also provide a relevant institutional base. The domestic contracting industry may be capable of undertaking rural road works with suitable types of technical assistance. Finally, community structures for making decisions and mobilizing resources may well offer models which can be incorporated into the design of rural road programs, particularly for maintenance.

Public vs. private sector

20. The growing shortage of financial resources for public investment in development projects, coupled with increasing pressure on governments to allocate such resources to services rather than to infrastructure, highlights the need to cut investment costs and to increase the effective use of existing infrastructure. The search for fiscal effectiveness has led many countries to consider changes that would allow greater participation in infrastructure development activities by the private sector (ref. 8). With respect to rural roads, these changes have to do mainly with the use of local contractors for construction and maintenance and with the deregulation of rural transport services.

21. The planning of road improvements in rural areas is clearly a matter of public concern which should involve some form of public participation. Private sector interests may contribute technical skills through consultancy arrangements and participation in public hearings. Specific development projects may also be served by roads built and controlled by private interests. In order to ensure a more efficient and equitable allocation of resources, however, it is necessary for rural road investment programs to be responsive to the needs of communities rather than particular individuals, and to be subject to the scrutiny of technical experts in the fields of economics and engineering. Nevertheless, such planning activities can easily envisage the mobilization of resources from the private sector for the construction, operation, and maintenance of rural road investments.

22. The use of local contractors for construction and maintenance activities appears to be particularly advantageous in situations where the capability to carry out rural road programs exists in the private sector but new institutions would have to be created in order to accomplish this in the public sector. Even when a public road-building agency exists, contractors can often provide the needed labor and equipment at lower cost and with less delay than the public agency. Obtaining services from contractors reduces the burden on public agencies to procure, operate and maintain equipment, to recruit and manage a large labor force, and to provide adequate pay scales and incentives for good performance (ref. 9).

23. The development of traffic on improved rural roads depends partly upon the arrangements made for the provision of transport services in rural areas. Such services are usually largely in the hands of the private sector, which responds quickly to rural road improvements wherever there is effective unmet demand and where the response is not limited by legal or regulatory constraints. Fuel tax and price policies, import restrictions on

vehicles and spare parts, and transport tariff restrictions, also play an indirect role in determining whether a rural road investment will be fully used and thus achieve its economic objectives.

Political commitment

24. A number of rural road programs have run into problems due to macro-economic forces over which they have little control. Currency devaluations, unanticipated price increases, and war and civil disorders may easily cause a shortfall in the resources available for project implementation. In such circumstances all public programs are forced to compete for the scarce resources available. Strong political support from the top is necessary for a rural roads program to survive in this situation.

25. Even when resources are theoretically available, rural roads programs may be severely hampered by cumbersome central disbursement and procurement procedures. These problems seem to be particularly acute when project funding runs through a central planning, financial, or political ministry rather than through a sectoral agency. Such problems make it more difficult for a line agency to carry out the program through force account, but they may be absolutely disastrous in the case of programs designed to be carried out by local contractors. Few contractors in developing countries have the means to meet the cash-flow requirements of carrying on a business when equipment or spare parts are not available or payments are delayed or even stopped.* Strong political support is required for this type of program to insure that "red tape" will be kept to a minimum and that funds will flow freely through the system.

26. Important sources of political support may be found in other sectoral agencies, in regional development authorities, and in beneficiary communities. All of these groups stand to benefit from the successul completion of a rural roads program. The key factor, however, is control over resources. Thus, local powers to levy taxes, to procure and manage equipment, and to mobilize labor, are key elements in the decentralization of responsibility for rural roads programs to local communities. Rural roads programs must be responsive to perceived needs, but they must also be tied in to the system through which resources are allocated to meet these needs. Recurrent budget allocations for rural roads operations and maintenance provide the ultimate test of political commitment and support for rural road programs.

* One way to solve this problem is to establish a revolving fund with an initial contribution from the government. The fund is then replenished on the basis of invoices for work accomplished and paid for. This system greatly reduces the debt burden on contractors and makes more efficient use of government funds.

Beneficiary participation

27. Central government support is essential to the success of any rural roads program, but it usually does not depend on the specific issues of project design and implementation. Beneficiary participation, on the other hand, not only serves to mobilize local resources but also provides a channel for feedback that permits tailoring project design to local needs and priorities. In order for this channel to work effectively, opportunities for beneficiary participation need to be built into the institutional design of a project from its very beginning.

28. Many national rural road programs now rely on formal or informal mechanisms through which the agenda for rural road improvements is generated at the local level. Usually such participatory planning mechanisms produce a "long list" of proposals whose implementation would exceed agency capacities. Thus, some form of screening is usually applied by the implementing agency, taking into account technical and financial constraints as well as political priorities and assessments of potential socio-economic impact. The highest priority improvements are then incorporated into the agency work program as far as foreseeable resources (funds, staff, and equipment) will allow. Rarely is there any feedback from the agency to the community at this stage, although this process could be helpful in making future proposals more suitable from a technical and financial point of view.

29. Because central government resources are often inadequate to meet needs, it is sometimes proposed that communities should participate directly in project implementation, either by raising some of the necessary funds, or by making a direct contribution of labor and/or local materials. Such an arrangement, of course, requires both community consent and active commitment to the project. The ability of the community to contribute depends heavily on the standards and timing of road construction activities. Areas with high underemployment rates, where road works can be timed to complement peak season agricultural activities, may be able to afford a major labor contribution but very little in the way of cash. On the other hand, areas which provide a source of migrant labor for better paying jobs in other places may be able to make important cash contributions, although they may not be in a position to provide labor inputs.

30. Beneficiary participation in construction activities requires the use of some social structure or institution to mobilize local resources (whether person-days of labor or contributions in cash or kind) and to direct them towards the achievement of community objectives. The organization and supervision of a volunteer work force is one of the major problems confronting any program based on community participation. Such activities require not only a high degree of technical and managerial competence on the part of the

supervising staff, but also a high degree of skill in interpersonal and intergroup relations. The management of community funds and payments to local workers also present potential areas of misunderstanding that will require skillful and sensitive handling on the part of agency staff.

31. Many developing countries find themselves in a position to provide resources, often through donor assistance, for programs of rural road rehabilitation and improvement. It appears to be more difficult for these countries to allocate resources for rural road maintenance, which must come largely out of local currency and must be anticipated in the annual recurrent budget. In this situation, rural road maintenance is often left to local communities. Yet few such communities are prepared to undertake this responsibility. Beneficiary participation in rural road maintenance depends upon a clear understanding of the tasks to be performed, the availability of needed tools and equipment, and the existence of an institutional structure through which the efforts of many different individuals can be coordinated. As with participation in construction activities, it also depends on the road design standards, on the timing of maintenance tasks in relation to agricultural activities, and on the availability of community funds.

Interagency coordination

32. Rural road improvements are not an end in themselves, but rather one of the means to accomplish the more general goals of rural development. Other means are also required to help accomplish these goals. In a few cases, lack of access may be the only constraint on development, and in such cases a spontaneous burst of social and economic activity could be expected to follow on rural road improvements. More frequently, however, the lagging development process in rural areas is due to factors even more fundamental than lack of access. Many may be attributed to a failure to provide adequate government services in rural areas, which is often closely linked to the transport constraint. Other obstacles, such as poor soils, lack of water, or declining fuel supplies, may be overcome only by the transport of physical inputs. Rural roads agencies play an important part in the rural development process in such cases, but they cannot succeed without the cooperation of other institutions at the national, regional, and local levels.

33. Traditionally, projects in the highway sector have involved little in the way of interagency coordination. Where external linkages have been sought, the road-building agency has taken the lead to solicit information from other sectoral planning agencies at the national level. There has been little feedback to these agencies and almost no attempt to coordinate the planning of service delivery in other sectors with the completion of rural road improvements.* In some countries, road-building agencies have attempted to expand their role to include leadership in all rural development planning activities, but these efforts have met with strong resistance from other sectoral agencies with vested interests in rural development.

34. A second model is the agricultural development project in which the rural road component is clearly designed to serve specific project objectives. This model includes plantation projects, land settlement schemes, and many irrigation and forestry projects. Here the responsibility for planning the rural road component rests primarily with the lead agency or institution. The component may be executed by a special project unit, by force account through contract arrangements with the traditional road-building agency, or by contractors. If responsibility for the road component is delegated outside the lead agency, it is usually confirmed through a written agreement or contract.

35. Rural roads executed as part of a multisector program for rural development necessarily involve some form of interagency coordination. Frequently an interagency coordinating committee is set up to ensure that this occurs. The effectiveness of such committees seems to be an inverse function of the organizational level at which they are established and a direct function of their resources in terms of technical staff and control over funding. Interministerial coordinating committees rarely exercise any real control over rural road programs, whereas coordinated planning at the project unit level may be extremely effective in the short run.

36. A review of Bank involvement in rural development programs with road components showed that the more far-reaching and innovative programs involved participation by several agencies in plans for service delivery in rural areas. Lateral linkages were established through coordinating committees, although these did not always function effectively. These programs also had stronger vertical linkages with the sources of national political power and with beneficiary groups. They were somewhat more likely than other types of projects to make use of labor-intensive methods and local contractors to carry out the road component.

37. Such projects often had more difficulty than others in executing planned road programs, due to the lesser degree of direct control exerted by the lead agency. However, they were more likely to be successful in improving the lives of large numbers of beneficiaries from the target population of the rural poor.

* There are several important exceptions to this rule, including Tunisia and Kenya. However, in such cases coordination is not effected "from the top" so much as at the provincial planning level.

This success may have been due in part to their greater flexibility in shifting resources between sectors in response to changing perceptions of needs and priorities.

Human resource development

38. Staffing and training are major institutional issues in the implementation of rural roads programs. Such programs require more innovative planning approaches, more supervisory manpower, and a larger proportion of labor inputs than do traditional highway projects. Often the agencies responsible for such programs are not in a position to provide the incentives needed to attract and retain a sufficient number of competent personnel. Staff seconded from other departments are not likely to view participation in the rural roads program as the key to professional success. The lack of qualified local staff has meant that many rural road programs in the past have depended heavily on the participation of expatriate staff.

39. These obstacles can only be overcome through time, by the development of well-designed recruitment, incentive, and training programs. Such programs should be designed on the assumption that a certain amount of trained manpower will ultimately flow into the private sector. The special skills required for the management of a rural road program should also be taken into account in planning for training activities. Training also presents an opportunity to develop a sense of common concern and an *esprit de corps* among the staff that will both improve their performance on the job and increase the probability that they will stay with the program over time.

40. Training concerns should not be limited to agency staff, particularly if a rural roads program is to be implemented by contractors or by community labor. Many small contractors can be assisted through short courses or seminars on the technical, financial, and managerial aspects of rural roads work. Communities often need assistance in understanding the criteria used for project selection, the technical requirements for community participation, and the ongoing needs for rural road maintenance. This educational function can be an important part of a good public relations program which also serves to build political support and to provide a channel for feedback from project beneficiaries.

CONCLUSIONS

41. Institutional requirements for the successful execution of a rural roads program depend partly on the nature of the program itself and partly on the environment within which the program is expected to function. It is apparent that no single institutional arrangement will work equally well in all situations. Different combinations of skills and resources are required to achieve different project objectives, and skills and resources are differently distributed in different project settings. The program designer's task is, first, to identify correctly the resources required to meet program objectives; second, to analyze the distribution of resources in the project setting; and third, to propose an institutional structure for the program that will maximize the probability that needed resources will be made available in a timely and appropriate way.

42. Ultimately, the objective of rural road institution-building should be that local communities develop the capability to assume increasing responsibility for rural roads. At the local decision-making level, power structures tend to be more integrated and sectoral issues less heavily institutionalized. Should local communities come to function efficiently and receive sufficient resources to take care of the rural roads within their jurisdictions, the institutional issues now surrounding rural roads programs would largely disappear.

43. It will, however, take a considerable time before such capacities are created. In the meantime, both national and provincial agencies dealing with rural roads need to be strengthened in parallel with efforts to improve local capabilities. Such institutional development objectives might include: decentralizing decision-making responsibility to the greatest extent feasible; expanding existing institutional mandates to include specific provisions for rural roads; developing linkages to the private sector; creating new mechanisms for beneficiary participation and political support; improving interagency coordination; and introducing staff training and incentive programs. The basic goal of all such activities should be to strengthen institutional capacities at the local level, with the leading role of the road-building agency clearly corresponding to a transitional stage in the rural development process. Institutional changes should be planned and phased with this goal in mind, but should retain sufficient flexibility so that the system can continue to function even if unexpected changes occur in the political environment.

44. The determination of an appropriate institutional design for a rural road program will take into account numerous locally specific socio-cultural factors, as well as technical, economic, and human resource constraints. Our experience indicates that it is difficult to introduce and succeed with institutional changes that are not perceived and accepted by the various participants as relevant and appropriate. A wide variety of institutional arrangements appear to work well under different circumstances. The people best able to make an appropriate choice are the local communities and agencies whose interests will be directly served by the project.

45. Once an appropriate institutional design for a program has been determined at the planning stage, several steps can be taken to strengthen the capacity of the selected institutions to carry out the project. This review has suggested several principles to be kept in mind in the design of institution-building components for rural road programs:

(a) Responsibility for the planning, construction, and maintenance of rural roads should be decentralized to the greatest extent feasible consistent with manpower, technical, and political constraints;

(b) Existing institutions, whether national or local, public or private, should be modified or strengthened to serve new needs, rather than creating new institutions;

(c) Opportunities to increase the involvement of the private sector should be sought, particularly in road construction and maintenance;

(d) Mechanisms should be developed to permit beneficiary community participation in the planning, construction, and maintenance of rural road improvements;

(e) Political commitment to the success of the rural road program, as expressed in support for the needed financial commitments, is a *sine qua non* for success;

(f) The lead agency implementing a rural roads program must possess the required authority and the skills needed to execute the work; it should also develop strong lateral linkages with other sectoral agencies involved in the rural development process;

(g) Explicit attention should be paid to any programs proposed for recruiting, training, rewarding and retaining skilled and motivated individuals within the implementing institutions, including not only the road-building agencies but also private contractors and community groups.

46. The goal of institution-building components in rural road programs should be to develop self-sustaining institutions which continue to function because they succeed in identifying and meeting community needs efficiently and economically. These institutions will be closely linked to the sources of political support, including the central government as well as the beneficiary communities. They will interact with other agencies involved in rural development in order to make the most effective use of all available resources. Institutional arrangements that work well for the planning, construction, and maintenance of rural roads may come to serve as a prototype for structures intended to mobilize resources in order to meet community needs in other sectors as part of the rural development process.

REFERENCES

1. WORLD BANK. Comparative Evaluation of Selected Highway Projects. Report No. 349, Operations Evaluation Department, 1974.

2. WORLD BANK. The World Bank and Institutional Development. Experience and Directions for Future Work. Projects Advisory Staff, May 1980.

3. *Ibid.*, 6-7

4. SMITH W. LETHEM F. THOOLEN B. The Design of Organizations for Rural Development Projects. World Bank Staff Working Paper No. 375, March 1980.

5. Ibid., 24-25

6. WORLD BANK. The Road Maintenance Problem. Transportation, Water and Telecommunications Department, December 1980.

7. MASON M. The Organization of Road Maintenance. World Bank Projects Advisory Staff and Transportation, Water and Telecommunications Department, June 1981.

8. WILLOUGHBY C. Infrastructure: Doing More With Less. Finance and Development, December 1981, 30-32.

9. ROBINSON R. Maintenance Management for District Engineers. Overseas Road Note No. 1, Transport and Road Research Laboratory, 1981.

10

The choice of appraisal techniques when resources are limited—a case study of the rural roads programme in Peninsular Malaysia

J. D. SMITH, MSc, DipTP, MRTPI, MIHE, University College London

Rural road projects are now seen as representing only one part of a wider development process in rural areas. The developmental role that some rural roads are expected to play has created the need for more realistic and complex models of rural transport behaviour. As a result appraisal techniques relying mainly on cost benefit analysis have been used, particularly where loan agencies have required detailed technical and economic feasibility studies. The trend towards more complicated appraisal methods in developing countries may not be practicable in situations where data and expertise are limited. Further the implementation problems resulting from a road programme in Peninsular Malaysia indicate that the use of several variables in the analysis stage tends to increase the potential errors inherent in forecasting future activities. This paper suggests that in such cases simplified procedures are needed to make the best use of limited resources.

INTRODUCTION

1. The evolution of appraisal techniques reflects the changing views of governments and aid agencies. These views, at present, relate to the integral role that roads play in rural development and in national socio-economic development (ref. 1, 2, 3). While appraisal techniques can, in theory, include physical, social, political and aesthetic assessments (ref. 4), current methodological improvements tend to concentrate on economic analyses. This is in part due to the difficulties in quantifying many of the non-economic benefits. A recurrent problem in the conduct of economic appraisals in developing countries has been the requirement for large-scale data inputs and expertise (ref. 5).

2. With the present trend to view rural road planning as part of a broader development effort, an understanding of that process, in particular the social and political context, has been advocated (ref. 6). Similarly in the UK following the presentation of the Leitch Report (ref. 7), the framework for the assessment of major road projects is being enlarged. This framework is now to include social, environmental and aesthetic factors in addition to economic and operational criteria. Impacts on different sections of the community are also to be assessed.

3. The use of appraisal techniques in decision-making is becoming more complex in developed and developing countries. In the latter, this has been largely due to loan agency requirements for more realistic models of rural transport behaviour. Even so the need to incorporate social and other criteria into assessment frameworks presents a formidable challenge to simple quantification. It is interesting to note therefore that in countries where expertise and accurate data is generally scarcer, a similar trend towards more complex techniques has been observed.

4. This paper suggests that while the comprehensive approach to project appraisal may be suitable in some cases, wherever resources are limited a simplified approach will have to be adopted. The paper is centered on the planning and implementation of a rural roads programme in Peninsular Malaysia. The first part of the paper provides a description of the main elements of the programme including objectives, expenditure patterns and planning techniques. The second part looks at some of the features and problems of the programme before going on to assess the implications for project appraisal methods. With reference to some important drawbacks in current approaches, the paper concludes by suggesting a simplified model based on reduced requirements for accurate data.

RURAL ROAD PROGRAMMES IN PENINSULAR MALAYSIA

5. There is a variety of methods by which rural road projects are implemented in Peninsular Malaysia. These include regional development programmes where roads are combined with other infrastructure, agriculture and community settlement projects. Single sector programmes are also undertaken where roads play a supportive role to the main aim of improving agriculture. The land settlement schemes of the Federal Land Development Agency (FELDA) are an example of the former, while the Ministry of Agriculture under its integrated agricultural development projects undertakes the latter.

6. Rural road projects are also carried out independently of other sectors by the public works department and the district offices of local government. There are two programmes:

(i) the Accelerated Rural Roads Programme (ARRP), and
(ii) the Village Road Construction and Improvement Programme.

The remainder of this part of the paper is directed towards a discussion of these two programmes.

7. The ARRP is implemented by the state public works departments (PWD) by way of grants from the federal government. Individual projects are identified at village level by development and security committees in liaison with the district officer. From the district level the projects are then submitted to the state government for technical appraisal prior to formal submission to the central government for financial approval.

Programme objectives

8. The objectives of the ARRP, which commenced in 1978 were set out in the Third Malaysia (national) Plan 1976-80:

(i) to raise the productivity and incomes of the rural people by:
- providing greater access to existing populations,
- improving marketing and distribution outlets for rural produce,

(ii) to provide the necessary physical infrastructure to meet the security needs of the country.

9. The ARRP outlines minimum design standards for new rural roads. These usually include laterite earth surfacing and reservation widths of 20 m. Some details are provided in Table 1. Maintenance grants are provided at a flat rate equivalent to M$ 3,850 per km per annum.

Table 1. Some minimum geometric design criteria for new roads in rural areas

Traffic		light
Group		01
Terrain		rolling
ADT Two ways	veh/day	<100
Design speed	km/hr	50
Surface width	m	4.5
Usable shoulder width	m	1.25
Formation width (a)	m	8.0
Reserve width	m	20.0

(a) includes 0.5 m each side for rounding
Source: Public Works Department, Kuala Lumpur

10. The second programme involves the construction and improvement of village roads and tracks by district offices of local government. The programme is designed to allow for easier access in the villages. Light machinery (tractor, tipping trailer and roller) is provided and an annual maintenance grant equivalent to M$ 310 is also given. The main intention of the programme is to upgrade existing tracks between small villages and their agricultural hinterlands. No design standards are set though construction is below 01 minimum PWD standard.

11. Under the Fourth Malaysia Plan 1981-85 the village road programme is to be expanded, with the following objectives:

(i) to facilitate marketing and processing services for agricultural produce, and

(ii) to provide better access to social amenities.

The expansion, which is in response to experience in the implementation of previous programmes, will take the form of drainage, culverting and surfacing works on existing alignments.

Development expenditure

12. Public finance is allocated for rural road expenditure under five-year economic plans. Table 2 gives details for the three periods covering 1971-85.

Table 2. Rural road programmes - public development expenditure allocations 1971-85

Programme	Development allocation (M$ million)		
	1971-75	1976-80	1981-85
Accelerated rural roads } rural roads	35.00	517.00	569.56
Village roads } rural roads	-	39.73	193.00
Sub total	35.00	556.73	762.56
% village roads to sub total	0.0	7.14	25.31
Total transport	1562.06	4462.99	4116.07
% rural roads to total transport	2.24	12.47	18.53

Note: June 1981 exchange rate £1 = M$ 4.65 .
Source: Fourth Malaysia Plan 1981-85, Government of Malaysia, Kuala Lumpur, 1981, p 334.

13. The village road construction and improvement programme did not commence until the Third Malaysia Plan (TMP) was implemented in 1978. Later the Mid Term Review of the TMP directed that increased emphasis was to be accorded to the improvement and extension of feeder, access and rural roads. A large injection of funds was granted with the ARRP making up 11.58 per cent of the total transport budget at that time.

14. However, this injection of funds gave rise to two serious bottlenecks. First, the number of proposals requiring detailed design was far in excess of state PWD capacity. Second, both public and private resources were stretched in the supervision and construction processes. In order to cope with these problems new packages of projects were awarded to contractors on a design and construct basis. A spillover of projects into the 1981-85 Plan period was however unavoidable.

15. Following the completion of the TMP period a switch of emphasis towards the village road programme was proposed. While this programme constituted only 7.14 per cent of the total rural roads allocation in 1976-80, under the fourth plan it increased to 25.31 per cent. In terms of the total transport budget, the combined rural roads programme had increased from 12.47 per cent to 18.53 per cent over the decade to 1985.

Levels of personal mobility

16. Over the ten year period 1971-80, 93.3 per cent of public expenditure on the two rural roads programmes was designed to accommodate four-wheel based traffic. Levels of private motorisation for the peninsula and for a

predominantly rural state are shown in Table 3. Data for a rapidly urbanising state is also provided for contrast.

Table 3. Levels of motorisation

	Registration[(a)] of private motor vehicles			
	Cars per 100 population		Motorcycles per 100 population	
	1970	1980	1970	1980
High income (urban) Selangor[(b)]	5.1	8.0	5.4	10.0
Low income (rural) Kedah	1.2	2.8	2.8	8.7
Peninsular Malaysia	2.7	5.0	4.1	10.0

(a) cumulative vehicle registrations
(b) includes Kuala Lumpur
Source: Fourth Malaysia Plan 1981-85, Government of Malaysia, Kuala Lumpur, 1981, p 107.

17. Both car and motorcycle ownership levels were lower in the rural state than in the urbanising state and the peninsula. Even in 1980 the level of car registrations in the rural state only just matched the 1970 peninsula levels. In contrast a substantial increase in motorcycle registrations has taken place in the rural state. Over 75 per cent of total private motorvehicles registered in the rural state in 1980 were motorcycles.

18. Vehicle ownership at the local level also reflects this two-wheel bias. Table 4 provides information for one rural district over the period 1970-78. While car ownership was less than one per 100 population in 1978, about one in 25 people owned motorcycles and one in 7 a bicycle.

Table 4. Vehicle ownership in Sik district, Kedah 1970 and 1978

	1970		1978	
Vehicle type	Total vehicles	Vehicles per 100 pop.	Total vehicles	Vehicles per 100 pop.
Cars	177	0.45	273	0.65
Motorcycles	653	1.67	1720	4.12
Bicycles	N/A		5834	13.97

Sources: 1970 Population and Housing Census of Peninsular Malaysia and Village Classification Survey, Prime Minister's Department, 1978.

Project identification and appraisal

19. The process of project identification outlined in paragraph 7 is carried out on the basis of guidelines laid down by federal government. The guidelines suggest that the proposed road should follow an existing village track alignment, connect a fishing village or agricultural scheme into an existing network. It should serve a minimum of 20 families and be not more than about 16 km long. Other factors such as reducing travel costs where an alternative route exists and providing socio-economic benefits to the rural people are also outlined. No quantification methods were provided for the latter.

20. The packaging of projects is undertaken by the state economic planning unit and state development office. Information on detailed alignment and costs are provided by the public works department and in some cases population data served by the route is also given by the district office.

21. Up until 1981 appraisal of projects was undertaken on the basis of cost and political acceptability in relation to the federal grant available. The selection of projects for implementation was coordinated by a state infrastructure planning committee headed by a senior state politician. Under the Third Malaysia Plan all projects selected at state level were eventually financed by federal government on a block grant basis.

SOME ASPECTS OF PROJECT PLANNING AND IMPLEMENTATION

Explicit social objectives

22. In both rural road programmes, social as well as economic objectives are expressed. The criteria of improving access to existing populations and to social amenities are common to both programmes. In the village road programme they are made explicit while in the ARRP productivity and income are to be raised implicitly, through increased access to health, education and other Government delivery services.

23. These objectives can be viewed as part of the wider socio-economic framework embodied in federal government policies. The New Economic Policy outlined in the national Malaysia Plans aims to eradicate poverty and redistribute income. As a large proportion of the poor are rural Malays, the implementation of rural roads programmes plays an important part in the attainment of national policies.

Roads and rural development

24. Public development expenditure allocations also reflect the role of rural roads in national soci-economic development. The significance of these programmes has increased since 1978, relative to other sub-sectors and to transport investment as a whole (ref. Table 2). The allocation of substantial new funds to accelerate the roads programmes signalled the government's intention to speed up rural development procedures.

Assessment techniques

25. The process of identification from the 'bottom up' means that most projects respond to local problems. Once identified at village and district level they cannot be altered at higher levels except on matters of technical detail such as alignment. An important role is therefore envisaged for state government officials whose task is programme coordination and appraisal. Unfortunately the number of departments involved (district office, state PWD, economic planning unit and development office) and their differing

responsibilities leads to problems of integration. Lack of qualified staff and of accurate and consistent data at state level aggrevated these problems.

26. The volume of projects put forward placed a heavy burden on the state PWD for detailed design and cost estimates. Under the Third Malaysia Plan, 747 projects had been approved for implementation in 11 states. With an average of 68 proposals per state and limited resources, detailed appraisals such as those based on producer surplus theory (ref. 8) were not practicable.

27. In any event, some of the projects were non-agriculture in their potential impacts. For example links were required from the main networks to new secondary schools, district hospitals and military camps. In these cases passenger rather than goods traffic would be generated in reflection of social objectives. It has been noted (ref. 9, 10) that current methods of appraisal are not wholly appropriate when applied to projects where social benefits are expected to be significant.

Design standards

28. Village road improvements are carried out by small local contractors or by community residents without adherance to design standards. In many cases such roads are too narrow for trucks and buses, though cars and taxis can be used. Taxis, and especially 'pirate' taxis (unregistered), often ply unsurfaced village roads though buses will not use such routes because of operating difficulties.

29. In contrast ARRP roads are designed with a minimum reserve of 20 m. Implementation of some projects gave rise to problems of severance of smallholding patterns. Initial delays were also created because of land acquisition complications caused by the large number of small farmers involved.

30. Part of the argument to justify such wide reserves in rural areas was based on the assumption of future traffic growth and related development. In addition a change of standards would lead to problems in the granting of maintenance funds.

31. In response to the severance problem, later designs planned alignments to bypass villages. While this approach reduced severance effects it created the need for new village access links. Of more fundamental importance, the bypassing of villages reduced perceived project benefits to rural voters and this was unpopular from the local political viewpoint.

32. The mismatch between design standards and levels of personal mobility over the 1970-80 period appeared as a problem of balance. For example the emphasis on high design standards in the ARRP to accommodate four-wheel traffic was in direct contrast to the ownership of two-wheeled vehicles commonly found in rural areas.

33. However, the provision of a 4.5 m wide paved surface allows for the introduction of truck transport (for agricultural inputs and produce outputs) and bus services. Both public and private agencies are involved in the provision of such services which meet local demands for longer trips.

Implementation and construction techniques

34. The integration of the road programmes into a cohesive network was made difficult because two different departments were responsible for implementation. Lack of qualified staff at local and state level meant that only rudimentary assessments of projects could be undertaken while wider frameworks such as road hierarchies linked to settlement patterns could not be formulated. One result was that truck and bus services could not penetrate many rural areas because of non-complementary project designs and surface treatments.

35. The state PWD has now been able to attract more staff from other states and is now establishing a decentralised operation at district level. It is hoped that this will improve project integration at the local level.

36. In addition to the problems of recruiting qualified staff to unfavourable areas, it was also found that local contracting firms were small and inexperienced. As part of government policy to stimulate local entrepeneurs many small rural road projects were awarded to small contractors. Lack of experience which in some cases was made worse where projects were located in difficult topographic areas, led to many uncompleted projects. Poor cost estimates meant cash-flow problems and on a cumulative scale resulted in state requests for extensions to block grants. Poor construction techniques often caused bridge failures, gully erosion and other surfacing problems.

37. Methods of construction also created environmental problems when implementation schedules were delayed. For example a common procedure was to construct to laterite standard first then upgrade with a premix surface at a later date. However, when upgrading funds were held up, dust pollution became a severe problem especially where distinct dry periods were experienced.

IMPLICATIONS FOR PROJECT APPRAISAL

38. From Malaysian experience there are three broad areas for concern for the appraisal of rural road projects where resources are limited:

(i) lack of accurate data and qualified expertise to carry out complex analyses,

(ii) difficulties in the measurement of certain benefits such as social impacts and generated passenger traffic, and

(iii) problems encountered in implementation which reflect on the practicability of using complicated methods where potential errors are high.

39. Multi-sectoral programmes have, however, been carried out successfully in Malaysia by a number of agencies including FELDA. In these circumstances, where finance and other resources

are available the use of producer surplus and other types of cost benefit appraisal has proved useful.

40. Data collected can be used for a number of different purposes, while cost benefit analysis assists in the understanding of local transport movements and input/output patterns. The existence of a specially-created authority to coordinate planning and implementation has also contributed to successful project completion.

Such a project has been successfully implemented in Kedah state under World Bank financing where double cropping of rice has been achieved. Substantial funds were allocated for roads and related infra-structure in the programme.

41. Where these resources are not available and where local government officers are expected to undertake planning and appraisal of rural road programmes, simplified techniques will be more appropriate. Such techniques should be based on:
(i) the level of data available,
(ii) the time available to carry out analyses,
(iii) the expertise of the staff,
(iv) the size and cost of the programme, and
(v) the objectives to be achieved.

42. It should also be stressed that where local level project identification and state level appraisal procedures exist, techniques should be standardised between different parts of the country.

THE USE OF A SIMPLIFIED APPROACH

43. Towards the end of the Third Malaysian Plan period, an economic evaluation study of the rural road programme was commissioned by the federal government (ref. 11). The study set out to rationalise planning and evaluation procedures on the basis of World Bank staff proposals (ref.12).

44. It was found that the level of detail required in those proposals was too time-consuming for the purposes of programme appraisal in the context of limited technical and financial resources. As a result a simplified method was adopted to calculate economic benefits based on ton-mile and passenger-mile cost savings for normal and generated traffic. Raw data was collected from a World Bank rural development project in one of the states to determine average cost savings. Three types of road improvement were analysed, good earth to laterite, poor earth to laterite and laterite to paved.

45. Due to time constraints, only 8 projects were eventually appraised using these simplified techniques, each analysis with local data collection taking approximately one man-week. Interestingly six appraisals produced greater cost savings for passenger traffic rather than freight. In aggregate the 66 miles of sampled roads were not found to be attractive in economic terms. Reasons for the low returns included relatively high construction costs, the availability of alternative access routes and the limited extent of future freight traffic increases. Population densities were also low in many areas. (E.g. the rural district outlined in Table 4 had a 1978 population density of 0.26 pph: the district capital had only 1800 residents.)

46. In order to rationalise assessment procedures in relation to resources available and the size of the programme, a screening procedure was devised (ref. 12). From the initial list of economic and social criteria, three were eventually recommended for use in the peninsula. These were construction cost and population served per km and the percentage of smallholdings to total cultivated area in the road's zone of influence. The latter was an attempt to measure income distribution.

Table 5. Illustration of simplified screening method.

Rural road: Kilometres:	A 8	B 6	C 7	D 10	E 12	F 5	Weight (%)
Population per km	1100	450	380	290	720	1000	40
% smallholdings/ total hectares	70	80	40	20	60	90	40
Construction cost per km ($000)	220	270	210	180	240	300	20

Adjusted for comparability (0-100), first column weighted rating, second column:

	A	B	C	D	E	F
Population per km	(100) 40	(41) 16	(35) 14	(26) 11	(65) 26	(91) 36
% smallholdings/ total hectares	(70) 28	(80) 32	(40) 16	(20) 8	(60) 24	(90) 36
Construction cost per km	(81) 16	(67) 14	(86) 18	(100) 20	(75) 15	(60) 12
Weighted rating	84	62	48	39	65	84
Rank	1	4	5	6	3	1

Source: ref. 11.

47. The criteria have now been incorporated into the Fourth Malaysia Plan process with the intention of ensuring better utilisation of resources. The choice of the criteria reflects the level of data and expertise available at state and district levels and the related requirement for accurate interpretation and analysis. Simple weights to reflect policy objectives were also recommended. Table 5 provides an illustration of the simplified screening method.

SOME REMAINING PROBLEMS

48. The use of simplified methods based on cost benefit analysis and screening criteria resulted in reduced requirements for data and time. However, many problems remained: assumptions on passenger and ton-mile cost savings were derived from different study areas. Benefits from increased internal travel, employment creation, supply of other infra-structure, social trips and quality of living improvements were not estimated.

49. Difficulties in measuring many of these benefits has led in the past to qualitative expressions of their importance (ref. 13). It is clear for example that making education and health facilities accessible for the first time has greater significance than can be reflected in passenger-mile cost savings.

50. In Malaysia, problems of urbanisation and rural deprivation are apparent. Policies to redistribute income and improve employment and living conditions include the location of government services and facilities in rural areas. Secondary schools, district hospitals and mosques for example, are not always sited in major settlements. In these cases access roads and passenger transport facilities are required. Techniques relying mainly on quantification of agricultural traffic benefits are inappropriate in such circumstances.

51. It was also found that at the project level, population and agricultural land use data was either not available or sufficiently accurate. Inter-district differences in data collection were also evident. Where such data is fundamental to forecasts of passenger and freight traffic, aggregations and the use of complex techniques are likely to increase margins of error.

52. In some cases the land use in the project area was under smallholdings with low yielding non transport-intensive crops. Where complementary programmes are not planned, and population densities static (farm saturation and outmigration balacing natural growth), future traffic flows are unlikely to be much higher than existing. In such circumstances neither road user nor producer surplus benefits may represent the true picture.

CONCLUSIONS

53. It has been suggested that based on experience from Peninsular Malaysia the use of complex appraisal techniques is not appropriate where resources are limited. Problems of implementation also indicate that the errors inherent in forecasting future activities in rural areas are likely to be compounded when many variables are included in the analysis.

54. Where a high level of data and expertise is available or technical and financial aid forthcoming the use of cost benefit techniques can provide valuable guidance in the assessment of alternative transport investments. The application of the Leitch Report framework and World Bank staff proposals to projects in the field will contribute important data and feedback. Such feedback will assist in the understanding of rural transport processes as well as in the evolution of future appraisal techniques for use in developing countries.

55. If data and expertise are principal constraints, alternative approaches based on simpler models will be more appropriate. For example a reduction in the number of variables to be used in the analysis would reduce potential errors in data manipulation. It would also allow for a concentration of effort on improving important base line data such as cost estimates and population forecasts from census data. The latter could be carried out as part of a wider and much needed effort to improve the data base for project planning as a whole. At present many departments in the peninsula use different population forecasts for use in a variety of projects.

56. Further advantages may be gained if the use of simpler techniques result in a reduction of time required for project preparation. A common complaint in Peninsular Malaysia concerned the lengthy procedures involved in project identification, analysis and approval. Project implementation may be speeded up if analysis can be reduced. Alternatively time saved could be spent in detailed design and improving implementation procedures as well as standardising data collection and updating methods.

57. A simplified analytical framework is more likely to be compatible with the level of skills available in state and local government departments. This is particularly important in countries such as Malaysia where projects are identified from the bottom up. Simpler techniques will be more readily understood by technical officers and politicians, improving the potential for inter-departmental cooperation and standardisation of forecasts.

58. The requirements for more appropriate models in these circumstances are conflicting. On the one hand they need to take account of social and other impacts, in addition to economic effects. On the other hand they need to be simple and easy to use in relation to the levels of data and skills available. They should also be in proportion to the total costs of the project - a more expensive project justifying more complex techniques.

59. Based on Malaysian experience, existing models using a small number of variables such as construction cost, length of project and population served (ref. 14) could be applied. Alternatively methods based on available census

data, including vehicle ownership patterns (ref. 15) are worth further investigation.

60. Where loan requirements direct, detailed technical and economic feasibility studies will continue to be made in developing countries, usually by consultants from developed countries (ref. 16). Where such requirements do not form part of funding arrangements, and other resources are limited, the development of simple models based on the most accurate available data will be appropriate for use in project appraisals. Work is now continuing in this Group and elsewhere (ref.17).

REFERENCES

1. ODIER L. The economic benefits of road construction and improvements. United Nations (ECAFE). Translated from the French Document 'Les intérêts économiques des travaux routiers' Undated, 105.
2. HINE J.L. The appraisal of rural feeder roads in developing countries. PTRC Seminar U. University of Warwick, July 1975, 145-150.
3. BEENHAKKER H L and CHAMMERI A. Identification and appraisal of rural road projects. World Bank Staff Paper no. 362. The World Bank Washington DC, October 1979, 1.
4. DEPARTMENT OF TRANSPORT. Report of the Advisory Committee on Trunk Road Assessment. (Chairman Sir George Leitch) HMSO, London October 1977, 17.
5. BOVILL D.I.N. Rural road appraisal methods for developing countries. Transport and Road Research Laboratory Supplementary Report 395. TRRL 1978, 19.
6. EDWARDS C. Some problems of evaluating investments in rural transport. PTRC Seminar F. University of Warwick, July 1978, 19.
7. DEPARTMENT OF TRANSPORT. Trunk road proposals - a comprehensive framework for appraisal. The Standing Advisory Committee on Trunk Road Assessment. HMSO London, October 1979, 5.
8. BOVILL D.I.N., HEGGIE I.G. and HINE J.L. A guide to transport planning within the roads sector: for developing countries. Ministry of Overseas Development. HMSO London, 1978, 33-34.
9. PLUMBE A.J. Implications of feeder road usage by the farming community of South East Thailand. PTRC Seminar F. University of Warwick, July 1978, 8.
10. CARNEMARK C.S. Some economic, social and technical aspects of rural roads. United Nations ESCAP Workshop on Rural Roads. Dacca, January 1979, 12.
11. STONIER C.E. Rural roads in Peninsular Malaysia. A preliminary economic evaluation. Government of Malaysia, Kuala Lumpur, December 1979.
12. CARNEMARK C., BIDERMAN J. and BOVET D. The economic analysis of rural road projects. World Bank Staff Working Paper No. 241. The World Bank, Washington DC, August 1976.
13. BRUSER A. Difficulties encountered in the economic evaluation of road projects in developing countries and outline of practicable solutions. PTRC Seminar H. University of Warwick, July 1979, 16.
14. JOHNSON G.P. and STEINER H.M. Evaluating social roads in Mexico. Journal of Transport Economics and Policy, VII(1), January 1973, 98-101.
15. HOWE J.D.G.F. and TENNANT B.S. Forecasting rural road travel in developing countries from studies of land use. Transport and Road Research Laboratory Report 754. TRRL 1977, 12.
16. LABAUGH W.C. Roads in developing countries. Transportation Research Board Special Report 160. Low-volume Roads, Washington DC, 1975, 25.
17. HARREL C.G. Transport research in the World Bank. Research News 1(2), The World Bank, May 1980, 5-7.

11

The development of priorities for rural roads

P. R. CORNWELL and J. M. THOMSON, Halcrow Fox and Associates

INTRODUCTION

1. The governments of many developing countries are trying to strengthen economic enterprise and production in rural areas, and to raise living standards there through improved health, educational and social services. Under favourable conditions, rural roads can contribute to these objectives by helping farmers to sell their produce profitably and by improving accessibility to social facilities.

2. Although global food production has increased in recent years at about the same rate as population, per capita food consumption has decreased in a number of countries, particularly in Sub-Saharan Africa (ref. 1), and distribution problems have aggravated the situation elsewhere. Some 26 countries in Sub-Saharan Africa with a population of more than 150 million are still reporting food shortages.

3. At the same time, population pressure and economic change have contributed to a continuing migration from rural to urban areas in many countries. The number of African cities with more than 500,000 inhabitants increased from three in 1960 to 28 in 1980 and the urban population is growing by around 6 percent per year (ref. 2). This phenomenon is well known also in Asia, South America and the Middle East. The resulting demand for urban housing, water supply, sewerage, transport, health facilities, schools and other social infrastructure places intolerable strain on both public and private resources and major efforts have been made over the past decade to discourage migration to primary urban areas by creating more productive jobs in rural areas and by improving rural living standards.

4. As the principal economic activity in rural areas, agriculture is the cornerstone of rural economic development. Major improvements have been made in many areas by the introduction of improved seeds, irrigation, the use of fertilizers and insecticides and improved cropping patterns, but much remains to be done to exploit fully the productive potential of these areas.

5. It is within this general context that the role of rural roads must be defined, whether to support schemes to improve agricultural production and marketing, to assist the deployment of better health and educational services or in a general way to encourage contact between relatively isolated rural people and modern ideas and techniques. Rural roads can be called upon to fulfill many different requirements:

- the POLITICIAN may want to promote a particular region in need of special assistance
- the ADMINISTRATOR may seek improved accessibility in and around his jurisdiction
- the ECONOMIST may aim to promote activities and investments to increase the productive potential of an area
- the SOCIAL WORKER may want to improve health and educational services.

6. A rural road may have to respond to some or all of these various requirements; there is usually no single objective and choices have to be made concerning the relative importance of broad policy objectives since they directly influence the priorities to be accorded to competing rural road investments.

7. Governments have insufficient resources to undertake all desirable schemes and a choice has to be made between competing schemes, taking due account of the relative importance of the objectives of different interest groups; the eventual rural road investment programme must be tailored to the budget available in such a way as to maximise the economic, social and political benefit from the resources invested. This requires the application of a consistent and comprehensive evaluation of alternative schemes so that:

a) worthwhile projects can be identified, prepared and selected, and

b) priorities can be established.

Highway investment in developing countries. Thomas Telford Ltd, London, 1983, 79–87

8. In the light of these introductory comments, the objectives of the paper are to:

- o clarify the broad objectives in attempting to evaluate rural roads;
- o focus attention on several practical problems which must be recognised when applying theoretical models;
- o identify criteria for the evaluation of rural roads;
- o propose a general evaluation framework incorporating the above criteria.

THE CHALLENGE OF RURAL ROAD EVALUATION

Characteristics of Rural Roads

9. Over the past twenty years, transport planners and economists have developed theories and methods for the analysis and evaluation of major highway schemes in complex urban areas and of interurban highways forming main links in national and regional networks. These methods, however, are not suitable for the evaluation of investment in rural roads (the term is used here to denote penetration or feeder roads).

10. The conventional road user surplus method of evaluation is appropriate for cases where common traffic (ie. common to the 'with project' and 'without project' situations) is substantial and generated traffic is relatively small; in such cases transport cost savings can be taken as a reliable measure of project benefits, and it is usually unnecessary to consider the underlying mechanisms of economic development. Rural road projects, by their nature, can give rise to relatively important economic and social development in the areas they serve and a large proportion, if not all of the traffic on the new or improved roads, may be associated with new development. Road user savings are not, in this case, a reliable measure of project benefits, and attention must be devoted to the underlying activities which create the associated development and its benefits. Furthermore, in the case of the typical rural road, common traffic is small and generated traffic depends upon development which may or may not materialise as rapidly and as extensively as hoped.

11. For these reasons, attention turned some years ago to the application of producer surplus methods to the evaluation of rural roads. The producers' surplus is simply the value of production, less all costs, including the cost of capital investment in equipment and infrastructure other than the road itself. The theoretical framework for the application of the producer surplus approach to the economic analysis of rural roads was published in 1976 by the World Bank (ref. 3). The method has since been applied in various countries and this conference provides an opportunity to discuss the practical lessons learned. This paper, while adhering to the producer surplus approach, does so in a very different way from most other studies, and the reasons for this are explained subsequently.

Some Practical Considerations

12. Whichever method of evaluation is used, it must have sound theoretical foundations and must respect a number of practical constraints. There are three related practical issues which are particularly acute in the case of rural roads in developing countries and which must be taken fully into account:

- o time lag
- o development constraints
- o risk and uncertainty.

Time Lag

13. A particular problem with rural roads is that development takes time - a gestation period is needed after a road is built and during which development takes place. The gestation period comprises two components: the technical gestation and the psychological gestation period. The technical gestation period consists of the minimum time necessary to bring an activity to full production; for agricultural activities it includes the preparation of land, and the planting and growth of new crops; in the case of intensive cattle rearing, for example, the elapsed time may be 10 years before full production is reached. The second component, the psychological gestation period, is the time it takes for farmers to decide and act to extend or intensify their activities; this period depends upon many cultural factors and is subject to considerable uncertainty.

14. Time lag is one of the most unpredictable aspects of the impact of a rural road and can have a profound effect upon the viability of a project. The effect of the gestation period on the net present value (NPV) of a hypothetical project is illustrated in TABLE 1. This shows the effect on NPV when annual income rises to $1000 over various gestation periods and it may be seen that NPV can be effectively halved if a 12 year gestation period is required.

TABLE 1: EFFECT OF GESTATION PERIOD ON NET PRESENT VALUE

Gestation Period (years)	Net Present Value ($)	% Loss of NPV due to gestation period
0	8060	-
3	6400	20,6
6	5430	32,2
9	4660	42,2
12	4040	49,9

NOTES: 1) Assumes 12 percent discount rate

2) NPV calculated over 30 years.

15. The rate of development associated with a rural road is therefore crucial to the economic viability of a project.

Development Constraints

16. Rural roads may or may not induce development, depending on the circumstances. It is not sufficient to hope that rural roads will in some way bring about favourable social and economic changes - it is necessary to have clear ideas about the mechanism for these changes and to incorporate them in the evaluation procedures; this requires technical knowledge related to specific development opportunities and imagination to foresee potential opportunities.

17. What are the factors to be considered? In general, there are four essential requirements for a road to induce development:

a) it must give access to areas of productive potential;

b) it must give access to a market where the output of the area can be sold;

c) the market prices must be sufficient to give an attractive profit over and above the costs of production and transport;

d) the size of the market must be sufficient to absorb the new source of supply without prices falling to an unprofitable level.

18. If these conditions are met, the opportunity exists for profitable development, and upon this profit the justification of the road rests. But, depending upon the type of production involved, other conditions may be necessary before the opportunity is exploited:

e) investment may be required, particularly if intensive production is anticipated;

f) there must be a sufficient supply of labour, whether indigenous or migrant;

g) adequate living conditions must be provided (housing, schools, medical facilities etc).

19. Transport is unlikely to be the only constraint. Even if it is the most critical one, once the transport situation has been improved, there may be others: market size, land tenure, finance. It may be necessary to overcome several constraints before the anticipated development can take place.

Risk and Uncertainty

20. In estimating the future economic and social development of an area, it is necessary to recognise the substantial uncertainty attached to many elements in the development process. For example, variations in weather and in international market prices can lead to substantial fluctuations in performance from year to year; individuals' responses to risk and innovation are largely unknown but can have a major influence on the success or failure of a scheme.

21. An evaluation methodology should try to take account of uncertainty and should recognise that it does vary from element to element. Thus for example, given the location and standard of a proposed route, construction costs can be estimated with a relatively high degree of certainty; on the other hand, future profit margins for cotton, for example, must be subject to considerable uncertainty. Implementation of a particular rural road scheme can be predicted with reasonable certainty by the agency directly responsible for constructing and maintaining rural roads, whereas implementation of the necessary complementary investments and actions by other agencies can be subject to considerable uncertainty.

22. Ultimately these factors are reflected in the estimated future traffic levels and economic benefits, and their related degree of accuracy.

23. The significance of this is shown in Fig.1, which depicts the probability distribution of alternative evaluation results. In Case A, a representative urban road, the rate of return on the project is estimated at, say, 12%, but of course there is some possibility of error. The analyst might concede that the true rate of return could conceivably turn out to be as low as 8% or as high as 16%. Curve A indicates his assessment of the probabilities of different results within this range. With the rural road, however, given the same 'best' estimate of 12%, the range of uncertainty is normally far greater, as shown in Case B. Indeed, although 12% may be the most likely result (rounding to the nearest whole number) the actual chances of this result occurring are relatively small, and the possibility of a very different result occurring is substantial.

24. In this situation it is rather misleading to say that the project is estimated to give a 12% return. The approach to be proposed is a probability approach; if the minimum required rate of return is 8%, say, one tries to assess the probability that the return will exceed 8%. Of course, this cannot be done in a rigorous, numerical fashion. Ultimately it requires a lot of judgement. But essentially one is asking <u>not</u> what is the rate of return? but rather what are the chances of the project yielding an adequate rate of return?

25. It is therefore suggested that the two principal questions for which evaluation criteria must be designed are:

i) Which of the alternative road projects are likely to induce at least sufficient benefits to make the road investment 'worthwhile' (ie which projects should proceed);

ii) of the projects which are considered 'worthwhile', what should be the order of priority and timing (ie how and when should limited resources be allocated).

26. We are concerned primarily with PROBABILITY of development and PRIORITIES for investment.

A-Urban
Probability
Rural-B
4
8
12
16
20
24
28
Rate of Return(%)

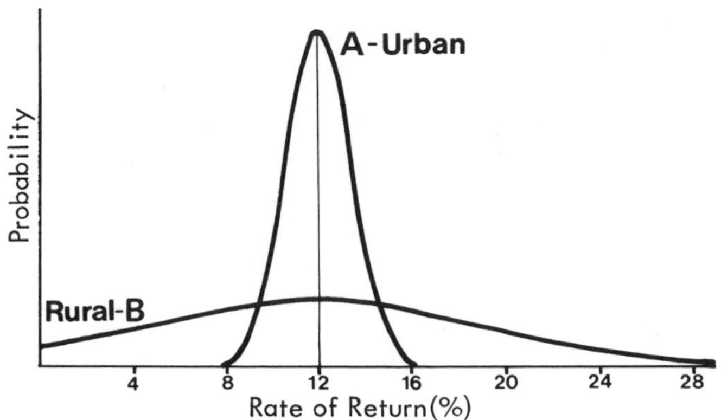

Fig. 1. Probability distributions of alternative evaluation results

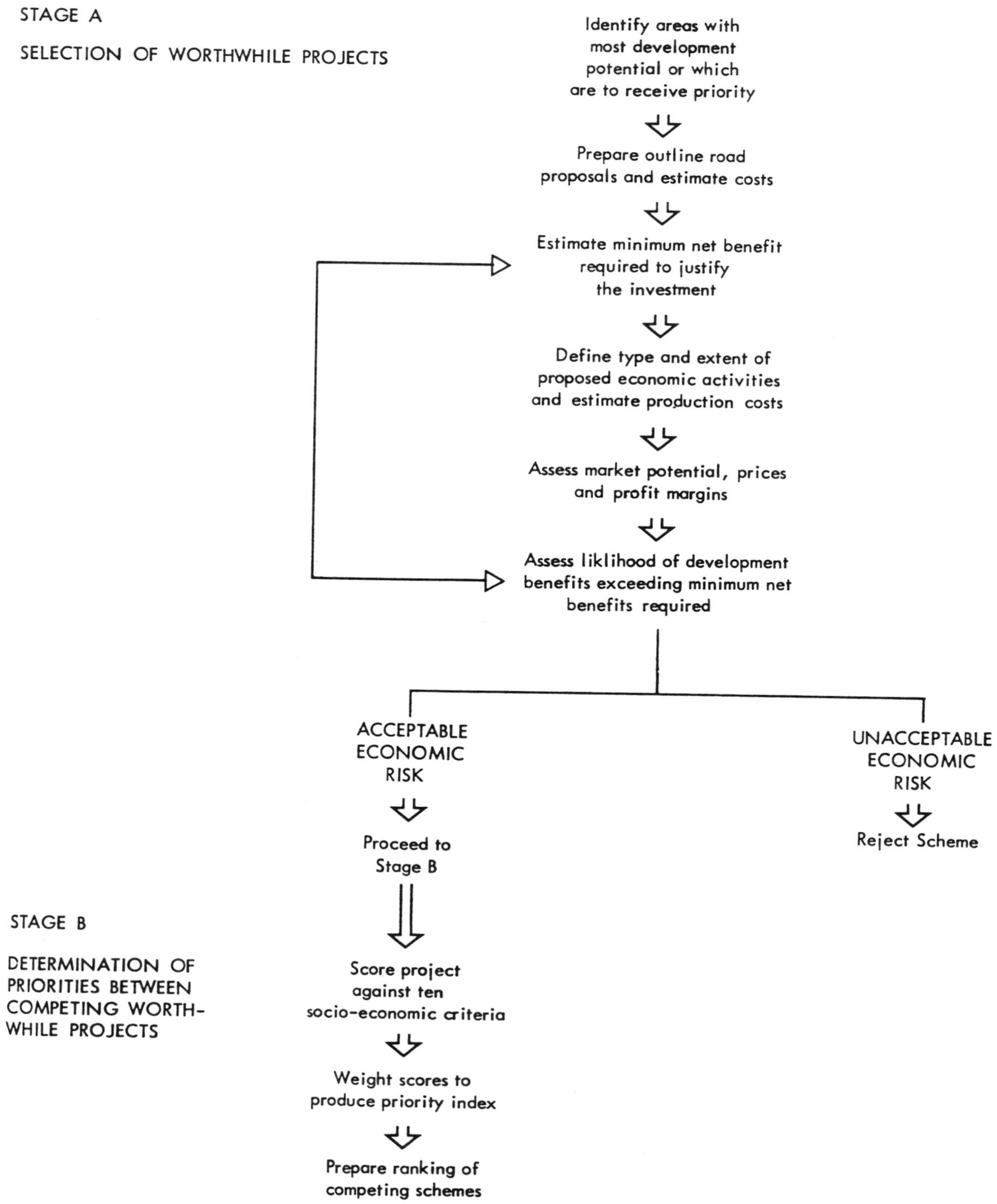

STAGE A
SELECTION OF WORTHWHILE PROJECTS
Identify areas with most development potential or which are to receive priority
Prepare outline road proposals and estimate costs
Estimate minimum net benefit required to justify the investment
Define type and extent of proposed economic activities and estimate production costs
Assess market potential, prices and profit margins
Assess liklihood of development benefits exceeding minimum net benefits required
ACCEPTABLE ECONOMIC RISK
Proceed to Stage B
UNACCEPTABLE ECONOMIC RISK
Reject Scheme
STAGE B
DETERMINATION OF PRIORITIES BETWEEN COMPETING WORTH-WHILE PROJECTS
Score project against ten socio-economic criteria
Weight scores to produce priority index
Prepare ranking of competing schemes

Fig. 2. Proposed evaluation framework

THE EVALUATION FRAMEWORK

Introduction

27. Because of the way departmental budgets are constructed and administered, and the way international loans are conceived and organised, the planner is usually required to:

a) identify candidate schemes for initial consideration;

b) select schemes for detailed evaluation;

c) determine which of the schemes are likely to be 'worthwhile';

d) determine priorities and timing.

28. He is concerned to obtain value for money and to satisfy particular social and political criteria; he must establish priorities WITHIN THE ROADS SECTOR and probably within the rural roads sector; he is NOT concerned in practice with demonstrating that a potential highway investment is more or less worthwhile than competing school or hospital investments. Such allocations are made at national level, and for social and political reasons, and not on a project by project basis.

29. The conventional method of choosing between a number of road proposals is to predict and evaluate their costs and benefits over a long period of time, and to express their net benefits as a rate of return on capital cost or, alternatively, as a net present value.

30. As indicated above, it would be misleading to predict the precise economic rate of return on a particular road, because there are too many uncertainties. One of the biggest uncertainties is the speed of development, which sharply affects the rate of return; market prices for primary commodities are notoriously unpredictable, and are often of critical importance to the rate of return; production and transport costs are relatively easy to predict but yields in virgin areas cannot be know reliably until farmers have actually started to grow crops there. The scale of development is also difficult to predict since it is influenced by non-economic factors like the supply of migrants and consideration of climate, culture and environment.

31. Consequently a less precise, two-stage approach is proposed. The general method has been applied in Bolivia and recommended for application in Upper Volta; conditions in these two countreis are very different and the application of the proposed method has highlighted difference in development requirements which can be associated with rural road projects.

32. Two main stages of analysis are proposed (see Fig. 2):

STAGE A: Selection of 'worthwhile' projects

STAGE B: Determination of priorities between competing 'worthwhile' projects.

33. The first stage of the analysis is concerned with establishing which of the competing road projects are likely to induce at least sufficient benefits to make the road investment 'worthwhile'. It is assumed here that projects are to be justified primarily on economic grounds, taking due account of social and other considerations; if roads are to be justified primarily on security, political or social grounds, they must be evaluated by other methods. When it is established that a rural road project is likely to induce benefits exceeding the investment anticipated, then it can be included on a list of projects which must compete in STAGE B of the analysis for the limited resources available.

34. The objective of STAGE B is to consider the economic and social case for each of the projects considered to be worthwhile on economic grounds, and to establish priorities for allocating budget and manpower resources. A scoring system is proposed, based on the application of ten socio-economic criteria, and results in a priority index for each road scheme, which directly determines its priority ranking. In order to obtain a balanced view of each project, it is considered essential to incorporate criteria which are not directly measurable, but whose importance may be reflected through the application of informed judgement and a scoring system. The proposed method is described below.

The Selection of Worthwhile Projects

35. The starting point is identification of areas to be developed, the elaboration of outline road proposals and estimation of the construction cost of the road, or of the first section of the road, together with feeder roads. This can be estimated quite accurately. Then the minimum benefit necessary to justify the investment can easily be calculated. The net benefit from the road consists of the net profits of the induced agricultural development, less road maintenance costs and any other net public expenditure. Given a knowledge of the crops to be produced, their production costs and market prices, one can translate the required benefits into the number of hectares of land that must be developed at an assumed rate per annum; and the slower the rate of development, the larger the number of hectares that must be developed. What is the probability that the scale and pace of development will equal or exceed these requirements? That is the key question to be answered. Also, given several proposals for penetration roads, which ones offer the highest probability of meeting their respective requirements.

36. Thus the problem is turned from an attempt to estimate a rate of return into a judgement of the probability that there will be a clearly defined minimum amount of development.

37. The first stage of the analysis proceeds by the following steps:

1) identify areas of undeveloped potential in the study area;

2) identify those areas included in 1) which are currently accessible by road, and explain the lack of development;

3) select those areas with most development and marketing potential or which it is required to promote for political reasons;

4) prepare outline proposals for penetration and feeder roads in these areas;

5) estimate investment costs;

6) identify development potential and economic activities anticipated;

7) estimate minimum net benefit required to justify the investment, taking account of the likely gestation period;

8) translate minimum net benefit into the number of hectares required to be developed and the corresponding required profit per hectare (different combinations may be reviewed at this stage);

9) consider the alternative production methods and yields in order to define expected scale and level of activity;

10) estimate production costs for the various crops and levels of technique anticipated;

11) note any physical or social constraints to realising the target development;

12) assess market potential and related transport costs;

13) estimate market prices and profit margins;

14) assess likelihood of attracting the required labour and capital;

15) assess the probability of development exceeding the minimum requirements.

38. Steps 1 to 3 are concerned with identifying areas where road investment should be considered in detail and may either be conducted by a central planning team or through regional development agencies, depending upon the size of the country and the administrative structure. The development aspirations of the country and specific development schemes provide the background and framework for identifying the favourable areas. Roads may be included for economic, political or social reasons at this stage (Step 4).

39. It is next necessary to consider how in practice sufficient benefits can potentially be generated in the area to justify the estimated capital investment in roads (Steps 5 and 6); natural resources, soil conditions, climate, water availability must all be considered in order to make a clear statement of the specific agricultural or processing activities anticipated. In particular, there must be an adequate market for the products proposed.

40. Now the total required annual benefits to justify the road can be calculated, in the light of the activities anticipated (Step 7), but at this stage the first major unknown factor must be introduced, namely the gestation period.

41. Given the total benefit to be generated, this can now be translated into the expected number of hectares to be developed, the intensity of development and the resulting benefit per hectare required to justify the road (Step 8). At this stage, the practical opportunities offered by a particular area must be considered carefully in order to ensure the realism of developing the target number of hectares at the required intensity. This means determining expected activity patterns (eg cropping patterns), technical means of production and expected yields (Step 9).

42. The costs of production can now be estimated (Step 10) from a knowledge of the techniques envisaged, the factors of production and expected unit prices. The inclusion of land and labour costs should not be neglected at this stage.

43. Before proceeding further, it is necessary to consider whether there are any particular physical or social constraints to bringing about the level and type of development anticipated. These can include environmental factors (disease, climate), cultural factors (eg tribal or linguistic differences which may inhibit migration) and social values (the importance or otherwise attached to material desires and entrepreneurial activity). If any such constraints do exist, then a judgement must be made concerning the likely effect on the proposed development (Step 11).

44. The economic case for justifying rural road investment rests upon increased production, either through the extension of the area of land being exploited or through the intensification of activities on currently utilised land. In either case, the production must be sold. Markets must be identified, their potential assessed (Step 12), and transport costs to market must be estimated.

45. Sooner or later, it is necessary to make a judgement concerning future market prices in order to estimate farmers' expected profit margins (Step 13). This step is fundamental, yet is subject to considerable uncertainty. Local markets frequently have only a limited capcity and a sudden increase in supply may lead to sharp downward movements in price. However, local markets sometimes serve as collection points for onward distribution to national and world markets; in these cases, market capacity may be less of a problem, but price can be subject to regulation and to numerous external factors. Whilst recognising these uncertainties, the analyst must make assumptions concerning the possibilities of selling produce and likely market prices. Once future market prices have been estimated, profit margins can be calculated by deducting production and transport costs. These profit margins can then

be compared initially with the targets established in Steps 7 and 8.

46. However, the final assessment of the likelihood of development potential exceeding that required to justify the road investment (Step 15) must take account of the requirements of labour and capital (Step 14). The opening up of virgin territory may require substantial migration, in which case careful consideration must be given to the potential availability of migrants, to the provision of housing and social infrastructure and to the likely social problems associated with such transmigration. Similarly, the successful realisation of capital intensive development must depend upon the availability of the required capital, with clear indications of its source. Thus the likely benefits of the anticipated development may be judged against the level required to justify the proposed road investment, in the light of explicit assumptions related to the economic activities anticipated and to the availability of specific market opportunities. If the scale and the profitability of the anticipated activities (Steps 9-13) are judged likely to induce benefits exceeding those required to justify the road investment (Steps 7 and 8), the road may be tentatively accepted for analysis of the relative priorities of competing projects in STAGE B.

Determination of Priorities for Investment

47. If the above procedures are applied consistently to a region or country, then a list of candidate projects may be established. The objective in STAGE B of the evaluation then becomes the determination of priorities.

48. The final assessment should not depend solely on the quantification of economic factors. Explicity or implicity, difficult judgements must be made of farmers' reactions, market prospects, the importance of social objectives and other factors. The method proposed for STAGE B of this analysis therefore centres on a scoring system incorporating the criteria deemed to be important.

49. A general framework is outlined below, but it is stressed that the formulation, numerical range and weighting may vary from country to country, according to circumstances.

50. Ten criteria are proposed for integration in the scoring system for each road and its associated development:

1) level of benefits required to justify the road investment;

2) level of profitability anticipated;

3) price variation;

4) price prospects;

5) market size;

6) labour requirements;

7) capital requirements;

8) climate;

9) socio-economic response;

10) accessibility.

51. In general, the lower level of benefits per hectare requried to justify the road investment, the higher the probability of achieving the required level. Thus criterion 1 is a measure of the scale of development required, and is expressed in money/hectare for the expected number of hectares.

52. Profit margins (Criterion 2), expressed in money/hectare and weighted for different crops, express in numerical terms the potential for achieving the required level of benefits. The higher the potential level of profitability, the higher the probability of achieving the benefit target and vice versa.

53. The prices of some commodities (Criterion 3) are highly variable, which introduces an additional risk into their production. World commodity prices since 1955 in constant dollars are available from World Bank statistics and the historical value of standard deviation to mean serves as an indicator of price variability.

54. Price prospects (Criterion 4) are, of course, fundamental to the viability of development. Local estimates of price prospects may be adopted or alternatively World Bank forecasts may be accepted. The suggested unit is forecast 1990 price divided by 1980 price (in constant money).

55. It is extremely difficult to estimate market size (Criterion 5), but an attempt must be made to judge the broad relationship between scale of production and markets in order to ascertain whether or not there is a reasonable possibility of selling the additional production at attractive prices. This judgement must be made, even if it is judged in subjective terms and scored on a coarse scale.

56. The availability of adequate manpower (Criterion 6) may be a potential constraint to development in some regions, particularly in opening up virgin areas where substantial migration may be required; in areas where adequate manpower is available, this criterion may be ignored. Requirements may be expressed in man-days per hectare.

57. Some types of development, particulary those with long gestation and those based on intensive production methods, may require capital to finance the early years when there is little or no income. In general, the more capital required per hectare (Criterion 7), the more difficult it will be to attract development. Once again, precise quantification is difficult, but subjective scoring of requirements on a 0-10 scale does give an indication of the relative importance of this factor.

TABLE 2: SAMPLE CALCULATION OF ROAD PRIORITY INDEX

Criterion	Score	Comment	Weighting	Value
1. Level of Benefits Required to Justify Investment	7	expressed in money/ha/year and scored on 0-10 scale	2,0	14
2. Level of Profitability Anticipated	8	expressed in money/ha/year	2,0	16
3. Price Variation	6	ratio of standard deviation to mean of prices over past 10 years; calculate and score on 0-10 scale	1,5	9
4. Price Prospects	7	ratio of predicted 1990 price to 1980 price; calculate and score on 0-10 scale	2,0	14
5. Market Size	4	judgement of market capacity in relation to expected additional production; score on 0-10 scale	1,5	6
6. Labour Requirements	10	ratio of manpower availability to manpower requirements; score on 0-10 scale	1,0	10
7. Capital Requirements	8	judgement of capital requirements and probability of securing; score on 0-10 scale	1,0	8
8. Climate	8	judgement of climatic constraints; score on 0-10 scale	0,5	4
9. Socio-Economic Response	7	judgement of likely reaction to development opportunities	1,0	7
10. Accessibility	4	accessibility index may be calculated to social infrastructure or urban centres, depending upon circumstances; calculate index and score on 0-10 scale	0,5	2

NOTES: 1) Scores are for hypothetical project

2) Weightings are indicative only

PRIORITY INDEX = 90

58. Climate (Criterion 8) may be a constraint in some instances, because of the attractiveness or oppression of temperature and/or humidity. This criterion will only be needed where schemes subject to different climatic regimes are to be compared. Scoring must again be subjective and on a 0-10 scale.

59. It may be observed that people from different social and ethnic backgrounds respond to economic opportunities in different ways. The above analysis implicitly assumes certain behavioural responses to opportunities created by road improvements, and in some circumstances it will be appropriate to take account of variations in response from one part of a country to another (Criterion 9). A subjective scoring on a 0-10 scale is again proposed.

60. Finally, accessibility is considered (Criterion 10). Depending upon the precise physical characteristics of the particular study area, an accessibility index can be constructed, as a measure of accessibility to centres of population and social infrastructure such as educational and medical facilities. Security considerations can also be included under this heading, if required. The index is calculated and transposed to a 0-10 scale.

61. Thus numerical scores may be attached to each one of the ten proposed criteria. Some criteria are clearly more important than others and a weighting should therefore be applied to each score. The choice of weighting is a matter of judgement and in the final analysis is for the decision maker to define. What the method does is to force the analyst and the decision maker to confront the complex criteria involved and to make explicit the importance to be attached to each factor, particulary risk and uncertainty, even when quantification requires the exercise of informed judgement.

62. Thus the numerical values attached to each of the ten criteria may be weighted to reflect political priorities, resulting in a priority index for each candidate road project (see example in Table 2). Each of the roads accepted in STAGE A is a worthwhile project and the broad priorities for implementation are given by the indices calculated in STAGE B. Within these broad priorities, the precise staging of each project must be established in the light of available construction resources and of the expected rate of development.

CONCLUDING REMARKS

63. The approach outlined above builds on the established theoretical framework and takes full account of the practical difficulties encountered in evaluating rural roads in developing countries. Firstly, it identifies those projects which are likely to be worthwhile in economic terms and, secondly, it establishes the relative priorities of candidate projects, taking due account of both economic and non-economic factors. The method recognises the uncertainties associated with development in rural areas and is deliberately flexible in order to adapt to local circumstances and values in a particular region or country.

64. By incorporating both numerical estimates of quantifiable factors and judgemental scores of other factors, the method blends the analytical skills of the professional engineer, planner and economist with the knowledge and values of the local decision maker.

65. What are the implications of adopting such an approach? To Government Agencies it means accepting the principle of applying a consistent, analytical method to determine the likely economic performance of alternative road investments and opens the way to including non-economic factors in an explicit manner in the determination of investment priorities; it means combining analytical techniques with local values; it means creating a Technical Unit in government, staffed by development economists, transport economists and engineers to conduct the identification, preparation and evaluation of rural road schemes.

66. To the Funding Agencies it means recognising that it is not realistic to attach precise numbers to economic rates of return, given the uncertainties surrounding rural road projects; encouragement must be forthcoming to broaden the evaluation process to include non-economic factors, whilst maintaining the criterion of likely economic viability for selected projects.

67. To Planners and Economists it means building on the present theoretical framework, presented in World Bank Paper No 241, to incorporate uncertainty and to introduce social and other non-economic factors in the development of rural road priorities.

68. To the People it means more chance of seeing the implementation of projects which will enhance human development, as well as stimulating economic opportunities.

REFERENCES

1. WORLD BANK, World Development Report 1981

2. WORLD BANK, Accelerated Development in Sub-Saharan Africa, 1981

3. WORLD BANK STAFF WORKING PAPER No 241, The Economic Analysis of Rural Road Projects

Discussion on Papers 9–11

Ms C.C. COOK *(Introduction to Paper 9)*: Paper 9 reflects the results of a review of some thirty completed and ongoing World Bank-assisted projects in the area of rural roads development, including both highway projects and road components in rural development projects. It also reflects the combined experience of the authors with rural infrastructure projects in their home countries - Finland, Australia, and the United States - and impressions gathered in our conversations with Bank staff, consultants, counterparts, and officials of other international development assistance organizations over the last ten years. We have identified some seven areas where rural roads project preparation requires decisions concerning the institutional framework within which projects will be planned, constructed and maintained. The Paper reflects our sense that the role of roads in rural development is somewhat different from the role of highways in national development, and that this difference implies the need for more decentralized, more flexible institutional structures, capable of mobilizing local resources and directing them towards locally defined development objectives.

Our principal conclusion is that there is no single answer to the question of appropriate institutional design for rural roads projects. In each situation, decisions must be made by taking into account the full range of available resources and the true nature of the task to be accomplished. In the long run, only local communities will possess the resources and the political will to integrate such projects in their programme of public responsibilities. It is essential, therefore, that communities become involved at every stage: in the selection of links to be improved and levels of improvement; in the choice of appropriate construction and maintenance technologies; in the mobilization of material, financial, and human resources for programme execution; and in the provision and use of transport services.

Mr I.G. HEGGIE *(Freelance Consultant, Oxford)*: Mr Smith in Paper 10 tackles an extremely important question: can we devise simple yet efficient methods of preselecting and screening rural roads? The purpose is to reduce the analytical effort involved and to make the methods simple enough for use by local provincial staff. Mr Smith says we can and suggests an index which in his case combines 'population per km', 'percent smallholdings to total area' and 'construction cost per km'. (Table 5, Paper 10).

I have some difficulty with this index. I would expect population, weighted by percentage of smallholdings, to contribute positively to an index of priority and for cost to contribute negatively to it. In his index, they are both added, and I suggest this may be incorrect. If, instead, you take weighted population and then calculate the ratio of cost to weighted population, you arrive at an index that gives the same priority ordering as Table 5 and you also have an index that corresponds to one of the two indices suggested in Carnemark's 1976 paper on rural roads (ref. 1). Carnemark suggested two screening criteria: the ratio of construction costs to population (this is comparable to Mr Smith's index) and the ratio of cost to the influence area of the road.

In a subsequent IBRD Mission Report on rural roads in Indonesia (ref. 2), these indices were compared with the actual rate of return on four roads, but the correlations were not impressive. The population index, at least, looked promising but the influence area index moved in the wrong direction.

What I find surprising is that indexes like this are proposed without any empirical justification and quickly become part of the conventional wisdom. I therefore decided to test them on material gathered by myself during a recent Asian Development Bank Mission in the Philippines. To do so I plotted the internal rate of return (IRR) for each scheme on the vertical y axis against Carnemark's two indexes on the horizontal x axis. The aim was to find a unique value of x which separated the acceptable IRRs from those that failed to provide a minimum rate of return. Neither of these indexes did that. Indeed there appeared to be no consistent relationship between them and the IRR.

My conclusion is thus that the indices being canvassed in Paper 10 are not efficient and cannot be accepted as a realistic alternative to current methods of evaluation. The approach adopted in Paper 11 by Cornwell and Thomson - which relies on permissive rather than acceptance criteria - seems more likely to succeed.

Mr A.J. PLUMBE *(Bradford University)*: It has been my experience in conducting a number of transport studies in rural areas of developing countries (of both on- and off-road movement), that passenger rather than goods movement forms the predominant component of total flows. Even within goods movement, the component of agricultural output from the area served by a feeder road is sometimes exceeded by that of construction materials and incoming food. Hence I would question the advisability of adopting in all cases the producer-surplus methodology advocated in Paper 10 for rural road appraisal, and it is not surprising that agricultural production in rural areas has proved to be a poor predictor of the level of generated traffic. The valuation of passenger travel in rural areas warrants urgent research in order to improve the forecasting, the choice of appropriate criteria and the appraisal methodology used in rural road assessments.

I find attractive the idea of changing from an attempt to estimate a rate of return to a judgement of the probability that there will be a clearly defined minimum amount of development. I envisage difficulties, however, in putting the concept into practice and would ask the Authors of Paper 11 whether it has been possible operationally to derive such probabilities measured in a rigorous manner?

The Authors of Paper 11 suggest that the minimum level of net benefits needed to justify a rural road project should be translated into a minimum number of hectares required to be cultivated at a given level of profit per hectare, and this should be compared with the actual cultivated hectarage which could be expected to occur with the investment. This implies forecasting future cultivated hectarage rather than agricultural output directly. Forecasting areas cultivated would be even more difficult than forecasting agricultural output owing to the lack of time series data, problems of measuring and defining cultivated areas, and the difficulty of translating cultivated areas into transport demand because of the fact that the intensity of cultivation is frequently adjusted during the cultivation cycle.

Mr H. VAN SMAALEN *(Agricultural University, The Netherlands)*: I agree with Mr Smith's simplified screening procedure (Paper 10), with the main weight on the criteria 'population per km' and '% smallholdings'. To underline the value of a method which does not entail the calculation of economic benefits, I would point out that I could guess with a nearly 100% certainty that all the delegates at this conference are living near a paved road and I am also sure that for none of these roads the economic benefits were calculated. The roads were built because we wanted them.

This raises the question: why should rural roads in developing countries be economically justified? My answer is: there is no reason. Roads serve people and the more they do the more they are justified. But money being scarce, we should combine Mr Smith's first criterion (population per km) with his third one (construction cost per km) into one criterion: population served per £1000. My final suggestion is that the present value of the maintenance costs should be added to the construction costs.

Mr N.D. LEA *(N.D. Lea & Associates Ltd, Vancouver)*: The method suggested in Paper 11 for developing rural road priorities uses judgemental scoring and weighting of ten indices. Until such methodology has been demonstrated as reliable through empirical economic data, practical experience in each region, such as reported in Paper 15, may be considered a more reliable guide for investment in rural roads. Proposals for investment in the upgrading of an existing road, even a seasonal road of poor quality, may be evaluated with respect to potential transport cost-savings.

Defining a penetration road as one which makes possible motor-vehicle access to a region which has previously been completely inaccessible to such vehicles, it is clear that penetration roads require different evaluation techniques. Because penetration roads have no history of motor-vehicle traffic, it is tempting to use a judgemental scoring technique in their evaluation. But even for this there may be better methods.

Parts of Canada are more poorly served by roads than the section of Ghana reported in Paper 15. Indeed, rather than the Ghana experience of only 2% of population farther than 2 km from a road, much of the Canadian north has only some 2% of the population closer than 2 km to a road. In Canada, a new penetration road may be essential to a new extractive industrial development such as a mine with the associated new mining town. In this case, for investment assessment the penetration road may be considered as a part of the total mining-project capital cost.

It is more difficult to evaluate penetration roads to existing communities which in Canada are usually fishing and hunting villages, and in tropical developing countries may be agriculturally-based communities. In both cases, however, some appreciation of the value of an all-weather road may be obtained either by monitoring present traffic, whatever the mode (aircraft or foot or bicycle), or by building a very low-cost road and evaluating further benefits through cost reduction. In northern Canada, such low-cost roads may be winter roads with a capital cost of practically zero and all costs are for maintenance and operating. These winter roads are only serviceable for about four months of the year, whereas in tropical developing countries very low-cost roads (say less than £500 per km) may be passable for 10 months. Even so, the traffic which develops on the low-cost roads is the sector of traffic which will benefit the most by an improved road. Thus building a low cost road is a good first step in most instances where a penetration road is proposed.

The low-cost roads may be expected to attract the more important truck traffic and, if adequate traffic does not develop, the low cost road is comparatively easy to abandon without serious political consequences.

The amount of person-travel on penetration roads is directly proportional to income and population and inversely proportional to a power of the distance to the next village or town (a situation similar to that found in Ghana as reported in Paper 15). This person-travel however is quite elastic to price, as can be seen by the much smaller amount of air travel. Thus care must be taken in justifying a high-cost penetration road through travel by persons who would not be willing to pay an appropriate share of the cost.

Mr E. KONSTANTAS *(Doxiadis Associates Int., Athens)*: I would like to describe the use and test of a process of evaluation of priorities similar to that put forward in Paper 11, adopted in Cross River State, Nigeria. The preparation of a road master plan in Cross River State and a 25-year programme for road construction or improvement was assigned to Doxiadis Associates in 1981. Cross River State occupies about 30,000 km^2 and has a population of about five million. The work was carried in three stages.

Stage 1. A detailed survey was conducted along the whole network; 2600 km of existing road network were monitored and the related infrastructural, operational, organizational and environmental problems were recorded.

Stage 2. A road network was proposed for the period up to 2005, consisting of about 4100 km of roads (classified in four categories). Population, population distribution, employment structure, social and economic development (agriculture, forestry and so on) were taken into consideration. The selection and establishment of roads cannot be made without such integrated regional development planning.

Stage 3. For the selection of the road projects to be executed initially the process of priority-evaluation was applied. The main steps of this process were:

(a) setting criteria for priorities
(b) identifying the measures of performance for each criterion
(c) weighting the importance of each criterion according to the measure of performance used
(d) weighing the importance of each criterion of priority against another
(e) identifying the performance of each road with respect to the measure of performance selected for each road
(f) multiplying the weight corresponding to the performance of each road with the weight of importance of each criterion for the road.

The criteria-setting for the evaluation of the priorities for road construction or improvement were:

(a) administrative status of town served by the road
(b) population estimated to be served in target year
(c) volume of traffic served
(d) present road pavement condition
(e) proposed hierarchy for roads
(f) present contractual status of road
(g) economic purpose of the road
(h) service provided to the whole state.

The weights of the criteria were established next. Having identified the weights of importance of each priority criterion, the performance of each and every road with respect to the measure of performance of each criterion was identified and then the sum of the products of the corresponding weight was calculated giving the priority ranking of all roads. This priority-ranking was adopted in priorities of construction or improvements of the proposed roads.

Mr J.L. HINE *(Overseas Unit, Transport and Road Research Laboratory)*: In Paper 10, Mr Smith indicates a need for simplified procedures for planning rural roads, and it has been suggested that conventional appraisals are too costly to be useful. I would like to suggest that conventional cost-benefit appraisal is a useful tool for planning rural roads. If necessary only a sample of rural roads need be appraised and the results grossed up.

Perhaps the most important component of a road appraisal is the change in transport costs which will be brought about by the road investment. Conventional transport user cost-savings suggest that normal traffic benefits are proportional to the change in transport costs. Induced traffic benefits (which may be regarded as a proxy for pure development benefits) can be seen to be a function of the square of the difference in transport costs. This can be seen from the simple example shown in Fig.1 of two instances in which transport costs are reduced from C_1 to C_2 and from C_1 to C_3, the latter being twice the former. Normal traffic benefits for traffic level Ot are equivalent to area C_1, ABC_2 for a change in transport costs from C_1 to C_2. Induced traffic benefits are equivalent to the area ABD for the same change in transport costs, the new traffic having risen to Ot_2.

Induced traffic benefits for a change in transport costs from C_1 to C_2 are equivalent to the area AEF (with traffic levels having risen to level Ot_3) which is equivalent to four times the induced traffic benefits of the previous case. This analysis is dependent on both a linear demand curve and on the usual assumptions of partial equilibrium analysis. However, I would suggest that the analysis is useful in drawing attention to the hypothesis that 'pure development' benefits are likely to be highly dependent on the forecast change in transport costs.

Another useful reason for identifying the change in transport costs is to help identify where social benefits from road construction might occur. It seems to me that for areas which have broadly even distributions of population and similar levels of accessibility, the social benefits from new road investment are likely to be highly dependent on the change in transport costs. It is difficult to believe that any substantial social benefits will arise from road projects which induce little or no change in transport costs.

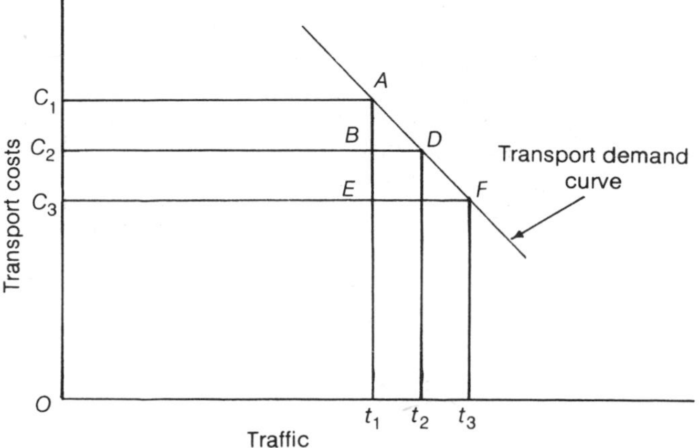

Fig.1

Mr P.W.D.H. ROBERTS *(Overseas Unit, Transport and Road Research Laboratory)*: There is an increasing recognition of the difficulties arising in the post-construction or operation phase of road projects. This is leading to an awareness of the importance of local institutional capacity. For the most part, the views expressed in Paper 9 are welcome in this respect, but I would add some warnings. It is important to distinguish between the following alternative approaches to institution development:

(a) the reorganization of existing institutions or the creation of new ones with the intention of establishing support for desired development plans
(b) the recognition and support of existing institutions (including some which are very informal, usually at the local level) with the adjustment of development plans to make best use of the capabilities of these institutions.

The second approach has been neglected too often in favour of the first. This is apparent from a number of maintenance projects in Africa. On the basis of this experience I should like to draw attention to two common and important problems.

In several cases the maintenance division of the national highway authority has been subject to extensive reorganization. This may have created a seemingly rational structure but has generally neglected the prevailing serious shortage of key resources and the very low productivity in most maintenance activities. Moreover, certain capabilities, such as the relative reliability of labour for routine roadside maintenance and the experience of unqualified supervisors and foremen, tend to be neglected or underrated and are therefore subsequently under-utilized.

A further error is commonly made where there is a shortage of technical ability at the district or rural level. This is the placing of responsibility for maintenance of local roads with the central government organization. As a result, the national maintenance organization is over-burdened and diverted from its main responsibilities which must be towards the national road system. Moreover this approach neglects the mechanisms which are effective in most local communities for identifying their own priorities and even managing or mobilizing resources to meet those priorities. The strategy for dealing with the present massive maintenance deficit in many countries must be to clearly place responsibility for local portions of the road network with the communities that they serve. Any pretence of meeting this responsibility from an inadequate central organization only undermines the commitment of the local institutions whilst failing to replace them with any effective alternative.

Dr J.B. METCALF *(Australian Road Research Board)*: Paper 10 gives a critique of cost/benefit appraisal methods when data and expertise is limited and outlines a simplified approach. In the simplified approach, clearly the weightings are crucial. Could the Author discuss their derivation and how they reflect policy objectives? Has any study been made of the relative ranking of project by the simplified method and cost/benefit analysis approaches, and does this lead us to a costing mechanism for policy objection?

Ms C.C. COOK *(World Bank, Washington)*: A great deal of thought has been given to the screening and simplified evaluation of rural roads for developing countries. The World Bank welcomes this opportunity for an exchange of views on a topic with which we have been concerned for some time. Currently, we have under review a number of projects on which different screening methods have been applied, in order to determine which parameters and relationships are most efficient in predicting the outcome of a subsequent economic evaluation. This work is not yet complete. However, our preliminary findings indicate that it is not efficient to mix social and political criteria together with economic criteria in a single screening procedure. We recommend that any socio-political prescreening procedure be carried out in advance of any analysis which would be expected to limit or replace a full economic evaluation. Related issues concerning the valuation of social benefits and the social distribution of both costs and benefits are being addressed by two other items in our research programme, one on rural mobility and one on rural transport services. We take this opportunity to solicit comments and contributions from the international engineering community on these topics.

Mr M. BOSTOCK *(Arup Economic Consultants)*: The work recently undertaken by Arup Economic Consultants for the Government of Botswana supports the main thrust of Paper 11. I share the Authors' sense of frustration over the state of the art. It seems to me, however, that their Paper does not go far enough in demonstrating their evaluation method. Paper 11 is confined to developing a procedure for selection and ranking. In my experience, it is necessary to separate social criteria from economic data at the initial stage of screening. This separation is particularly important because bilateral aid-donor agencies use different criteria in selecting rural roads and different emphasis is placed on the social aspects of development. Moreover, with a mixed social/economic ranking, there is no guarantee that the roads given the highest priority in the screening process will be those which demonstrate the highest internal rates of return. Certainly, it seems that, given the present evaluation procedures of most multilateral agencies, we cannot get away from appraising roads based on internal rates of return.

Ms COOK *(Paper 9)*: We certainly do not disagree with Mr Roberts' recommendations; indeed, we feel that they are strongly supported by the arguments presented in Paper 9. The Bank is continually trying to find ways in which we can support or strengthen local institutions through international development assistance. In some cases, we have found that a central government organization can become an effective intermediary for the development of community capabilities. In other cases, our efforts in this direction have failed. At present, we do not advocate any particular approach, but urge that institutional project design take into account all the factors present in each specific situation, in order to serve the long-range objective of developing local institutions to the point where they can assume full responsibility for rural road maintenance.

Mr SMITH *(Paper 10)*: In response to Mr Heggie's interesting views I would make two observations. Firstly, the construction cost criterion, as he rightly points out, is a negative contribution to the index and has therefore been applied inversely, that is, the higher the cost the lower the points awarded. Secondly, the index as referred to in Table 5 is an illustration and is to be further tested and modified as recommended in the study (ref. 11 of Paper 10).

I agree with Mr Plumbe about the importance of rural passenger transport and am at present carrying out research in this area (see paragraph 27). The producer-surplus approach is thus only one of a number of tools for use in project appraisals.

In answer to Mr Van Smaalen's question, I would suggest that the problem is not so much the economical justification of rural roads as the justification of the amount of expenditure *vis-à-vis* other sectors of public expenditure. I agree that maintenance costs should be included in the cost criterion.

Cost/benefit analysis, as Mr Hine points out, is a useful technique, amongst others, for planning rural roads and aids in the understanding of rural travel demand (see paragraph 54). The difficulties in quantification of some social benefits have in the past, however, meant that economic issues have been given more importance. In rural development programmes where social improvements are required, greater emphasis needs to be placed on the assessment of social rather than economic benefits.

The derivation of weights used in the simplified approach should reflect policy objectives, as Dr Metcalf points out. The weightings illustrated in Table 5 are only preliminary and reflect perceived differences of importance. They are derived from national and regional, socio-economic and political objectives in response to regional and ethnic imbalances. (See paragraphs 22-24.)

Mr CORNWELL and Mr THOMSON *(Paper 11)*: Mr Plumbe mentioned the importance of passenger travel in rural areas of developing countries and suggested that the producer's surplus method of evaluation ignored passenger traffic. This is not necessarily true: the producer's surplus method applies in principle as much to passenger as to goods traffic. In Upper Volta we tried hard to study the impact of better roads on the generation and distribution of passenger trips. But in some cases where a penetration road is built into an uninhabited area the object is to attract migrants in order to stimulate production. Passenger trips arise, of course, but they do not mean much in themselves; one has to consider the living conditions of the migrants in comparison with what they would have had if they had not migrated. The essence of the producers' surplus method in spite of its unfortunate name - is that it looks at the activities served by trips rather than at the trips themselves, and this applies to both passenger and goods trips.

Mr Plumbe suggested that it might be difficult in practice to estimate the probability of obtaining a satisfactory rate of return. So it is, but the point is that the conventional method can be so misleading, when applied to rural road projects, that one must try to do something better. Using the conventional method we may conclude 'Our best estimate of the rate of return is 12%,' but we do not add, 'However, the probability of getting 12% - to the nearest whole number - is only 5%; moreover, for all we know, there may be other results, like 7% or 15%, which actually are more probable than our best estimate, because two or more less-than-best estimates could give the same result. Hence, although our calculations, which contain an embarrassing number of guesses and assumptions, yield a figure of 12%, we could not honestly advise anyone to make a decision on this information alone.' In fact, rural road schemes are inevitably speculative, if they depend on development impacts, and the only sensible course is to treat them as such and try to assess, in general terms, the probability of success. But to do this rigorously is no easier with a road scheme than it is with a stock market speculation.

We do not agree with Mr Plumbe that it is more difficult to predict the area of cultivation than the output, although we would agree that both are difficult.

Mr Van Smaalen speaks like an anti-economist. There is no reason, he says, why rural roads should be economically justified. Yet he immediately concedes that money is scarce and starts to suggest ways of deciding which roads should get the money. So he is not really an anti-economist, just a bad economist.

We acknowledge the other comments bearing on our Paper and are broadly in agreement with the views expressed.

REFERENCES

1. CARNEMARK K. The economic analysis of rural road projects. International Bank for Reconstruction and Development (of the United Nations Organization), Washington D.C., 1976, Staff working paper no. 241.
2. INTERNATIONAL BANK FOR RECONSTRUCTION AND DEVELOPMENT (of the United Nations Organization). Mission report on Indonesian rural roads. IBRD, Washington D.C., 1981. Unpublished.

12

The economics of overlay design and road rehabilitation

R. L MITCHELL, MLM, BSc(Eng), BA, MPhil, FICE, FZweIE, University of Zimbabwe

In Zimbabwe the Ministry of Roads and Rail Traffic has long experience of rehabilitation. It has been found that strengthening by granular is usually cheaper than by premix overlay, is more flexible, and results in a better ride. The economics of design are analysed and rehabilitation costs given.

INTRODUCTION

1. It is North America and Western Europe that lead the advance of knowledge in highway engineering. However in these countries premix surfaces are the norm and traffic counts are high, the cost of overlay being insignificant in terms of traffic usage. In developing countries income from the road user is small compared to the essential expenditure on roads, and it is often impossible to find the necessary capital to follow overlay methods developed in the advanced countries. Moreover the traffic warrant seldom justifies the use of premix.

2. Zimbabwe is more advanced, and indeed, more wealthy per capita than most developing countries, and has for more than three decades maintained a small road research organisation whilst continuing with the construction of its national highway programme. Since its road system has been designed for twenty years, and much of the main primary road network has reached its design life, it has developed methods of pavement rehabilitation which are adequate but cheaper than those pioneered overseas. The techniques are discussed in terms of 1980 costs, converted to US dollars.

3. In the thirties, strip roads were constructed along the then main road skeleton of Southern Rhodesia, but this system proved to be uneconomic after Worldl War II with the increased pace of development, and a programme of two lane black top roads was commenced until, thirty five years later, a network of some 6000 km existed in the newly independent state of Zimbabwe. However, even with excellent maintenance, many kilometres needed rehabilitation a decade ago, and ten years experience has been gained in the art of strengthening and rehabilitation, which is now the major main road problem.

4. At that time the only non-destructive testing records were those of road roughness, as measured continuously with a bump integrator, and of Benkelman beam deflections. Traffic records in terms of both vehicle counts and axle loading were available, and typical road rehabilitation carried about 500 v.p.d per lane with 15% vehicles over 3 tonnes. Allowing for growth design for the following 20 years usually was for one to three million E80 axle repetitions.

5. Typical formation of the early period was a surfacing width of 6,7m on a gravel width of 9,6m with shoulder slopes of 1 in 2, the F.R.L. being some 500mm above ground level, whilst current new construction is for 7m black top on a 10m base, with 1 in 4 shoulder slope. Figs 1 and 2 show typical roads, whilst Figs 3(a) and (b) illustrate alternative rehabilitation techniques.

Premix Overlay

6. Since overseas techniques were to strengthen by premix overlay this method, which called for no deviations, was initially adopted. The traditional American approach of "Try three(inches)and see" was patently inapplicable and reference was made to a series of curves giving overlay thickness and future traffic v. existing deflection. (Millard and Lister (ref 1)). This design applied to thickness of gap-graded premix overlay to similar existing premix surfaces in the British climatic environment, and not to asphaltic concrete overlay on surfaced dressed gravel bases where black top temperatures of 70°C are not atypic, but no other design techniques data were available.

Premix Overlays and Design Experience

7. Although deflection data was measured yearly at 30m spacing, scatter was excessive. To isolate sections for different treatment recourse was made to a moving mean plot of 5 deflections - Fig 4 shows typical data - to isolate stretches for different treatment. The next problem was selection of design risk. Mean values are patently inacceptable, and a design risk from the mean deflection plus one or two standard deviations is chosen. Mitchell (ref 2) has illustrated the effect of cost on risk. He showed that, for given funds,

Fig. 1. Typical road (but in good condition) before rehabilitation

Fig. 2. Typical (new) construction showing finish achieved after granular overlay

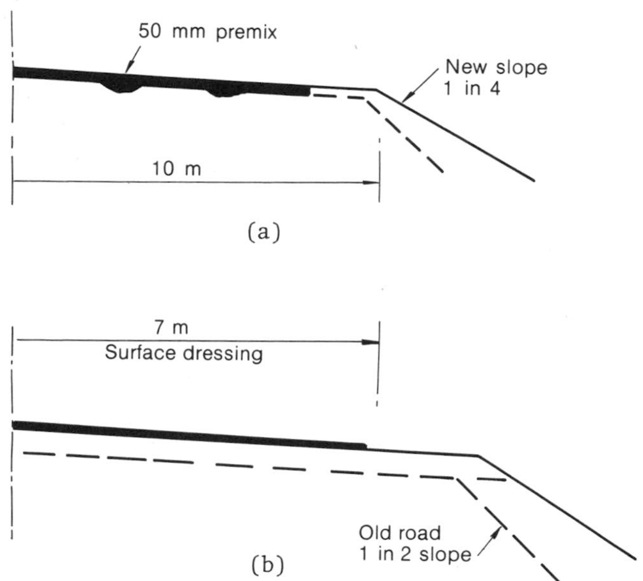

Fig. 3. Rehabilitation: (a) by premix overlay; (b) by 150 mm granular overlay

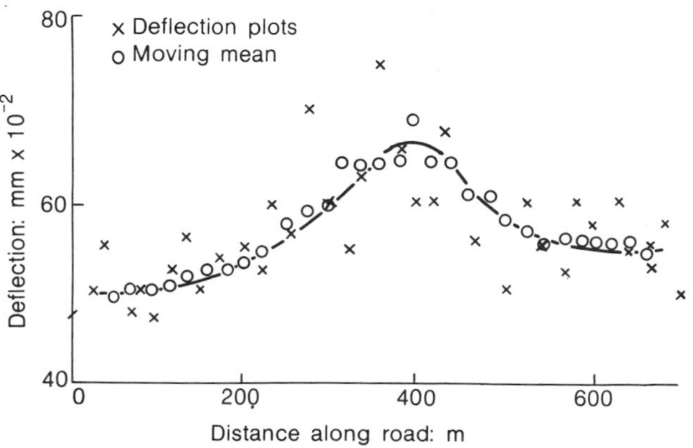

Fig. 4. Analysis of deflection data

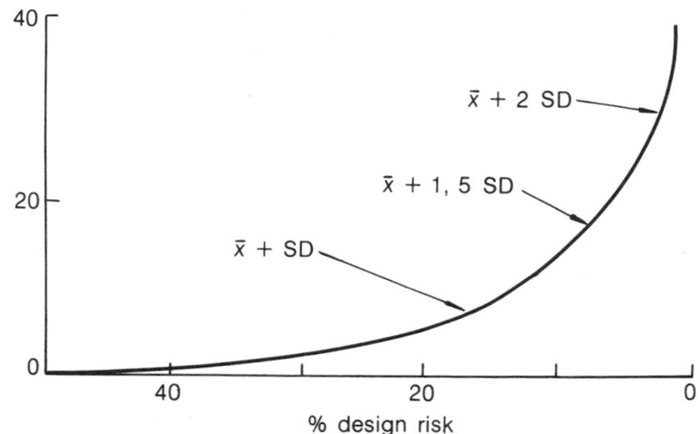

Fig. 5. Premix overlay costs v. risk (after Mitchell, reference 2)

100 km could be strengthened for 50% risk or only 76 km for a 2% risk of premature failure. Patently risk had to be accepted, and, from the shape of the curve in Fig 5 this should be about 10%.

8. In Fig 5 a practical decision might be to vary the asphaltic overlay between 0-250m, 250-500m and 500-700m. The computer can be programmed to give the moving mean incorporating any risk via the standard deviation. If the design traffic were for 3 million E80 axles and the chosen statistical deflection risk results in 60, 80, and 70mm x 10^{-2} deflections respectively, and using the chart of Millard and Lister (ref 1) overlays of 50, 90, and 65mm would be required in the three sections. This is feasible, but subsequent strength will still not be uniform.

9. Usually, however, very high deflections result over vlei (valley) crossings, where the roadbed soil is expansive, and it proved to be economically impossible to add sufficient premix to meet the cover requirements. Indeed, costs of premix at some $150/m³ (1980 prices) restricted overlay thickness to 50 mm save for short stretches where this would patently be inadequate.

10. Again sections occur where deflections indicate that premix is unnecessary, but it is not feasible to carry out short stretches of surface dressing, and not feasible to reduce premix thickness to less than 40mm, due to aggregate size.

11. Much base laid two decades ago was found to be too plastic even to meet today's sub-base standards. In some cases such material was patently too soft or cracked to risk an expensive bituminous overlay; in others it was holding up, thanks to rich bleeding surface treatments which waterproofed it. Experience has usually shown that, when given a 50 mm premix overlay, such bases still perform well and only occasional patching becomes necessary. However doubts do remain.

12. Finally the universal aggregate in Zimbabwe is granite, which has a high propensity to stripping. Whereas gap graded premix - suitable aggregate gradations are not locally available - may not readily strip, asphaltic concrete material tends to, irrespective of precautions taken, and the premix itself often does not last for even ten years. New additives of long chain amines may cure this, but have not yet stood the test of time.

Practical Problems with Premix Overlays

13. Costs proved to be considerably in excess of estimates, largely due to the need to restore shape. Cross sections showed settlement in wheel paths, as indicated in Fig 3(a), and irregular longitudinal depressions. This necessitated a levelling course being placed and compacted before the overlay, and 65mm of premix proved to be necessary to ensure 50mm of cover over high spots.

14. Further problems resulted from the need to add extra material to restore the profile to the shoulders - such material had to be mixed and compacted with existing material after the overlay had been given. Again it was undesirable to leave shoulder slopes at 1 in 2, and this too necessitated placing extra material, and concomitant plant.

15. A further problem was the need to treat very weak sections which called for reconstruction, with its attendant construction plant and deviation problems, before the overlay could be laid. Similarly whilst most original horizontal alignment was found to be adequate, vertical alignment changes,both to improve sight distance and to reduce fuel consumption were often necessary, and in other cases the pavement had to be widened to permit of a heavy vehicle climbing lane, necessitating a construction unit.

16. Thus not only did premix overlay costs exceed estimates but some reconstruction usually proved necessary. Costs of premix too, rose on rural roads where haulage distance was high. This cost rise led to a change to granular overlays.

Granular Overlays

17. Granular overlays, which are sealed like new construction with a surface dressing, involve the construction of a traffic deviation and the use of normal construction plant. The technique involves

(a) The construction, to new sub base level, of all new work such as improvement to horizontal and vertical alignment and of climbing lanes, etc.
(b) The reconstruction by excavation and replacement of material in weak areas e.g. over expansive soil, antheaps, old rock bars, etc.
(c) The flattening of shoulders by importation of new material - on occasion culverts have to be widened, but this too applies with bituminous overlays.
(d) Ripping and spreading existing base and shoulders, stabilising with 2% cement or lime as necessary, and recompacting as sub base.
(e) Importing and compacting a new 150mm base, and
(f) Applying a 2 coat surfacing dressing to the traffic lanes, and a single coat seal to the shoulders, together with centre line and edge marking.

Experience shows that the ride is often better - a PSI of the order of 3,5, as opposed to 3,0 with a premix overlay.

Design of Granular Overlay

18. If only deflection data is available, it is possible to approximate granular overlay thickness by doubling the premix thickness required. However there is no reason to assume that asphaltic concrete, of 40-60mm

thickness is more efficient than granular material when its stiffness is reduced by temperatures of 60°-70°C. Ideally the existing road should be sampled and evaluated in terms of a design for new construction, but this is an inconvenient and expensive procedure. Sampling can be considerably reduced, (and testing may be restricted to gradation and plasticity), if it is undertaken in conjunction with a deflection survey at occasional stretches of higher deflection values. However CBR values can be quickly estimated by dynamic cone penetrometer tests (ref 3) to 400mm depth in as many seconds.

19. In the event, variations in base overlay thickness - as long as only one layer is required - will result in only marginal cost changes. Layers of less than 120mm (allowing for practical variations) will tend to shear under the roller, and of more than 150mm cannot readily be compacted. Bearing in mind that level will be lost in shaping and widening existing base and shoulders as sub base, it becomes advisable to standardise overlay thick-ness at 150mm. Indeed, if more than 150mm of granular overlay is needed, the equivalent premix overlay is likely to exceed 80mm which makes it economically unviable in developing countries. In such circumstances, the existing road is likely to have completely failed and have lost shape, in which case reconstruction, rather than overlay, will be the only remedy. Details of this argument are listed in Appendix 1.

20. The only variation in granular overlay specifications, then, is the necessity or otherwise to stabilise the existing base as sub base. Here the material must be checked for plasticity, and, if it fails present standards, it will need 2% of cement or lime. No further testing is needed since material for which this treatment is inadequate could not have performed as base under past generations of traffic.

Analysis

21. A cost analysis of 1980 Zimbabwe costs in US dollars for rehabilitating a 7m on 10m road, is given in Appendices 2 and 3. It can be seen that to rehabilitate a 50mm minimum thickness overlay costs about US $87 000 per km, whilst to overlay with 150mm of cement stabilised base costs about US $78 000. Moreover, if the existing base is sound (as should be the case if a premix overlay were the alternative) and the new base also needs no stabilisation, this cost will reduce to about US $66 000. It is interesting to note that these costs are about half of those for new construction.

22. A major item of granular overlay costs is the need to construct and maintain a deviation. The figure of US $16 000 quoted is typical, but it depends on the length of time in use, the amount of traffic, and the soils and construction materials and patently can be reduced or materially increased with planning; for example construction during the dry season.

23. Exact foreign currency data is not available, but it is likely to be of the order of 60% for the granular overlay, and 90% for the bituminous overlay method, so if the costs are shadow priced, the discrepancy in economic costs increases considerably.

24. Cost savings resulting from the use of granular, as opposed to asphaltic overlay are thus a not significant 20%. However, important as this may be, financial restrictions usually apply in developing countries, the availability of funds being the overriding control. This being so, an extra 20% of the network can be rehabilitated by the granular method for given funds - an important consideration.

25. Again, less foreign currency being required, the multiplier comes into effect, and more labour can be employed, a desirable feature in developing countries.

26. Finally, based on more than ten years experience, it is suggested that a 150mm granular overlay is more satisfactory than a 50mm asphaltic concrete overlay in distribution of stresses in conditions where black top temperatures exceed 60°C.

Conclusions

1. In developing countries asphaltic concrete overlays are more expensive than granular overlays three times as thick.
2. For short sections of repair asphaltic concrete is more convenient as no construction unit is required.
3. When no deviation is possible asphaltic concrete overlays constitute the only solution.
4. Where future traffic exceeds the potential of surface dressings asphaltic concrete is the economic solution; however surface dressings in Zimbabwe have carried 3 million E80 axles, and the upper limit could be as high as 10 million, in which case stage construction with surface dressing would be economic.
5. In all other cases, which are the norm in developing countries, standardising on 150mm granular overlays can well result in better structural improvement and a 20% cost saving.
6. Granular overlay should be used in all cases where a full construction unit is required, that is when *any* realignment, vertical or horizontal or *any* improvement to geometrics is necessary. Such improvement is likely to be increasingly warranted with the energy crisis.
7. Cost patterns will vary between countries, but, although the conclusions are drawn from Zimbabwean experience, granular overlays should be considered as possible alternatives to asphaltic before design is finalised.

ACKNOWLEDGMENTS

The author is indebted to Eng P Mainwaring BSc(Eng) MICE MZweIE, Secretary for Roads & Road Traffic, for permission to quote recent unit costs.

APPENDIX 1 JUSTIFICATION OF 150mm GRANULAR OVERLAYS

Mitchell (ref 4) has determined that, for the average requirements of SR1 (now replaced by MPD1), RN31, or draft TRH4, (ref 5, 6, 7 & 8) the necessary cover, T_n, for N_{E80} repetitions of the E80 axle is given by the formula

$$Tn = (565-330 \log_{10}CBR)\ (-0{,}92+0{,}38 \log_{10}N_{E80})$$

In Zimbabwe, where the average CBR is 7, this reduces to

$$T_n = 286\ (-0{,}92+0{,}38 \log_{10}N_{E80})$$

This may be tabulated, using 1980 construction costs,as follows:

Repetitions of E80 axle (N_{E80}) in millions	0,1	0,3	1,0	3,0	10,0
Pavement thickness (T_n)	280	332	389	441	498
% increase in thickness compared to that for 10^6 repetitions	-28	-15	0	13	28
1980 Costs of typical construction (US $1 000)	150	170	189	237	275
% change in cost as compared to that for 10^6 E80 axles	-21	-10	0	26	46%***

It can be seen that the pavement thickness, as compared to that for 10^6 E80 axles, changes by only 28% for a tenfold increase or reduction of traffic. Alternatively if costs are reduced by 21%, life will be reduced by 90% from 20 to 2 years, whereas, if the investment is increased by 46% life will be increased by 1000% from 20 to 200 years (assuming constant traffic).
Again, 120mm of extra base will increase life more than tenfold; thus rehabilitation should not require greater granular overlay thickness than this, but the increase to 150mm will add little cost but give a higher factor of safety.*** This cost is extrapolated, on the conservative assumption that such traffic requires 50mm of premix.

APPENDIX 2 COST SCHEDULE, 1980 PRICES, US $ FOR 1km OF ROAD

50mm asphaltic concrete overlay, 7m wide, single seal on shoulder to 10m width.

Item	Description	Quantity	Unit rate	Total cost
1	Widen shoulder slopes from 1 in 2 to 1 in 4	$1300m^3$	3,00	3900
	Tack coat	$7000m^2$	0,37	2590
	Levelling course, avg 15mm	$105m^3$	151	15860
	AC overlay, 50mm	$350m^3$	144	50400
	EO for premix haulage (say 15 km)	$6825m^3$-km	0,93	6350
	Compacted gravel to shoulders	$225m^3$	12,00	2700
	Prime to shoulders	$3000m^2$	0,37	1100
	Single seal to shoulders	$3000m^2$	0,60	1800
	Widen culverts		lump sum	2000
Total cost/km				$86700

APPENDIX 3 COST SCHEDULE, 1980 PRICES, US $ for 1 km OF ROAD

150mm granular overlay with 2 cost surface treatment, 7m wide, and sealed shoulders to 10m width

Description	Quantity	Unit rate	Total cost
Widen shoulder slopes from 1 in 2 to 1 in 4	$1500m^3$	3,00	4500
Spread and recompact existing base as subbase	$1260m^3$	2,40	3020
Add 2% lime as required (E.O.)		3,15	(3970)
New granular base, 150mm thick	$1590m^3$	11,37	18080
Add 2% cement as required (E.O.)		4,79	(7620)
Prime coat	$10000m^2$	0,37	3700
2 coat surface dressing	$7000m^2$	2,47	17290
Single seal on shoulder	$3000m^2$	0,60	1800
Widen culverts		lump sum	2000
Grade and maintain deviation		lump sum	16000
Total cost/km	$66400 - $78000 (with stabilisation of base and subbase)		

REFERENCES

1. MILLARD R.S. & LISTER N.W. "The assessment of maintenance needs for road pavements". Proceedings of the Institution of Civil Engineers, v.48, 1976.

2. MITCHELL R.L. Author's reply to "The overloading and strengthening of flexible pavements with special reference to airfields". Rhodesian Engineer v.6, N9, 1971.

3. STANDARDS ASSOCIATION OF CENTRAL AFRICA "Methods of testing soils for civil engineering purposes", CAS A43, Part 2 "Strength Tests" Salisbury, 1974.

4. MITCHELL R.L. "Economic justification of highway pavement design...", M.Phil thesis University of Rhodesia, Salisbury, 1980.

5. MITCHELL R.L., van der MERWE C.P. & GEEL H.K. SR1"Standardised flexible pavement design for rural roads with light to medium traffic". Stationery Office, Salisbury (out of print), 1975.
6. MITCHELL R.L. MPD1 "Highway pavement design procedure", (in preparation), 1981.
7. ROAD NOTE 31, "Guide to the structural design of bitumen surfaced roads in tropical and sub-tropical countries", HMSO London, 1977.
8. DRAFT TRH4, "Structural design of road pavements", National Institute of Transport and Road Research, Pretoria, 1979.

Discussion on Paper 12

Eng R.L. MITCHELL *(Introduction to Paper 12)*: In Zimbabwe we have had long experience of strengthening pavements and it has been proved that surface dressing is adequate for possibly five million E80 axles - sufficient for all rural roads in Zimbabwe where axle loading was enforced. Thus a viable alternative to premix is a granular overlay, which is easier to lay and which needs minimal thickness design. It can result in a better finish and, above all, is about 20% cheaper, and requires much less foreign currency. Recent laboratory studies have shown that rutting is more or less than proportional to the log of repetition at more than 10^5 repetitions of load, which helps to explain the proven economic practice of dispensing with premix in rehabilitation.

Mr J. REICHERT *(Belgian Road Research Centre)* and Mr V. VEVERKA: Road strengthening fits into the general pattern of policies aimed at the progressive improvement of the road network. It is possible to distinguish between strengthening programmes which affect the entire road network or part of it, and those which envisage the strengthening of an individual route. The strategy used when strengthening an entire road network (or part of it) attempts to evaluate the needs for strengthening in relation to the overall quality of the network, establish the choice of priorities and put forward a co-ordinated strengthening programme in accordance with the priorities and the available financial resources. Strengthening strategy for an individual route is aimed at eliminating the causes of deficiencies, adapting the bearing capacity of the road to meet future traffic requirements, and keeping costs to the road authorities and the road user to a minimum.

All strengthening programmes for the road network must be based on an assessment of its overall quality. There are various methods of assessing the quality of a road network based either on the use of a single parameter (deflexion, roughness) or on the use of several parameters. The first approach, using a single parameter, runs the risk of leading to somewhat sketchy conclusions, whereas the second type, using several parameters, permits a better analysis of the physical phenomena which give rise to deficiencies. A method is recommended which makes use of three different parameters: deflexion, the pavement distress as revealed by visual inspection and the structure (thicknesses and types of the courses).

The quality of a route is characterized by its suitability for passenger and freight transport at a given speed and with a predetermined degree of safety. These requirements come down, in engineering terms, to the road's traffic capacity, its bearing capacity, the pavement condition and its skid resistance. In developing countries the problems arising from the traffic capacity and skid resistance do not appear to be the most pressing; at present the accent should be on problems of bearing capacity and pavement condition.

The bearing capacity is considered to be adequate when deflexion is below the limit value for deflexion, calculated on the basis of a deflexion criterion in terms of the cumulative number of standard axles which have travelled the road in its service life (Fig. 1) The condition is considered satisfactory when deficiencies affect only a small percentage of the wheel tracks; occasionally quantitative assessment criteria may be used.

The comparison of bearing capacity and pavement condition allows a first classification of the roads with a view to proposing a choice of priorities. The priorities must take account of various factors of an economic, financial, and social character, such as: the importance of the route in relation to the national network; the

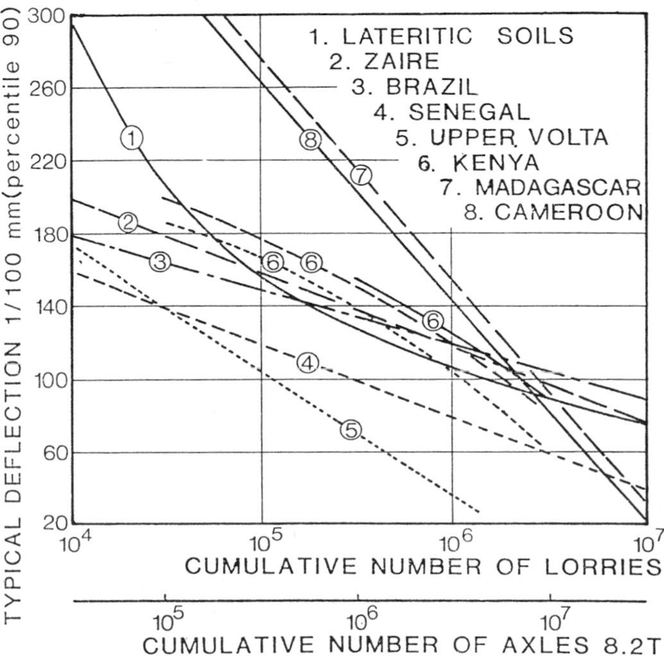

Fig. 1. Maximum permissible deflexion under a load of 13 t.

present and projected traffic volume; the cost of strengthening and, possibly, the evolution of maintenance and user costs concerning the strengthened road and the road in its present condition. The choice of priorities and decision-making can be facilitated by using technical and economic models (e.g., the HDM model).

Individual project strategies are used for routes which are selected for strengthening; the main operations involved are the diagnosis and the choice of solutions. The diagnosis is formulated on the basis of the results yielded by the analysis of file data (survey), thorough visual and instrumented inspection. (Generally speaking, the variation in bearing capacity over the route and data on the structure are regarded as being indispensable). The causes of deficiencies must be specified, and decisions must be taken as to whether any widening work is necessary and whether further scanning is required.

The choice of solutions generally concerns

(a) the application of one or more courses (e.g., bituminous overlays)
(b) the replacement of one or more layers of the existing structure after removing the damaged layers
(c) the modification of the hydrological conditions of the structure (drainage, vertical capillary screens, and so on)
(d) the reworking or treatment of the material in situ (on unpaved roads)

Calculations used for determining the thicknesses of strengthening layers resort to either empirical formulae or design models based on multi-layer systems theory which take into account the mechanical characteristics of the subgrade and other materials as well as their seasonal variations.

Mr J.B. COX *(N.D. Lea and Associates Ltd, Jakarta)*: Paper 12 (and Paper 18) recommends gravel overlays in preference to hot-mix overlays for rehabilitating existing roads. This is exactly the opposite conclusion to the one that N.D. Lea and Associates reached in Indonesia on technical grounds. This difference of opinion may be due in part to the different environmental conditions (as the drier conditions in Africa, or in Australia, may favour surface dressings on crushed rock bases) which are in contrast to the situation in Indonesia where conditions are much wetter. However, I do believe there are some basic technical differences in opinion leading to these opposite conclusions.

In Indonesia we have taken the emphasis off structural pavement design in favour of an economic systems-approach, emphasizing vehicle-operation costs which make up 70-90% of the total capitalized costs of construction, maintenance and vehicle-operating costs over the design life of the pavement. In fact, this statement is too weak - we have, in fact, rejected the overlay design techniques used by Mr Mitchell (Paper 12), such as Lister's Benkelman Beam deflexion/axle-loading relationships, as too conservative and grossly uneconomic. I have never understood why the Lister curves, which gave axle loadings to give rut depths of 10 mm on new roads, should be used for overlays on existing roads where most pavement compaction has already been completed.

Moreover, why do we use rut depths as a pavement design criterion when most rutting does not affect pavement roughness and therefore vehicle-operating costs? Our experience in Indonesia indicates that the major reasons for pavement failures are other than strength - durability of surfacing, drainage problems, material problems, and so on - and it is therefore senseless to overdesign on strength. Similar results have come from Kenya and Paper 19 shows that initial deflexions and axle loadings show no correlation with future pavement roughnesses and pavement life. We have therefore relaxed deflexion criterion - mainly using failure rather than critical conditions - and are happy if we can reduce deflexions under an 18 000 lb axle load to 1.5 mm - about double that recommended by Mr Mitchell.

We would emphasize the properties of the surfacing as performance trials on actual field pavements have indicated that initial pavement roughness and its increase with time is more as a result of the type of surfacing rather than pavement strength *per se*. For example, in Thailand, a pavement roughness survey has indicated approximate average pavement roughnesses of 2000 mm/km for hot mix, 3500 mm/km for surface treatment and 5000 mm/km for penetration macadam surfaces. The reason for the low roughnesses of hot mix is that this surfacing is more durable and, even when badly cracked, the roughness does not exceed 3000 mm/km. Economic analyses using these roughness figures have indicated that the most economic solution, including vehicle-operation costs over the pavement design life, is the use of non-structural overlays (such as sand sheets or 3 cm of hot-rolled asphalt) even though structural criteria were not satisfied and we expect a badly cracked pavement at the end of the design life. I would therefore like to see Mr Mitchell carry out a total economic analysis using two pavement overlays - crushed rock/surface treatment and 3 cm of hot-rolled asphalt.

Our general pavement design philosophy is to accept the economic benefits of non-structural overlays and design the material properties to withstand whatever deflexions we have with these thin overlays. In general, we are promoting high-bitumen-content surfacings such as sand sheets or hot-rolled asphalt which are more flexible and durable.

Mr R.J. FREER-HEWISH *(University of Birmingham)*: The most successful pavement design methods have been developed by strong national road organizations to suit the environmental conditions of their own country. Europe, North America, South Africa and Australia are examples of this philosophy. This development has taken place over a long period of time. There has been continuity of thought, experience and honing of the system to bring it to its present state, and it is very difficult to argue against systems developed in such a way, which appear to give at least as good a result as any other. This 'hard' side of engineering, is still, whether we like it or not, primarily empirical and we have a long way to go in the analytical approaches.

The success of empirical design methods in the beforementioned countries is due to the common denominator of a strong infrastructure within their highway organizations. In the developing countries, infrastructures vary considerably: a wise investment would be to improve the weaker of these infrastructures. These remarks apply equally to highway maintenance.

How do we improve these organizations? It has to be by training: training in concepts (such as is being carried out in Birmingham), in organization and in techniques. The latter two aspects are the difficult areas to tackle and the most important. To introduce these into a training programme, two distinct stages are desirable: implementation and operation.

When little or no organization exists, then expertise from outside the country should be used to form an organization which, in conjunction with local input, is considered to be capable of carrying out highway work. The key to the success of forming a competent organization is not so much in the implementation stage but in the following operational stage. Unless the seed of motivation has been sown within the national highway organization to continue to develop and hone the system set up initially, then the scheme will collapse. The successful methods of today were not built overnight and have generally been developed from within the organizations themselves. This latter aspect must be the objective of training activities in the future.

Dr J.B. METCALF *(Australian Road Research Board)*: Mr Mitchell has presented a very interesting paper in which he claims a life of 3 x 10^6 ESA for a sealed road - clearly supporting Australian work on this form of surfacing. Does he recommend a tolerable level of deflexion before overlay or does he use roughness or rut depth as the criterion? If the latter, how is overlay thickness decided? I would also question the techniques of overlaying flexible bases with high deflexion levels; even with a stiff overlay, fatigue failure is likely. Could he comment in the context of criteria for investment on criteria for deflexion, roughness and rut depth with respect to type/volume of traffic, route classification and so on?

Mr D.P. POWELL *(T.P. O'Sullivan & Partners)* and Mr G.R. KIRKPATRICK *(Ministry of Construction, Jamaica)*: Mr Cox's comment that Paper 19 shows that initial deflexions and axle loadings show no correlation with future pavement roughnesses and pavement life is inaccurate. What we stated in our Paper was that in Jamaica, where pavement structures are strong and a large proportion of the subgrades are very strong (almost 80% of 440 miles of the trunk road network pavement tested before overlaying gave deflexions of 0.020 in. or less under a 14,000 lb axle load) and where traffic is relatively light, the deterioration in surface irregularity observed appears to be more dependent on time than on traffic loading. Contrary to Mr Cox's apparent belief we feel sure that there is a correlation between initial deflexions and axle loadings and future pavement roughness and pavement life. However, what is quite apparent from our observations in Jamaica is that this correlation is almost completely masked by the effects of weather with time.

Eng MITCHELL: I strongly support all that has been said by Mr Reichert and Mr Veverka. However, in Zimbabwe at least, many development roads are built on final horizontal, but on only a temporary vertical alignment. Thus, on improvement or rehabilitation cut and fill is involved, which calls for new design and granular construction which, in turn, is more compatible with granular overlays than with premix.

When Mr Cox states that premix overlays are more economic in Indonesia, I wonder whether he has included shadow pricing - or does Indonesia perhaps have the advantage of a favourable balance-of-payments position?

Mr Cox may well be correct in considering the different climatological conditions, although asphaltic concrete performs badly in the wetter areas of Zimbabwe as the granite aggregate has a high propensity for stripping.

I agree with Mr Cox on the economic systems approach. In my thesis (ref.4 of Paper 12) a copy of which was sent to consultants in Indonesia, I showed that whilst vehicle-operating costs were a very small part of rural transport costs in low count roads (say 10^5 E80 axles in 20 years), the infrastructural costs were miniscule compared with vehicle costs on high count roads of, say, 10^7 E80 axles. However, Mr Cox has misread my concepts on the Millard-Lister deflexion curves. As Dr Millard has often said, these apply to gap-graded overlays on gap-graded premix roads in the UK. Indeed, in my paper to the June 1982 Trondheim conference (ref. 1), I have showed that deflexion reduces with age on roads with gravel bases and surface treatment! Certainly surface durability, matèrials, and particularly drainage - this latter can never be overstressed - are more significant than deflexion. However, where deflexion under a 40 kN twin-wheel load exceeds 1 mm, possibly less, surface cracks will result; surface dressing being more pliable than premix, as Hveem has shown (ref. 2), is a better paliative. I would never specify premix overlays in conditions of high deflexions as Mr Cox's 30 mm over-rich premix may, of course, act more like surface dressing than a rigid structural layer in this regard as excess binder could seal the cracks. Nonetheless the problem would be controlling the excess bitumen to ensure that some air voids remain, as bleeding must be avoided. Without voids the stability is lost and strengthening would not result.

It seems likely then, that Mr Cox is really referring to maintenance, whilst I was talking of strengthening for future traffic increase - I can, of course, see no problem with a granular overlay sealed with 30 mm of rich premix in lieu of 10 mm of surface dressing. Indeed does Mr Cox not say this in his last paragraph?

Mr Freer-Hewish confines his remarks to training - with which I cannot disagree. Nevertheless, although the training of professionals is important - I have been doing this for seven years - as is that of technicians - the training of craftsmen is possibly more important and is overlooked in many developing countries. Surface dressing can last 15 years if laid by a craftsman - 15 months if not - and surface dressing will certainly remain the most important surface technique in developing countries not only in Zimbabwe and Australia, as rightly mentioned by Dr Metcalf.

In reply to Dr Metcalf, I cannot quote limiting deflexion figures, as, when writing, I have no notes available. However, on surface-dressed gravel-based pavements I regard the Benkelman beam as a qualitative, not a quantitative tool. I use it to identify sections of potential problems - not to design overlays. Anything giving more than 0.8 mm deflexion (under a 40 kN load) is potentially troublesome - and could indicate inadequate subgrade cover (rectifiable by drainage or a granular overlay), or a plastic base (rectifiable by stabilization). Rut depth is possibly more significant - unless it is due to inadequate initial compaction when recompacting the base is the cheapest solution - as ruts hold water which, in time, percolates through the pavement and triggers off failure.

I agree completely that stiff overlays will crack and fail over sections of high deflexion - hence the need for stabilization and/or additional granular overlays. Design must be based on in situ testing (the dynamic cone penetrometer is ideal for this) and applying principles of design for new construction such as MPD1 (ref. 3). However, in practice 150 mm of gravel is usually adequate - savings by reducing this are negligible as can be calculated using the formula in Appendices 1-3. Technological design gives way to economic logic.

Dr Metcalf in his last sentence, asks for investment criteria. I think this apparently innocuous request requires a book in reply. I regret that I have no experience on the economics and technology with different route classification Perhaps, however I can avoid the issue by noting that in South African heavy vehicle simulator studies, and in research in the University of Zimbabwe, possibly half of ultimate rutting or settlement seems to occur under a few initial passes on well-designed pavements - this reshaping and recompaction should materially prolong a road's life. It is all too easy for engineers to bury their mistakes under premix - it is easier to ignore drainage and complicated but more economic solutions than to explore them. Surely this is rather akin to the adage that a doctor buries his mistakes?

REFERENCES

1. MITCHELL R.L. Repetitive laboratory plate-loadings and implications on pavement performance and overlay design. Bearing Capacity of Roads and Airfields. Vol 1, Norwegian Institute of Technology, Trondheim, 1982.
2. HVEEM F.N. and CARMANY R.M. The factors underlying the rational design of pavements. Proc. Highw. Res. Bd, 1948, 101-136.
3. MITCHELL R.L. Flexible pavement design in Zimbabwe. Proc. Instn Civ. Engrs, Part 1, 1982, Vol. 72, Aug., 333-354.

13

The building of tracks and secondary roads

A. BERRENS, Directorate-General for Development, Commission of the European Communities

1. The building or improvement of tracks (to open up areas, service tracks, agricultural tracks) is a special case among the road infrastructure activities in which the European Development Fund engages.

   These tracks do not have to come up to the same technical standards as roads that carry a heavy and fast traffic load.

   The main objective is to ensure an all-weather means of communication. The technical requirements are often reduced to finding an alignment which follows the lie of the land as closely as possible, keeping the track clear of water, providing adequate drainage and making a surface.

2. These tracks could be built using the two traditional methods, either by private contractor or by direct labour. However, both methods have major disadvantages:-

(i) to employ a private contractor, following an invitation to tender, prior studies have to be carried out (design plans, quantities, bills of quantities, technical details and so on) to enable the contractor to make a valid offer and to avoid as far as possible the risk of claims arising during the works. Generally speaking, these studies are too costly in relation to the work to be done and the consultancy firm carrying out the studies usually produces a solution that is too sophisticated and hence too expensive. Also, the traditional contract is too rigid an instrument to be used for the kind of adaptable work programme often needed for this type of project;

(ii) using direct labour permits this large deal of flexibility in carrying out the works. The competent services of the Administration concerned can dispose freely of staff and equipment. The public works formula should even enable the work to be carried out more cheaply than if a private firm were employed since the expenditure does not include the profit that the firm has the right to add to its cost prices (1)

But experience has shown that the output from a direct labour operation is quite often far inferior to that of a contractor. One of the main reasons for this is the trouble that the Services experience in servicing and maintaining in a good state of repair the equipment needed for the work. So the low output generally cancels out the advantages of the lower costs that a well organised direct labour operation could offer.

3. There is a third formula that combines the following advantages:-

   - the invitation to competitive tender
   - the use of contractors (2)
   - the high output of a contractor
   - the flexibility of direct labour

This formula can be called the "public/private formula". It means that a contractor is chosen after tendering and it then places at the disposal of the Administration the work force (management staff and workmen), equipment and vehicles to carry out the work. For its part, the Administration appoints a Works Manager who then directs the works using the facilities provided by the firm.

The firm is paid, on the basis of a price-list, according to the number of hours the machinery has operated and the number of kilometres usefully covered by the vehicles on site.

This formula has been used and perfected on the following projects financed by the European Development Fund:-

- cattle tracks in the Central African Republic

---

(1) A frequent error committed in comparing the cost of works carried out by a contractor with that of works done by direct labour is that the fact that the firm has to pay taxes is not taken into account

(2) It is often advantageous to keep a firm on the spot using this formula, even if there is no other large scale work to be done

Highway investment in developing countries. Thomas Telford Ltd, London, 1983, 105–107

- tracks in the Basse Kotto Region (Central African Republic)
- tracks in the plateaux region (Togolese Republic)

An invitation to tender, for this method of execution of a project, has been established. It is useful to make certain remarks and reflexions thereto:-

(i) The first comment concerns the composition of the mechanical unit(s) to be used according to the work to be done (choice of machinery).
A typical unit could consist of:-

- 2 crawler tractors of approximately 150 HP, with angledozer and ripper
- 2 graders of approximately 150 HP
- 1 shovel loader of approximately 150 HP
- 4 8-ton dump lorries
- 1 8-ton transport lorry
- 2 6 000-litre tankers
- 1 20-ton pneumatic tyred compactor
- 1 motor driven pump, capacity approximately 80 cubic metres/minute
- 1 compressor with pick hammer
- 1 tracked machine transporter
- 1 mechanics workshop, static or mobile

A clause enabling the contractor to offer variations in the composition of the mechanical unit is useful.

In certain cases, it can also be useful to make provision for the setting-up of a laboratory with basic equipment. It must be stressed, however, that such equipment will be used only to make it easier for the Works Manager to take decisions and to tell the contractor how the machines are to be employed (e.g. the number of times the compactor has to pass).

(ii) An estimate must then be made of the number of kilometres the lorries should cover.

It is obvious that these estimates will always contain a margin of error - given that there is no prior study - but that efforts will nonetheless have to be made to reduce this margin as much as possible.

(iii) There is no advantage in laying down a time schedule in the contract. The time taken to do the work depends first of all on the competence of the Works Manager and the volume of work involved. It is desirable, however, to give some indication of the length of time required.

(iv) In order to prevent the tenderer from covering himself for certain risks - for neither the time schedule nor the volume of work required can be fixed in advance with any degree of precision - it is necessary to take account of certain items on the price list and estimate:-

a. bringing the technical unit to a given point and withdrawing it. The contractor must be made to give a precise breakdown of the price of this item and it must be checked that this corresponds to the reality;

b. the management staff of the contractor is to be paid a monthly flat rate;

c. so that the contractor is not penalised by temporary breakdowns - which are bound to happen - of certain equipment or vehicles, two prices must be worked out for the use of this equipment: the first for a day on site, but out of action through no fault of the contractor, and the second for hours worked or kilometres covered;

d. regarding the drainage works, the price list can stipulate either that the contractor is both to supply the equipment and lay it, or that the contractor is simply to lay the equipment supplied by the Administration.

5. As a rule, the Conditions of Contract should remain as simple and flexible as possible in view of the special nature of the contract.

In the absence of any study and, hence, of any technical details fixed in advance, the project has to be "invented" by the engineer as it progresses.

For example, the typical cross section in the dossier (carriageway 5 metres to 6 metres wide - pavement 3 metres to 3.5 metres) can state the crossfall and the shape of the side ditches but it cannot say how thick or how compact the surface is to be, since these details depend on the nature of the soil encountered and are, hence, the engineer's (Works Manager's) responsibility.

Surfacing should normally be provided for, moreover, since it gives an added guarantee of the longer life of the track and helps save on future maintenance costs.

6. Regarding the management staff of the contractor, experience has shown that one works foreman, representing the firm, and one mechanic responsible for maintenance and repairs to the equipment are sufficient.

7. It is obvious that the success of such an operation depends very largely on the professional competence, the sense of organisation and the qualities of the engineer chosen by the Administration to manage the works.

   His task is to know how to organise the work - setting out of the road alignment and structures, finding borrow material - sufficiently well in advance so that the works are not interrupted, to adapt quickly to unforeseen occurences and exercise authority and firmness in dealing with the contractor.

   In the following European Development Fund projects the national administrations concerned had recourse to expatriate Works Managers of consultancy firms, whose salaries were paid out of the appropriations for the project itself.

8. As I have already mentioned, this public/private formula has been used successfully several times: the construction periods have been short and the costs very reasonable.

- Cattle tracks in the Central African Republic
  - Length : 112 km
  - Cost per km : CFAF 658 000
  - Duration : $5\frac{1}{2}$ months
  - Year : 1972/3

- Tracks in the Basse Kotto Region (CAR)
  - Length : 150 km
  - Cost per km : CFAF 1 200 000
  - Duration : $5\frac{1}{2}$ months
  - Year : 1973/4

- Tracks in Togo
  - Length : 50 km
  - Cost per km : CFAF 1 700 000
  - Duration : 6 months
  - Year : 1974/5

- At present, other works are underway using this formula

Account must be taken when comparing these figures of the volume of work, which differed greatly from one project to another.

It is also obvious that the costs per kilometre would have been lower if sufficient funds had been available to build more kilometres of track, since certain fixed costs would thus have been more easily amortised.

Identification of priority schemes—the Sudan example

S. P. GOODWIN, MA, and R. THURLOW, BA(Econ), Economist Intelligence Unit Ltd, London

The study carried out in the Sudan by Norconsult AS and the Economist Intelligence Unit Ltd in 1979 defined a national feeder road construction programme for a six-year planning period. The starting point comprised a complete absence of engineered feeder roads and a very limited number of other roads. Consequently the study had to cover the full range of identification of potential roads and the determination of their priorities. The three separate stages in this process were; firstly, identification and screening of areas with potential for feeder road development; secondly, identification and preliminary evaluation of feeder road projects and finally, a detailed evaluation of selected projects.

INTRODUCTION

In 1978 Norconsult AS of Oslo and the Economist Intelligence Unit Ltd of London were selected to prepare a feeder road construction masterplan for the Government of the Democratic Republic of the Sudan. The assignment was carried out over a period of 12 months, beginning in early 1979. The work was divided into two phases; the first involved the production of a long term masterplan for the construction of feeder roads, and the second a detailed evaluation of priority projects for inclusion in the Sudan's current development programme. As the second phase of the study was a conventional economic and engineering evaluation this paper concentrates on the first phase, describing the unusual conditions which confronted the consultants and the method used to complete their task.

The methodology was developed rapidly in the early stages of the study to meet the particular requirements of the assignment. There will be specific requirements unique to individual assignments even where similar objectives are shared and it is inconceivable therefore to present a blueprint for universal application. However with appropriate modifications and refinements, the method of approach that is described in this paper can be applied in contexts other than that of this one case study.

The terms of reference for the study required the consultants to identify potential feeder roads throughout the Sudan and to rank them in order of priority for construction. The evaluation technique to be used in the appraisal and ranking of priority projects was specified in the terms of reference; it was to be a conventional cost benefit approach; that is a discounted cash flow analysis of construction and maintenance costs, vehicle operating costs and passenger time savings, and development benefits from induced agricultural and other production. However the approach to be used for the initial identification of projects was not specified, except for the requirement that .... "the identification shall be based on appropriate economic analysis".

THE SUDAN CONTEXT

Before describing the methodology adopted for the first phase of the study it is necessary to give a brief summary of certain features of the physical and economic geography of the Sudan to put the consultants' task into perspective.

The Sudan is the largest country in Africa, with an area of 2,506,000 km^2, roughly equal to that of western Europe, but with a population of only 16 million. The country can be divided into three regions. The northernmost third of the country is characterised by barren desert with most of the population and economic activity concentrated in the narrow strip of land along the banks of the Nile. The central belt forms the eastern part of the Sahel region of Africa with areas of sandy desert and of fertile black cotton soil. The main economically active areas of the Sudan, including the capital Khartoum, are in this central zone with the population widely distributed throughout it. The southern part of the country is tropical, very undeveloped and almost isolated from the rest of the Sudan. Land transport to the north is greatly restricted by the swamplands surrounding the White Nile and all routes are impassable in the rainy season. This southern part of the country forms the semi-autonomous Southern Region, giving it some political independence from Khartoum.

The main transport network of the country, which the feeder roads were to link into, consists of a very limited system of primary roads, the railway network and river transport on two reaches of the River Nile. The road network in 1979 consisted of less than 1,000 km of paved highway almost all of which were incorporated in various sections of the Khartoum - Port Sudan road which was then nearing completion. Several other roads were under construction or planned to extend the main road network over a wide area of the eastern half of the central belt of the Sudan. In the south there were some 6,000 km of gravel roads of varying quality. In addition to these roads an additional 12,000 km of roads were officially recognised but in fact were no more than tracks defined by the passage of vehicles. They were not engineered in any way and receive no maintenance, in some cases a number of quite separate parallel tracks defined the route. Many sections of these routes were impassable to cars and other conventional vehicles throughout the year and to all vehicles in the rainy season. In the dry season it is possible to drive almost anywhere across the Sudan by lorry or four-wheel drive vehicle. Where a track is used frequently it becomes sufficiently well defined to become a 'road'. 'In other countries roads are made for vehicles but in the Sudan roads are made by vehicles'. As a consequence the total length of routes in use by vehicles in the Sudan is far greater than the 12,000 km of officially recognised roads.

From information available in Khartoum it was impossible to locate many of these roads. The most detailed maps available, a 1:250,000 series covering the whole of the Sudan, were based on field surveys carried out in the early years of this century or the end of the last century with only limited updating since then.

Although the main road network was very limited the rail and river transport systems extended the main transport network to most areas of the country, thus greatly increasing the number of potential feeder roads. The railway system consists of almost 5,500 km of narrow gauge single track with its principal function the haulage of freight between Port Sudan and major inland centres. The main services of the river transport system are on the White Nile and connect the southern part of the country with central and northern Sudan. These services operate on the section between Kosti and Juba, a distance of over 1,400 km, but the total length of navigable river sections on which services are operated or planned is approximately 3,000 km.

IDENTIFICATION AND SCREENING

Unlike a main road network which normally can be considered as a single 'central' network a national feeder road system consists of a large number of roads distributed throughout the country and serving the wide range of needs of particular communities, each with their own individual characteristics. These features often render difficult the identification of potential feeder road schemes let alone the formulation of a national construction programme based on a consistent set of priorities. The method described here to do this was chosen because it ensured a consistent approach. It is also a logical first step before the evaluation of selected road projects since it requires similar input data to a road evaluation. An alternative method of preliminary selection is a compound ranking procedure with assessment by a rating given according to how well various economic social and political criteria are met. Such an approach can require as much information as a preliminary economic evaluation and weighting of the various criteria is a very subjective process. The consultants were required to select roads which would improve the communications of economically active areas inadequately served by roads and to provide access to underdeveloped areas of economic potential. Some guidelines for the definition of a feeder road were agreed. In particular the feeder roads should:

- link into the existing and planned main transport network, including road, rail and river systems;

- connect the focal point of an area to the main transport networks;

but should not:

- form part of the envisaged main road network; that is their main function should not be to carry through traffic;

- be internal to a development scheme and which would be financed and implemented as part of that scheme;

- be short links (less than 10 km) from urban areas to the main transport network.

Although many of the existing tracks would obviously form part of the future main road network many thousands of kilometres fell within the definition of feeder road used in the study and therefore had to be considered. There was, of course, no information on traffic levels on these roads. (During the course of the study it was discovered that they ranged from 1 or 2 vehicles a day to more than 800 vehicles per day). From Khartoum it was impossible to gain an accurate picture of the circumstances in each part of the country while the size of the country and poor communications made it impossible to visit all areas. It was therefore decided that before the main task of the first phase of the study could be carried out, that is the identification and preliminary evaluation of feeder road projects to give a priority ranking, there had to be a screening and identification of areas with a potential for feasible feeder road development.

The identification and screening process also gives an indication of which areas, if any, should receive priority if there is found to be a time or other resource constraint affecting the completion of subsequent work.

The process involves the following steps:

- establishment of a zoning system;

- quantification of potential feeder road traffic generating activity;

- determination of threshold traffic levels below which no feeder road project could be justified on economic criteria;

- identification of areas with a strong possibility of significant development benefits generated by the provision of feeder road schemes.

The zoning system

The zoning process, as the name implies, requires that the entire subject area is divided into zones of reasonably homogeneous characteristics using as broad a data base as is available at the national level. In the case of the Sudan 80 such zones were identified on the basis mainly of a recent livestock and natural resource inventory. This inventory provided systematic information on the absolute and relative levels of population, cultivation and livestock which could be grouped into areas of relative homogeneity reflecting soils, vegetation, land use, population and cropping densities. Whilst it is always preferable to have precise data with which to work, at this stage a reasonable picture of the relative distribution of key variables is a more important objective.

Having drawn up appropriate 'patchwork' maps of the relevant factors the main transport network, to which the feeder roads will connect, is superimposed. The zones can be then compared on the basis of criteria appropriate to a judgement of their potential to support one (or more) feeder roads. Ideally a single criterion should be aimed at. That chosen was the estimate of whether or not each zone was likely to generate enough traffic to justify a single feeder road. To calculate this the following parameters were used:

- zone area;
- zone population (urban and rural);
- irrigated and mechanised farming scheme crop production and inputs tonnages;
- 'traditional' sector crop production;
- per capita consumption of goods 'imported' into zones;
- freight traffic growth rates based on population growth and income growth;
- per capita crop retentions in the 'traditional' sector of crop production;
- urban consumption of local rural production.

In addition a series of general parameters were required:

- road construction and maintenance costs per kilometre for each soil type and terrain characteristic;
- vehicle operating costs for the standard freight vehicle in use on bitumen roads, gravel roads and on unmade tracks (for the range of soil types).

From the above it was possible to estimate the threshold level of present day traffic which, allowing for growth, might justify the construction of a feeder road. Values assumed for construction costs and for vehicle operating costs were deliberately set to prevent any zone failing to meet the criteria solely because of inaccuracy in the data. The result of the screening process was that 40 out of the 80 zones met the minimum traffic threshold criterion. A further 10 zones failed when traffic was reduced to reflect the proportion of each zone that a single feeder road might influence. Some other zones were excluded from further analysis because their proximity to the main transport network obviated the need for feeder roads.

It must be emphasised that at this early stage the assumptions must be kept sufficiently conservative to prevent the exclusion of areas with potential for feeder roads that might prove economically feasible as a result of a more refined analysis. To further eliminate this possibility the theoretical screening exercise was complemented in a number of ways.

Firstly suggestions for feeder roads were obtained from appropriate departments of central government and other public and private sector organisations. Secondly, a pilot field study of one area was undertaken to test the validity of the theoretical exercise and other short visits to different regions of the country were made. Thirdly, ideas and lists of perceived priorities were sought from regional administrators.

Those zones which met the minimum traffic threshold criterion were grouped into those of high, medium or low priority in planning field work to identify specific road projects. This assessment of priority included considerations of such additional factors as:

- density of traffic per km^2;
- special projects not specifically allowed for in the traffic estimation exercise;
- strong possibility of significant development benefits in the traditional sector of agriculture.

The zone screening exercise identified a high priority area of 338,000 km^2 (16 zones) and a medium priority area of 71,000 km^2 (7 zones). The high priority group of zones included 2,500 km of suggested road projects and the medium priority group a further 600 km.

THE PRELIMINARY RANKING OF PROJECTS

From the results of the zone screening exercise the areas to be subjected to further study and a list of road projects, albeit on the basis of qualified suggestions, can be drawn up.

As a result of visits to each of these areas the list of road projects will be amended by the elimination of poor candidates and the inclusion of others which analysis based on locally acquired data reveals as being more worthy of study. In the Sudan study for example, 76 projects totalling

4,500 km were evaluated at this stage. These projects had a wide range of characteristics. In Darfur Province, in the far west of the Sudan, feeder roads were identified which connected with the railway system but which were more than 1,000 km away from any constructed road. In the area immediately south of Khartoum feeder roads were obviously required in the huge Gezira irrigation scheme. This is an area which although roughly the size of England and surrounded by sections of the main transport network contains no constructed roads despite high traffic levels. Further south in Blue Nile Province other areas of very high agricultural development potential and with the first stages of development underway appeared to justify feeder road development but have virtually no traffic at present.

In order to compare so many projects with this range of characteristics it was decided that a standard economic cost-benefit analysis was the most appropriate method to assess each project so that they could be given a priority ranking. The benefits included both road user savings and induced agricultural production. However whilst the results can be presented in a similar manner to that required in a detailed feasibility study the number of projects is too great for the input detail to be sufficiently accurate for the results to be regarded as final.

DETAILED EVALUATION AND FORMULATION OF AN IMPLEMENTATION PROGRAMME

The detailed evaluation should be carried out in the form of a standard cost-benefit analysis exercise but using input values calculated on the basis of more precise data. Thus the difference between the preliminary ranking and the final evaluation is not one of methodology but rather one of increased accuracy in the input parameters. In the Sudan study, seven projects were selected out of the best 16 on the basis of the preliminary ranking. At the request of the Government of the Sudan an eighth project, 46th in the preliminary ranking, was also included in the final evaluation. These eight projects totalled just over 600 km in road length and were located in 6 different provinces of the country. They exhibited a diverse range of characteristics; their length ranged from 6 km to 157 km, the area of influence population ranged from 20,000 to 500,000, and the communities served varied from modern farming based, to those based on traditional farming, mining and forestry. The final evaluation produced a revised set of net values and internal rates of return and a slight change in the rank order of the projects. In order to establish an implementation programme the optimum year of opening was calculated for the eight projects plus, to give some idea of a longer term implementation programme, the other projects which the preliminary evaluation indicated should be accorded a high or medium priority in road construction planning.

Finally an implementation schedule for feeder road construction can be drawn up. Ideally each project should be completed in accordance with the results of the optimum timing analysis results. However there are likely to be practical and or financial constraints which prevent this. In the case of the Sudan example it was found that construction of six projects should be started immediately (ie 1980), a seventh in 1985 and the eighth was not justified before the year 2000. With the two constraints noted being an annual total construction rate of 100 km per year and no more than two projects under way simultaneously it was clear that not all the projects could be undertaken in accordance with the theoretical optimum.

The criterion used for determining the project implementation schedule was the value of net benefits foregone as a result of delaying each project. For a delay of one year this value equals the benefits associated with the year of opening, using an annualised value for generated benefits, minus the opportunity cost of the capital required for the project. This procedure is represented for subsequent years of delayed implementation. The two projects with the greatest loss of benefits as a result of delay should clearly be implemented first. As each project is completed the accumulated loss of benefits for each remaining project is then reviewed and again that with the greatest loss should be commenced.

A further adjustment in the schedule could have been necessary in the Sudan example, although in the event it was not. It is clearly desirable that the feeder road should not be built before the associated main road is completed. Many of the feeder road projects under study were to link in with parts of the main road network which was planned rather than under construction or completed.

CONCLUSION

The Sudan example, we believe, demonstrates well how a national feeder road construction programme can be formulated from a starting point at which there are no existing feeder roads and a limited network of main roads. From a starting point with only general geographical and economic information available it is possible to undertake a study which progresses rapidly from the general to the particular in a series of logical and consistent steps.

Following the results of the study the Sudan feeder road construction programme commenced with the final design of two of the recommended projects.

15

The impact of feeder road investment on accessibility and agricultural development in Ghana

J. L. HINE, BA, MA, MCIT, Transport and Road Research Laboratory, and
J. D. N. RIVERSON, BSc, MSc, MGhIE, AMITE, Building and Road Research Institute

In a cross-sectional study of 33 villages in The Ashanti Region of Ghana, little evidence was found to suggest that agriculture was adversely affected by inaccessibility, apart from some difficulty in obtaining loan finance in the more remote areas. The more accessible villages were observed to have a higher proportion of people employed outside agriculture. The improvement of existing road surfaces was estimated to have a negligible impact on prices paid to the former. However, connecting a village to a road head by converting a footpath to a vehicle track was calculated to have a gross beneficial effect in the order of a hundred times greater than improving the same distance of earth track to good gravel road.

INTRODUCTION

1. Many case studies carried out in different parts of the world have often pointed to significant development benefits stemming from rural road investment. These studies however, have largely been carried out in untypical isolated locations where the road investment has brought about large changes in transport costs often arising from a change in transport mode from perhaps headloading to vehicle transport. More usually rural road planning is concerned with less dramatic projects to improve existing roads and tracks where no change in transport mode is envisaged.

2. In order to help with road investment planning in a more typical environment a study of the impact of feeder roads was carried out in the Ashanti Region of Ghana by the Building and Road Research Institute (Kumasi) in cooperation with the Transport and Road Research Laboratory. The study was carried out for the Ghana Highway Authority as part of its Second Highway Project and was supported by the World Bank.

3. The purpose of the study was to determine how parameters of rural development (particularly agricultural practises, costs and prices) varied with accessibility within the region. From this it was hoped to infer how rural development would change if access were improved through road investment, and hence lead to better methods of planning rural roads in Ghana and elsewhere.

4. In this paper only a brief resumé of the findings of the survey will be given. More extensive coverage of details will be published by the Transport and Road Research Laboratory and Building and Road Research Institute in due course.

SURVEY BACKGROUND

The region

5. Ashanti Region has an area of 24,000 km^2. The capital Kumasi has a population of over 400,000 which is many times larger than the combined population of all other urban centres of the region. Over a million people live in 2,500 small rural towns and villages dispersed widely over the region, except for the uninhabited Afram plains in the north east.

The road network

6. Besides being the major administrative centre Kumasi is also the major market, transport and distribution centre of central southern Ghana and all major roads in the region radiate from there. Excluding Kumasi and the Afram plains, ie in 70 per cent of the region, there are 4,400 km of roads and motorable tracks. Ninety-eight per cent of the rural population lives less than 2 km from a road or motorable track but only 0.3 per cent lives more than 5 km from a road or track. Thirty-one per cent of the land area of the region lies more than 2 km from vehicle access but only 3.3 per cent lies further than 5 km from a road or track.

The rural economy

7. Food crop cultivation and cocoa farming are the major sources of livelihood for most of the population. This is supplemented by the rearing of poultry, sheep or goats. Marketing, the provision of services, rural industry and handcrafts provide additional sources of income to a small proportion of the rural population.

8. Because land is relatively plentiful, shifting cultivation remains the dominant pattern of food farming. A plot of land is cropped for up to 3 years and then left to bush fallow for up to 10 years to regenerate the fertility of the soil. When the area is to be used again the land is cleared by fire. Large trees and tree stumps are left standing, and the open patches of land are cultivated with hand hoes. Machinery is largely inappropriate to this type of farming; labour (and working capital to hire additional labour) is a more critical factor of production, although modern inputs such as cocoa insecticide and fertilizers are widely used.

SURVEY METHOD

Definitions and sampling frame

9. Throughout this paper the term holder is used to denote an individual who manages a family farm holding. One holding may represent several dispersed fields or farms but in general totalling less than 20 acres (8 hectares). Data was collected on a holding basis.

10. Ministry of Agriculture enumerators collected cross-sectional socio economic data for the study from 491 holders in 33 villages. The sampling frame for the normal Ministry small holders survey was used to keep the data set conformable with other Ghanaian statistics. All but two of the villages in the sample had vehicle access and were between 8 and 102 km by road from Kumasi, lying in the cocoa growing forest zone (except for two villages in the savanna to the north of the region). Figure 1 shows the location of the survey villages.

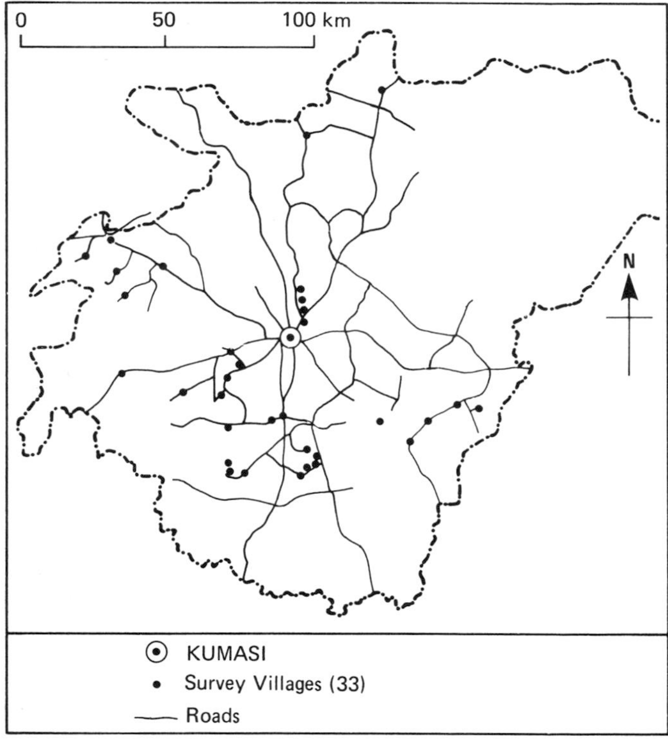

Fig. 1. Ashanti region showing location of survey villages

Parameters of agricultural development

11. A number of parameters such as farm area, cocoa production, cocoa and maize yields, labour input, finance, crop sales and the use of modern inputs were used to indicate agricultural development. Individually none of these parameters would be adequate but considered together they could provide a good overall indication of the pattern and nature of change of agriculture in the region.

12. Additional data on the local population density, soil fertility, crop diseases and rainfall were all collected as parameters of the region and were used to test alternative development hypotheses. The Soil Research Institute (Kwadaso) assisted by analysing soil samples collected from each village. The Ghana Meteorological Services Department confirmed that rainfall could be excluded as an explanatory variable of agricultural production in Ashanti region in 1979, because rainfall in the main crop season was adequate for all crop requirements.

Parameters of accessiblity

13. Two key parameters of accessibility were used in this study. These were:-

(i) the transport charges of moving a unit of produce equivalent to a headload of produce from each village to Kumasi,

(ii) the transport charges of moving a unit of produce equivalent to a headload of produce from each village to its district centre.

The transport charges of moving produce from field to Kumasi and from field to district centre were also used as subsidiary parameters of accessibility. The transport charges were found to vary directly with distance travelled.

Data analysis

14. In order to determine whether agricultural development can be explained by accessibility a cross-sectional framework of analysis was used. Survey data from each holder was collected and averaged within each of the 33 villages. Using this data the parameters of accessibility were tested as explanatory variables of the parameters of agricultural development by regression analysis.

SURVEY RESULTS: THE RELATIONSHIP BETWEEN DEVELOPMENT AND ACCESSIBILITY

General characteristics

15. Over half the holders interviewed were female; the proportion of male holders in each village increased the further the village was from Kumasi. Nearly sixty per cent of holders were over 40 years old, this proportion also increasing with distance from Kumasi.

16. The average household size in the survey was found to be 4.66 people, and the average holding size 4.2 acres (1.68 hectares). Total farm area per holder was found to increase with inaccessibility. This applied to both cocoa and non cocoa holdings.

Income

17. 59 per cent of holders reported that their major source of livelihood came from food farming, a further 28 per cent claimed this to be cocoa, and the balance looked to remittances and paid employment. In terms of the geographical distribution of holders between these sources of livelihood food farming was relatively more important in the more accessible villages and cocoa was more important in the more remote villages. Non farming jobs were more commonly reported in the more accessible villages.

18. By way of confirming this last observation household labour input into farming was found to rise with inaccessibility when measured either in terms of days worked per person or in terms of days worked per holding. The labour input per hectare tended to fall with inaccessibility although this probably reflects the smaller labour demands of the cocoa crop.

Modern inputs

19. More holders used fertilizers and other modern inputs in the savanna than in the forest zone villages. One remotely located village in the far north of the region was found to have 32 per cent of the reported extension contact, 65 per cent of the machinery hire and 75 per cent of the fertilizer used in the whole study. Even if the two savanna villages are excluded from the analysis there is still no evidence to show that inaccessibility prevented the use of fertilizer, machinery or insecticide or that it prevented contact with extension workers. Overall it appears that the pattern of extension contact is more dependent on the local management and enthusiasm of individual extension workers than on the problems posed by inaccessibility even though the latter may well hinder directly or indirectly the overall efficiency of each extension organisation.

Finance

20. The proportion of holders in a village that applied for loan finance was found to rise with inaccessibility and age.

21. A different picture emerges with success in obtaining loan finance. Farmers in the more remote locations experienced greater difficulty in securing finance. Loan applications were more successfully made the more accessible the village.

Crop production

22. The survey found that the proportion of holders growing cocoa increased with inaccessibility. Both the average cocoa crop area per holder and the proportion of farmed area covered by cocoa increased with inaccessibility. No significant relationship was found between accessibility and cocoa sales per grower or cocoa sales per hectare. The data does suggest that women holders are more successful in maintaining higher cocoa yields.

23. No evidence was found to suggest that maize yields or food crop husbandry varied with accessibility. However data relating to crop yields are notoriously difficult to interpret where multiple intercropping is practised as widely as in Ashanti.

Animal husbandry

24. Small numbers of poultry were kept by a large proportion of holders in the survey, however data collected on this topic was too unreliable to be analysed in detail. Other sources of information suggest that commercial scale poultry farming is concentrated in and nearby the major towns in the Region. The major towns provide a market and are also major distribution centres of chicken feed concentrate which has been in short supply for some time. In these circumstances a remote location would put the commercial poultry farmer at a distinct disadvantage.

25. Nearly 600 sheep and goats were kept by the 491 holders in the survey. No evidence was found of any significant relationship between accessibility and the ownership of sheep and goats; although one of the most inaccessible villages accounted for a quarter of all sheep and goats recorded in the survey.

THE RELATIONSHIP BETWEEN ACCESSIBILITY, TRANSPORT AND MARKETING

The initial movement and location of sale of crops

26. The average distance between field and village was found to be 3.9 km; most of this consisted of footpaths. In over 90 per cent of the households surveyed the principal means of carrying goods from the field was by headload. Tractors were used occasionally in the savanna villages.

27. Fifty-seven per cent of holders sold the dominant proportion of their food produce at their house. A further 24 per cent sold their food principally at the local village market. Cocoa was sold at the village buying posts of the Cocoa Marketing Board at a fixed price set for the whole country. Food is mainly sold to travelling wholesalers at the village who arrange for its transport and onward sale in urban markets. It is expensive for the farmer to arrange to sell his own produce in urban markets because not only must he pay his own return fare but transporters charge two to three times as much for individual loads (such as a bag of maize) than they would charge for movement of goods in wholesale quantities.

Accessibility and food sales

28. Less than 5 per cent of holders identified road conditions which would disrupt the movement of vehicles as a contributory cause of their produce becoming rotten before it could be sold. As farmers were referring to particular instances they remembered over the last few years, only a minute fraction of produce was effectively lost through poor road conditions.

29. Overall it appears that accessibility does not easily explain the proportion of farmers in a village selling food crops. Although the proportion of farmers selling more than 30 per cent of any crop (including cocoa) does apparently increase with inaccessibility this may reflect the indirect influence of other factors such as farm size and the use of labour which probably vary more directly with accessibility. The level of inaccessibility in the more remote areas of the survey was insufficient to hinder food crop sales. Over 55 per cent of all holders surveyed reported selling maize, 36 per cent cocoa, 17 per cent cassava and 13 per cent plantain. No significant relationship was found between the sale of maize and accessibility but cassava was sold relatively more frequently in the more accessible villages. By contrast plantain was sold more frequently in inaccessible locations. This is probably because plantain tends to be grown as a cover crop for cocoa.

Social mobility and migration

30. The level of trip making per holder was found to vary greatly with proximity to urban centres. As might be expected, the most accessible villages demonstrated much higher levels of mobility than the more inaccessible villages. For example one village very close to

Kumasi reported a trip rate to Kumasi of 84 journeys per holder per year. By contrast the most inaccessible villages were found to have trip rates to Kumasi of only one journey per holder per year. The average trip rate of Kumasi for all villages was 19 journeys per holder per year.

31. The percentage of holders having migrated to the region was found to be closely associated with the sex of the holder and accessibility. The least accessible areas in the region are now attracting the most migrants. Male holders are now much more likely to migrate and establish new farms than female holders.

The impact of accessibility on farm gate prices

32. The impact of accessibility on farm gate prices was estimated using Ministry of Agriculture data. Regression analysis confirmed that transport charges were closely related to travel distance. If it is assumed that one third of the Kumasi market price covers wholesale and retail margins and that all producers' prices are set in relation to the Kumasi market price, then it can be calculated that farmers located 100 km from Kumasi would receive 6.7 per cent less for their maize than those selling direct to wholesalers at Kumasi market. The calculated decline in farmers prices was little different for yam (6.5 per cent) or for plantain (5.2 per cent) at the same distance from Kumasi.

ROAD INVESTMENT AND FARMERS' PRICES

Improvement from earth road to good gravel surface

33. In order to assess the relative change in farmers' prices following road investment it is necessary to estimate the proportionate change in transport costs to the transporter following an improvement in the road surface. Unfortunately because of the difficulty in quantifying the engineering standards of motorable tracks and earth roads an exact figure cannot be given and so two separate estimates of the change in vehicle operating costs were used to calculate reduced transport charges following road investment. Scott Wilson and the Economist Intelligence Unit (ref.1) have suggested a 32 per cent reduction in transport costs between an earth and gravel road for a mammy wagon in Ghana. The Transport and Road Research Laboratory (ref.2) have suggested (for somewhat different circumstances) a change of only 6 per cent for a light goods vehicle.

34. By using two alternative methods, which varied in their treatment of standing charges, of calculating the impact on wholesale transport charges, coupled with the two alternative estimates in the reduction in vehicle operating costs, four estimates of the reduced transport charges for each commodity were made for each considered improvement of an earth road to gravel road standard. In this way increases in farm gate prices were predicted for different road lengths and different commodities from these estimated reduced transport charges. Averages of the four different increases in farm gate prices following road improvement are shown in Table 1.

Table 1. Potential improvement in farm gate prices following a road upgrading from earth to gravel surface

Length of improvement	Average percentage increase in farm gate price		
	Maize	Yam	Plantain
5 km	0.08	0.11	0.09
20 km	0.29	0.3	0.24
50 km	0.67	0.5	0.37

35. These figures demonstrate the very small increase in prices that can be expected from a road improvement. All the figures here assumed that the transport cost savings would be fully passed on to the farmer, and that none of the benefits from the road investment would go to the final consumers or to the wholesalers, retailers or transporters.

An improvement from pathway to basic earth road

36. Headloading is many times more expensive than vehicle transport, the survey found that the average charge to a farmer for moving one headload of produce from farm to village was ₵2.9 for 3.9 km. The impact on farm gate prices of converting a footpath from the village to the road head to the most basic vehicle track can be substantial. Nevertheless though large they might not justify the costs of the construction and maintenance. Although a majority of holders preferred to use domestic labour for this purpose, 40 per cent of the holders did hire labour when necessary.

37. If it is assumed that it costs ₵0.5 to move a standard 40 kg headload one kilometre then the costs of moving a 100 kg bag of maize would be ₵1.25 per km. Assuming that the farmer is able to sell his produce to a travelling wholesaler at the village after the construction of vehicle access, the calculated proportionate increase in farmers' maize prices following the conversion of a footpath to an earth road is shown in Table 2.

Table 2. Potential improvement in farm gate maize prices following the conversion of footpath to an earth road

	Length of footpath to be changed to vehicle access		
	2 km	5 km	20 km
Improvement in farm gate maize prices	4.3%	11.4%	70.6%

These estimates suggest that it is in the order of one hundred and forty (140) times more beneficial to the farmer to have vehicle access brough 5 km nearer to his village (where the alternative is headloading) than to improve 5 km of existing earth roads and motorable tracks up to a good gravel standard.

CONCLUSIONS

38. Within the range of accessibility considered in the study little evidence was found to suggest that agriculture was adversely affected by inaccessibility. It appears that the more inaccessible villages concentrate more on agriculture than the more accessible villages. The latter have the advantage of their position to concentrate their efforts on non agricultural sources of income such as marketing, rural industry and the provision of services. Accessibility was also shown to influence strongly the level of passenger trip making.

39. The only important drawback of inaccessibility identified was difficulty in obtaining loan finance. The provision of other modern inputs to agriculture were not observed to be adversely affected by inaccessibility. The pattern of extension contact was more dependent on the local management and enthusiasm of individual extension workers than on the problems posed by inaccessibility, even though the latter may well hinder directly or indirectly the efficiency of each extension organisation.

40. The study found that the improvement of short lengths of roads and tracks would have a negligible effect on the prices paid to the farmer. However replacing a 5 km footpath between a village and the roadhead by a vehicle track may benefit the farmer through increased farm gate prices by over one hundred times more than improving the same length of poor quality road surface to a good quality gravel road. However these benefits would have to be carefully weighed against the cost of construction.

41. Overall the figures indicate the advantages of ensuring that all villages have direct vehicle access. The quality of the road surface is of minor importance. From the points of view of agriculture, investment in bridging, minor drainage work and other small scale remedial work to extend vehicle access and keep routes open to vehicle traffic probably represent the best use of scarce of engineering resources.

ACKNOWLEDGEMENTS

42. This work forms part of a joint research programme between the Overseas Unit (Unit Head: Mr J N Bulman) of the Transport and Road Research Laboratory and the Building and Road Research Institute, Kumasi, Ghana (Acting Director: Dr Gidigasu). The work was undertaken for the Ghana Highway Authority as part of their Highway Research Programme.

REFERENCES

1. SCOTT WILSON KIRKPATRICK AND PARTNERS, ECONOMIST INTELLIGENCE UNIT, Road vehicle operating cost manual. Ghana Highway Authority, Accra, 1975.

2. ABAYNAYAKA S W, H HIDE, G MOROSIUK and R ROBINSON. Tables for estimating vehicle operating costs on rural roads in developing countries. Department of the Environment, Transport and Road Research Laboratory. Crowthorne, 1976. TRRL Laboratory Report 723.

16

Rural transport policy in developing countries

G. A. EDMONDS, BSc, MSc, PhD, MICE, MInstHE, AMBIM, International Labour Office

The objectives of rural road transport planning can be, and often are, varied and diverse. In a strict planning sense the objective can be seen as purely to extend the road network, to provide the final link in a chain which is seen to start at the port or a major city. More rationally, rural roads can be planned to meet various development objectives, be they strictly economic or socio-economic. On the other hand, they may be prepared to meet political objectives related to integrating the rural population or strategic/defence objectives.

INTRODUCTION

In recent years, there has been a major shift in emphasis in road planning away from the primary network, which in most countries has already been constructed, towards secondary and feeder roads. More attention has therefore been paid to rural planning in general and the definition of selection and evaluation criteria in particular.

There is justified concern of governments and aid agencies to invest limited funds in the most effective manner. This has lead to useful developments in the way in which transport planners view and analyse rural roads. In the first place, they are now seen as an integral part of the road network. Second, it is accepted that they merit detailed assessment in their identification and selection.

However welcome these developments may have been, it is time to consider, perhaps, whether they are not valid responses to the wrong questions.

In the first place, seen from the centre, rural roads are the end of the chain, consequently they are, by definition, less important. In fact, seen from the farm they are already a long way down a chain which starts <u>on the farm</u>. Moreover, in economic terms, rural roads are the most important elements of a road network. Without them any idea of improving the country's economy through the exploitation of agriculture is merely a pious hope.

Secondly, even when their importance has been recognised it is unfortunate that their role has been perceived only in terms of their economic or strategic/political role, rarely in their potential for social development.

In the following sections, these assumptions regarding rural transport planning are discussed and it is suggested that there are more basic questions which remain to be answered.

Rural road transport planning

The first illusion that we have to divest ourselves of is that rural transport planning is synonymous with roads. The road, as most people understand it, is a luxury to which the majority of the rural population in developing countries do not have access. For example, in Mexico, before the start of a major rural roads programme in 1971, 80,000 of the 97,500 communities had no direct road access; in other words, 60 per cent of the rural population were not served (1). Although matters have now improved in India, the previous state of affairs was typical of that still existing in many developing countries. Thus, prior to the initiation of the third development plan, one out of every three of India's half-million villages was still more than 8 km from a dependable road connection. A study in Maharashtra (2) found that 27,000 out of 34,000 villages were "off the main road system, and that half had no approach roads linking them to the main road". In Kanpur, 37 per cent of the villages had no road access and 20 per cent of these were at least 16 km from any road access. In Kenya, more than 60 per cent of rural holdings are more than 4 miles from a government secondary school, 50 per cent more than 4 miles from a health centre and 35 per cent more than 4 miles from a recognised road (3).

The standard answer to this serious lack of road access has been to pump money into the rural roads sector. Thus, whereas in 1965 only 35 per cent of roads built under the World Bank's lending programme were rural, accounting for 12 per cent of highway loans, by 1977 the figure had risen to 94 per cent, taking up 37 per cent of the loan budget (4). However, the objective of providing a network which gives access to the majority of

Highway investment in developing countries. Thomas Telford Ltd, London, 1983, 119–124

the population is a long way from being achieved, and indeed it seems unlikely that it ever will be. One further example helps to illustrate the point. Let us assume that it is planned to double the size of the road network in, say, Tanzania, Malawi, Kenya and Zambia over the next 15 years, a not unreasonable objective given the present level of access. At present-day prices, this would mean a massive increase in foreign exchange outlay, requiring an average increase of 3.5 per cent in export earnings (5).

This appears to be a fairly gloomy picture. However, it must be recognised that because there is no road, it does not mean that transport does not take place. In fact the bulk of rural transport takes place off the road. Studies have shown that the demand is for the transportation of relatively small loads (less than 150 kg) over relatively short distances (up to 15 kms) (6).

Our natural disappointment when seeing the figures presented above, may therefore be based on a false conception. If we see a rural road as being the major transport mode in the rural areas then our disappointment is justified. However, if we recognise that it is merely the end of a chain which starts on the farm then it serves to focus our attention on the actual transport demand in the rural areas.

Road selection

As suggested above the emphasis on rural roads has lead to a concerted effort to devise effective selection criteria for rural roads. By and large these have concentrated on purely economic criteria. There are several reasons for this not least of which is the pressure from funding agencies to justify, with a number, their investment. Moreover, it must be added that, unlike social benefits, it is actually possible to quantify economic benefits.

120 years ago Henry Law wrote "in applying ourselves to the formation of roads in new countries, we have peculiar mathematical resources to guide us in the selection of these lines which are most direct" (7) (author's emphasis). He was, inadvertently, perfectly correct. The selection of rural roads for construction is now accepted as being only one element in a range of activities which together will lead to the development of that area or region. Thus the provision of extension services, of health centres, schools, public transport, all these and more must be considered simultaneously. Given the variety of factors to be co-ordinated we should accept that we cannot predict with accuracy the actual outcome.

Unfortunately, however, we are in an age when it is considered necessary to justify every decision with a number. To produce that number it is necessary to have a "scientific" formula. Naturally, it is perfectly feasible, in theory, to assess which group of roads and activities will maximise the value added of a region. However, in practice, the data required to carry out the evaluation would probably require more man-days than that required to build the roads themselves, using labour-intensive methods! It seems that we have been intent on developing formulae for solving a problem which is incompletely defined and poorly understood and depends for its solution on data which can be no more than informed guesswork.

In the selection of roads therefore we suffer from two problems. First, an over-zealous application of scientific method and the assumption that we can and must quantify all the variables. Second the erroneous assumption that we are dealing with a choice between which group of roads should be built whereas in fact it is a question of choosing, for budgetary reasons, which group of roads we should best afford to build given that there is demand for all of them.

The major cause for concern here is not that developed country academics are wasting their time developing extremely convoluted selection criteria. That is their choice. What is worrying is that these are then inflicted on developing country engineers and planners as a necessary prerequisite for securing loans for road construction. The fact that the World Bank has increased its rural road lending from $40 million in 1969 to $238 million in 1977 shows the massive shift in emphasis from trunk roads to rural roads. However this shift in emphasis seems to have been accompanied by the development of a whole spate of evaluation techniques designed for the most part to justify this shift. For instance, it is argued that in rural areas where there is already a high level of economic activity the traditional road user savings methods would be used as there is already a substantial demand for transportation. That all sounds very well until one questions the causal relationships assumed. First, there is little evidence to show that movements related to economic activity are the major element of transport in the rural areas. In fact the little evidence there is suggests that it is not (8). Moreover, the assumption of a low level of price elasticity of demand ignores the monopolistic nature of transport in the rural areas of most developing countries. The supply of goods to be transported may be high and thus the demand for transport, but the supply of transport vehicles is low. Consequently, the benefits attributed to the new or improved road may or may not be passed on to the users of the available transport facilities.

The emphasis on economic analysis would, of course, be justified if the roads thus chosen actually responded to the demand. Recent evaluation studies (9), (10), (11) suggest however that this is the exception rather than the rule. To quote only one "The roads were not economically justified, improved cultivation was and is dependent upon the availability of inputs and a better marketing system, not better roads" (12).

Rural transport vehicles

The over-emphasis on the selection criteria has obscured the other more fundamental issues of rural road planning. We have already suggested that we are fairly ignorant of the actual transport demand in the rural areas. If we knew more, it would be possible to respond effectively to the question of "for which type of vehicle should we be designing the rural transport system?".

By any standards of appropriateness the European vehicle is inappropriate to the needs of the rural population. They are extremely costly; they are not available to the rural population; they require a particular type of track which is expensive to build and maintain; their technology is far removed from that used in the rural areas; they must be either imported or manufactured in mass-production factories; and they need to transport large loads to justify investment in them. If no alternative existed to vehicles of this type one would, sadly, have to accept their inappropriateness. However, as the World Bank (13) and UNIDO (14) have recently pointed out, there is a large range of others at present in use, which could be more effective if serious attention were paid to improving them. Animal-drawn carts, the bicycle and its numerous derivatives and simple motorised transport are prevalent in most developing countries. However, they are seen as quaint, second-rate substitutes for the car or pick-up. They should, however, be regarded as a real solution to the lack of mobility of rural people. They meet the criteria of affordability, availability, utility and compatibility with local conditions which the European motor vehicle does not and, moreover, they could be manufactured locally.

For those who find it difficult to believe that any form of transport is a luxury to the majority of people, the following figures from a study made in the northern region of India (15) provide food for thought: during the reference period (July 1977-June 1978) 73 per cent of rural households did not own any type of vehicle, nor did 89 per cent of households having holdings of less than 5 hectares. Of the total quantity of goods transported from the village to a recognised road, 92 per cent were taken by head load or animal cart. Even on farm-to-market access roads as much as 63 per cent of goods were transported by cart.

It may be therefore that the major effort in rural transport planning is being directed at a perceived demand for a form of transport (i.e. road and vehicle) with which we are familiar. In fact this effort should, perhaps, be directed at an actual demand for forms of transport with which we are not familiar but, in fact, responds to the real needs of the people concerned.

Whether the emphasis can be changed or not, it is clear that investment in rural roads will continue for some years to come at a relatively high level. Consequently, given our present state of knowledge, it is important that thought is given to two major aspects. First the techniques to be used and second the eventual maintenance of the roads.

Road construction

In the case of construction, it is now clear that labour-based techniques are economically and technically viable as far as the construction of rural roads are concerned. The detailed research that has been carried out has proven the validity of this conclusion up to wage rates well in excess of the agricultural wage rates in the great majority of developing countries. Naturally we are talking of efficient labour-based techniques, not of subsidised employment creation schemes.

The problems associated with the substitution of labour-based for equipment techniques are not technical or economic. They are problems of attitude, of training and of organisation and management. The use of labour-based techniques implies various investments. In training - because the responsibility is devolved further down the chain of command and one is dealing with the management of men and not machines; in administration - because the existing systems are often incapable of ensuring that large groups of labourers are paid on time or that tools of good design and quality are procured; in planning and reporting - for it is important that tight control is kept over productivity and performance. For most engineers, this investment is a risk, as they are not conversant with labour-based techniques. Moreover, there is a feeling that these techniques are backward, second-hand, out-of-date. Their benefits are seen to be socio-political rather than financial and there is, therefore, a reluctance to accept them. Consequently, it is in the reduction of this risk that technical assistance efforts should be directed. Moreover, as the number of successful pilot projects increases, the fear of the using of these techniques will diminish.

Road maintenance

Whatever technology is used for the construction of rural roads it is in their maintenance that any hope of realising their potential benefits lie (i.e. it is rather strange that so little attention has been paid to maintenance. However meticulously we assess the benefits that might accrue, whatever rate of return we place on the investment in a road will have absolutely no meaning whatsoever unless the road is maintained. In fact, in most countries, the effective size of the network is diminishing due to lack of maintenance. An analysis carried out by the author on the level of maintenance in various developing countries is summarised in table 1.

Maintenance costs have been the subject of many learned papers. However, whilst few agree on the actual figures, most experts recognise that there are two elements to the cost: a fixed cost, which is irrespective of the traffic and related to vegetation and drainage clearing; and an element which relates to the level of traffic expressed in average daily flows. A cost formula should include the average total cost of maintaining a road over its life. Thus, the paved road formulae, for instance, should include an average reconstruction figure expressed as a yearly figure.

It should be recognised that maintenance expenditure figures do not provide any indication of maintenance quality. The evidence suggests that a large part of maintenance expenditure is lost in overheads or in badly executed work.

The second point is that the estimated maintenance costs are rather consistent at about $1,000 per km per annum. This, however, is an average cost for paved, gravel and earth roads; the range being from about $2,000 down to $400-$500.

Further, actual maintenance expenditures are generally made on the primary network. Thus, not only is the actual maintenance expenditure insufficient for the needs of the network, it is only used to maintain a relatively small amount (in length) of the network. The fact that most of the maintenance expenditure goes on the primary road network should not be surprising because:

(a) those who travel on the primary network are often those who are in the best position to apply pressure for better maintenance; by and large, those travelling on the earth roads are not powerful members of the society;

(b) the calculable economic returns are much greater for the maintenance of paved roads;

(c) the maintenance of secondary and feeder roads is often the responsibility of local administrations; their maintenance budget is usually taken from a general recurrent budget and may be used for "higher priority" activities;

(d) local administrations do not have the administrative or technical capacity to deal with road maintenance;

(e) donor agencies provide training and equipment for primary road maintenance activities.

Furthermore, the small amount of money that is spent on road maintenance is often not spent very efficiently. This is also a reflection of the lack of attention that is paid to maintenance. To many, the emphasis on new construction and the lack of road maintenance are inter-dependent. Much of

Table 1. Maintenance expenditures

Country	Road density km/100 sq km	km/head of population	Average maintenance expenditure /km	Estimated average maintenance cost
Benin	6.4	2.9	$ 400	$ 935
Bolivia	3.5	7.1	$ 260	-
Botswana	2.0	23.0	$ 220	-
Burundi	11.0	2.3	$ 242	$ 900
Cameroon	6.0	4.7	$ 450	$ 800
CAE	1.7	-	$ 180	-
Honduras	7.0	2.0	$ 887	$1630
Guinea	5.4	2.5	$ 170	$ 950
Ecuador	8.3	3.3	$ 436	$1000
Guatemala	13.0	2.7	$ 650	$ 950
Lesotho	11.5	2.9	$ 490	-
Mali	1.0	2.4	$ 260	$ 700
Mauritania	1.0	5.6	$ 230	-
Nepal	3.1	-	$ 450	$ 900
Paraguay	-	2.6	$ 400	-
Senegal	6.7	3.4	$ 530	-
Sudan	0.7	1.0	$ 50	$2000[1]
Swaziland	15.0	5.5	$ 460	-
Togo	13.0	3.2	-	$ 950
Upper Volta	6.0	2.9	$ 230	-
Malawi	11.0	2.2	$ 222	-
Tanzania	5.0	3.0	$ 470[2]	$1100[2]
Zambia	4.5	7.0	$ 430	-

[1] For paved roads. [2] Trunk roads only.

the finance for new construction is external, whilst maintenance finance comes from government recurrent expenditures. From the psychological point of view, road construction is much more visible to the people and status-bearing for the engineers than road maintenance.

The logical conclusion of this argument is, therefore, that one should stop spending money on new construction and direct the resources into road maintenance. Or, put more brutally, until developing country governments spend more on road maintenance, donors will not finance new road construction. It has also been suggested that any new roads that are built should be to the highest possible standard to limit the level of maintenance required.

This emphasis on an either/or solution, though economically justifiable, clouds the real issue. This is, "how, with limited resources can a developing country provide a level of <u>transport</u> service which provides maximum benefit to the largest number of people?"

Whilst it is true that there should be less emphasis on new construction, it is for the broader policy reason that it is inconceivable that most developing countries will be able to develop a similar type of road network to that of the developed countries.

Similarly, even if the recurrent budget for road maintenance in many developing countries were tripled - an unlikely event - it would merely mean that it would be possible to maintain the paved road network to the required standard. Given the various financial, administrative and technical restraints, it is clear that the best one can hope for as far as the road network is concerned is a <u>minimum level of maintainable service</u>.

There is no doubt whatever that as far as section improvement and routine maintenance of rural roads are concerned, labour-based methods, in general, provide the least cost alternative.

In Kenya, for example, the old lengthman system of maintaining roads has been utilised in an effective manner for the maintenance of the Rural Access Roads Programme. The details of the system are described elsewhere (16). The major advantages of the labour-based system are:

i. it is cheap (see below);

ii. the skills required in the construction of the road can be utilised in its maintenance, thus limiting the need for training;

iii. administration is simplified by having a contract which specifies the output expected and the payment to be paid;

iv. no equipment is involved and therefore spare parts and skilled labour are not problems;

v. each contractor has a section of road which means that he takes pride in it and, moreover, the local people know who to exert pressure on if the road is not maintained.

The system is based on payment by results. Each contractor is responsible for a certain length of road, payment being made after a satisfactory monthly inspection. The system is extremely simple but depends entirely on effective supervision. It is estimated that one supervisor/roads inspector assisted by a clerical officer (for payment) can effectively supervise 300 km of road.

With equipment-intensive maintenance, it is usually fuel and spare parts which restrict operations. In a labour-based programme, it is the timely payment and close supervision which ensure proper maintenance.

In Kenya, routine maintenance using this system will cost in the order of $150 per km which represents 2.5 per cent of the construction cost. Wages represent about 65 per cent of the total cost; (payment is made on the basis of 12 days effective work at $1 per day which, on a monthly basis, represents a daily wage of $0.5).

The system is, therefore, ideal for implementation by local authorities which lack the sophisticated support services required for machine-based maintenance. This type of system is known to be used in various other countries such as Burundi, Uganda, Papua New Guinea and Mexico.

SUMMARY

In a necessarily cursory discussion of rural transport, it is inevitable that the individual problems have only been touched upon. The article, however, has been written to stimulate discussion on issues that are often ignored even though in terms of the needs of the mass of the population they are crucial.

REFERENCES

1. For a description of the programme see EDMONDS G.A. The roads and labour programme, Mexico, in G.A. Edmonds and J.D.G.F. Howe (eds.): Roads and resources. Intermediate Technology Publications, London, 1980.
2. GOVERNMENT OF MAHARASHTRA, Finance Dept. Report on regional transport survey of Maharashtra State (Bombay, 1966), referred to in Owen W. Distance and development. Brookings Institution, Washington, 1968.
3. Republic of Kenya. Integrated rural survey 1974- 5: Basic report. Central Bureau of Statistics, n.d., Nairobi.
4. World Bank. Note on rural road lending. Washington, 1977.

5. EDMONDS G.A. Labour-based road maintenance: policy and prospects. Paper presented at the PTRC Summer Annual Meeting. Warwick, United Kingdom, 1980.
6. BARWELL I.J. and Howe J.D.G.F. Appropriate transport facilities for the rural sector in developing countries. ILO, Geneva, 1979.
7. LAW H. Construction of common roads circa 1850. John Weale, London, 1855.
8. HOWE J.D.G.F. Surface transport in Africa - the future. Journal of the Royal Society of Arts, London, Aug. 1975.
9. BLAIKIE P. et al. The effects of roads in West-Central Nepal. A report to ESCOR, Ministry of Overseas Development, London, 1976.
10. HOWE, J.D.G.F. The impact of rural roads on poverty alleviation. Working Paper No. 106. ILO, Geneva, Nov. 1981.
11. TENDLER, J. New directions rural roads. USAID, March 1979.
12. USAID. Jamaica feeder roads: an evaluation. AID project impact evaluation report No. 11. Agency for International Development, Washington, 1980.
13. World Bank. Appropriate technology in rural development: vehicles designed for on and off farm operations. Washington, 1978.
14. UNIDO. Appropriate industrial technology for low-cost transport for rural areas. Monographs on appropriate industrial technology, No. 2. New York, 1979.
15. India, National Council of Applied Economic Research. Transport technology for the rural areas: India. Mimeographed World Employment Programme research working paper; restricted. ILO, Geneva, 1981.
16. de VEEN J.J. The rural access roads programme: appropriate technology in Kenya. ILO, Geneva, 1979.

Discussion on Papers 13–16

Mr W.R.G. EAKIN *(Ward Ashcroft & Parkman, Liverpool)*: I agree with Mr Berren (Paper 13) that for some very low-grade road projects the conventional procedure of awarding the construction of the road to a contractor who has tendered on a fully designed and documented project may not be appropriate. However, the disadvantage of the method he describes is that the profit motive does not exist. There is no incentive to carry out the work as quickly and efficiently as possible and consequently the final cost may be high. A 'target-cost' type of contract may be better.

Mr Berren's main objective to the conventional procedure of awarding work to a private contractor is that it means that a full design is done in advance and this design is often too sophisticated and expensive for the purpose. I agree that designs have sometimes been of an unnecessarily high standard, but this is a fault of the design, not of the method. The design parameters adopted were not appropriate to the situation.

I would agree that for some projects it may be more economical to carry out the detailed design just ahead of construction. However, it seems to me that, before construction starts, a preliminary engineering design, at least, is absolutely necessary. The preliminary study would

(a) select the most suitable corridor, taking account of both engineering aspects and the potential for development in the region traversed by the road
(b) define the most appropriate geometric standards
(c) define the type and thickness of pavement
(d) prepare a specification (if detail design is to be done during construction)
(e) identify any major structures
(f) prepare a realistic cost-estimate

Sometimes budget cost-estimates are prepared simply by taking costs/km for a similar type of road in another country. This can be very misleading because costs vary widely, depending on many factors including and especially soil conditions and availability of materials. To illustrate this point I would refer to the study by Cross in Papua New Guinea (Paper 7) where it was found that the cost of pavement construction in the coastal area was approximately half the cost of pavement construction in hilly terrain.

In a study of Northern Somalia on which I was engaged I estimated the reverse to be the case. On the flat coastal plain there are no suitable materials for pavement construction and materials have to be hauled some considerable distance from the hills lying inland and this is expensive as no haul roads exist, whereas in the hilly terrain there are suitable materials all along the route. Further, the thickness of pavement can be reduced because the sub-grades are stronger there than in the coastal region. Thus in the Somalian study, pavement construction on the coast will be much higher than in hilly terrain inland.

I am sure that there are very good reasons for the pavement-cost differences given in Paper 7 and this example illustrates how very careful one has to be in taking costs from one project and applying them to another project to a different country. The preparation of an accurate cost estimate is one of the objectives of a preliminary study.

Mr O. BENNETT *(Touche Ross & Co.)*: Paper 14 raises issues similar to those described in Papers 10 and 11. I was involved with Ove Arup and Partners in Botswana in a two-stage study similar to that in the Sudan, but in our case we took account of non-economic factors in Phase 1. We had 4500 km of road in Phase 1, reduced to 1500 km in Phase 2. As in the Sudan, the terms of reference referred to a conventional cost-benefit analysis in the second phase, but not in the first. We therefore had the problem that roads which ranked highly in our first phase might not rank so highly when subjected to a conventional cost-benefit analysis.

We therefore adopted a system where we had two separate rankings in our first phase, one based on conventional economic criteria and one relating to less easily quantifiable social criteria. We also had a combined ranking. In Sudan, the approach was primarily economic, but I should like to ask the Authors of Paper 14 if they are confident that, of the 76 projects which came out of their preliminary ranking, others than the final seven would not have ranked higher, given a detailed cost-benefit analysis?

Paper 16 has made a useful contribution to the debate on rural road transport planning. Some members of an aid institution recently drew my attention to a clear preference for feeder roads (which in this particular instance were tracks put to satisfactory use by animal-drawn sledges) over roads which were the final mechanism of transporting produce to market.

Mr P.J. MACKIE *(University of Leeds)*: We know that in the UK travel-time savings comprise some 70% of total user benefits on COBA-type schemes. We also know that the values used are based on shaky foundations, i.e. some research work done 15 years ago, together with some economic theory of doubtful validity. I support the use of cost-benefit analysis as an appraisal tool, but the results, in terms of numbers of projects justified, and their priority rankings must depend critically on the values of time used. I would like to hear more from the Authors of Papers 14 and 15 about where they got their values from, and if their values are really based on consumers' preferences.

Mr L. ODIER *(Bureau Central d'études pour les Equipements d'Outre-Mer (BCEOM) Paris)*: I confirm the fact, underlined by Mr Hine in Paper 15, that there exists, in developing areas, cases where road construction or improvement has no effect or very little effect on agricultural development. I give the example of a study of rural road projects made by the BCEOM in Tunisia. Roads were classified into three groups. One such group included the roads for which there was a very low development/accessibility relationship. Such roads were found, for example, in agricultural areas devoted to olive production, where the lack of access is not an obstacle to development, but where the growth of output is hindered by other problems (land reconstruction, technical progress). In the evaluation of the project, these roads were not retained, but it was recommended that they should be included, eventually, in any adequate agricultural development project.

Mr C.K. KAIRA *(Karlsruhe University, West Germany)*: Articulated road networks in rural areas of developing countries consist of

(a) trunk roads integrating all regions
(b) arterial roads linking to trunk roads joining market towns with principal towns
(c) feeder roads traversed by trucks and buses to link villages to the nearest town
(d) rural access roads joining farmsteads and hamlets with the market and collection/buying centres located on the feeder roads
(e) footpaths, animal-cart and bicycle-tracks, forming the local transport system, linking homesteads with market and collection/buying sub-centres located on rural access roads.

As Dr Edmonds asserted in Paper 16, the bulk of rural transport takes place along the local transport system. However, planners do not seem to be interested in local transport systems, neither the network nor the transport. Usually, improvement of transport in rural areas stops at rural access roads as far as the network is concerned and at conventional motor vehicles as far as transport is concerned. Consequently, failure on the part of the government and transport planners to recognize the existence and function of the local transport system has been responsible for a development of a sub-optimal transport system which is characteristic of rural transport in developing countries. Indeed there cannot be significant improvements in rural transport unless the local transport system in the influence area of a feeder road is incorporated into the feeder-road project.

Improvement or development of the local transport system is necessary for the following reasons.

(a) Transport of people or goods start and end from or at dwellings and farms and not at bus-stops or collection or buying centres located on roads.
(b) Due to the massive funds needed to construct rural roads, it is not possible to bring all households within 1 km of a road.
(c) Increased agricultural output needs efficient transport methods for transporting inputs or products around the farm or to market and collection or buying centres.
(d) Transport is a constraint to rural development as the majority still use primitive and inefficient methods of transport to travel and/or to move loads.

Improvement of the local transport system can be achieved through the intensive use of intermediate-technology transport and aids and community action in constructing and maintaining all-weather tracks and paths. Intermediate-technology vehicles are the best alternative as they meet the criteria of affordability, availability, utility and compatibility with local conditions. Although acquisition, maintenance and operation costs of such vehicles and aids are still not low enough for every household to own one, households can have access to a variety of vehicles at very low hire-charges through village co-operatives.

The research I am conducting at Karlsruhe University has revealed that a household which owns an animal-cart spends about US$ 170 annually on operating and maintenance costs compared with between US$ 60 and 90 paid by a household that hires an animal-cart for four hours a day for 150 days in the year. Furthermore, a village co-operative serving a community of 150 households, requiring a fleet of 50 bicycles, 30 bicycle-trailers, 20 bicycle-sidecars, 30 animal-carts, 10 handcarts and 30 wheelbarrows needs US$ 22,300 which is the lowest construction cost of 1 km of a feeder road or of buying two saloon cars. Therefore provision of village co-operatives with fleets of vehicles would not significantly raise the costs of a feeder road project. Moreover, the costs are likely to be offset by the benefits.

As construction of roads using labour-intensive methods is gaining momentum in rural areas, the population is likely to gain skills in construction and maintenance techniques. Therefore, through good co-ordination and organization at national and local level, communities can provide free labour for construction and maintenance of tracks and paths, especially during non-peak agricultural-activity periods. However, the government should provide supervision and technical services as well as special tools, materials, equipment, culverts and so on. Furthermore, land-tenure laws should allow for tracks and paths between land tracts, and through legislation the use of tracks and paths should be limited to one-ton motor vehicles, thus excluding heavy trucks which normally destroy carriageways.

An improvement of the local transport system based on the above maxims is likely to result in less feeder-road kilometres and an integrated, efficient and self-reliant system in the rural areas of developing countries.

MR I. HAMILTON *(Organisation for Rehabilitation through Training, London)*: The ORT has been in the vocational training business for one hundred years and has trained around one million people. During the past twenty years the Technical Assistance Department of ORT has carried out over 80 training projects in 42 countries in Africa, Asia, South America, Europe and the Pacific. ORT has carried out road-maintenance training projects in 12 African countries and studies in half-a-dozen more.

Although we are very happy to see that the training for road maintenance is now being recognized on all sides as a very significant factor in investment-decision on highway systems, I would like to make one or two qualifying comments. Firstly, training in isolation is not a remedy for all problems. Taken in isolation, without being provided in a proper management context, training may be successful. If the Maintenance Department is starved of funds, spare parts or fuel, no amount of training will ensure satisfactory road maintenance. Secondly, it is absolutely essential that manpower planning, leading to the design of sensible training programmes, is included at the start of the projects and not tacked on at the end. In short, alongside the traffic and economic models prepared at the beginning of the project, there would also be a manpower-planning model.

Finally, the training should be appropriate. If we are providing or enhancing skills, then on-the-job training would be the most effective way of achieving the required aims. If the aim is to transmit knowledge, then a classroom system using modern audio-visual methods is required. In most training programmes a sensible, planned combination is required.

Mr P.M. HARTLEY *(Trevor Crocker & Partners, Mitcham)*: The economic benefits of increased accessibility for a rural population by the provision of feeder roads have been acknowledged, but I feel that we need to keep the wider implications of such accessibility in sight. In Paper 14 the Authors acknowledge the possibility of using economic/social/political criteria for their preliminary sieving of possible schemes, yet they rejected these criteria in their work in the Sudan, in favour of more tangible criteria of each zone in terms of traffic/population/agricultural production criteria. These tangible factors are indeed much easier to handle; they are also re-used in the subsequent stages of the project.

I would like to sound a note of caution about our assessment of the benefits of improved accessibility: in addition to providing improved transport links for agricultural produce, better communications can also encourage migration of the rural population to urban areas. This is perhaps one of the main factors in the very rapid growth of poverty/unemployment/social problems in the growing urban areas of the developing world. I feel that we should perhaps be more aware of this factor when evaluating feeder-road projects at the preliminary stage and that greater weighting should be given to these sociological factors. The approach suggested in Paper 16 might also be given greater weight.

Mr W.J. WYLEY *(Howard Humphreys & Sons, Leatherhead)*: My comments relate to the subject of national and provincial roads and to Paper 13. Although I would not decry the considerable resources allocated to planning and design, comparable attention has not always been given in developing countries to ensure that cost estimates are achieved in practice - thus enabling the criteria established for investment to be fulfilled. I therefore welcome Mr Berren's Paper on this subject. For undesigned tracks, he suggests a 'public/private' formula involving the management skills for a contractor on a cost-plus basis.

Flexibility in design and construction methods and the removal of risk from the contractor are very important elements in reducing costs. However, I believe that there is another fundamental element about which Mr Berren is less explicit, and that is incentive. Incentive must be provided firstly to the contractor so that he commits personnel of the requisite calibre to the project and secondly for him to reduce the cost to the client.

For a road in Tanzania funded by the Overseas Development Administration, Howard Humphreys and Sons have developed and are using - with considerable success - a target-cost management form of contract - the objective of which, as defined in the Conditions of Contract, is the construction of the road to the required standard at the minimum cost. Savings from the tendered target are shared between client and contractor. Details of the form of contract are available to anyone interested.

Finally I should add that the general approach of incorporating flexibility and cost-effectiveness are equally applicable to the activities of rehabilitation and maintenance.

Mr D. BOVILL *(Dar Al-Handasah Consultants, London)*: I would like to emphasize the significance of fashion in project appraisal. Over the past 20 years we have gone through a gamut of fashions. To name but a few, we have had major inter-urban roads, the 'trickle-down' effect, multipliers and linkages, complex computer models, rural and feeder roads, roads as a 'stimulus' to development or simply 'permissive', integrated rural development, institution-building, distribution of benefits, maintenance and energy conservation. Woe betide any analyst not taking account of the fashion that is prevalent at the time of his approval! At the height of the feeder road fever some five years ago, a French colleague swore that at a recent astrological conference he attended, there was a paper on the impact of feeder roads on Venus. He could possibly have been lying, but the point to bear in mind is that fashions do influence the method of appraisal and it is often difficult for the analyst to put the prevailing fashion into perspective.

On a similar note, there is something to be gained from a historical view of the way in which economic appraisal has progressed over the years. Initially we simply looked at resource-savings to normal traffic, (i.e., savings in tyres, petrol, spare parts and so on). Then, when we could not get enough benefits to justify investment, we introduced commercial drivers' time, then generated traffic. Then, when this was not enough, we brought in value of leisure time (just in urban areas, though, because that is where they were needed). Then, for justification of rural and feeder roads where benefits were inadequate, we thought about value added to the agricultural sector. Then, when this proved inadequate, we discovered the distribution of benefits, social benefits and multi-criteria weighting.

It is difficult to imagine where we will go next, but I am sure we will think of something to satisfy the needs of the aid donors (in fact, we could give an explicit weighting to those needs in our appraisal).

In summary, any analyst must have, at least in the back of his mind, an awareness that prevailing fashions can bias approval and that there is always a tendency to find enough benefits to justify investment. The best that can said about any appraisal is that, in the opinion of the author, a project is no good, marginal or good.

The significance of the magnitude of user cost-savings is discussed in Paper 15 and I would like to suggest that, before carrying out any detailed investigation, a preliminary analysis should be carried out as to the magnitude of potential user cost-savings per vehicle for the alternatives under investigation and the impact that these changes might have on the possible beneficiaries. Let us consider two extremes i.e. feeder roads and paved-road improvements.

Feeder roads

The transport cost-savings often exceed 50% of the 'without project' case. This seems significant but how does it effect the farmer? Are the prices of his produce fixed by government? Are transport costs fixed or will the transporters reap most of the benefit? Even if the benefits are passed directly to the farmer, will even a 50% reduction in costs stimulate a farmer into changing his pattern of production, given that transport costs are seldom more than 10% of the selling price of his product? (Thus a 50% reduction in transport costs will only increase profits by 5%.) Given that variations in market prices, even throughout one season (let alone on a year-by-year basis), usually vary by considerably more than 5%, how will the farmer react?

Focusing on such issues at the initial stages of a study will help to establish, prima facie, whether there are any reasons why, even on simplistic economic grounds, there is any likelihood of significant changes in agricultural production. Such preliminary investigations should also include an appreciation of farmers' risk-aversion, possible influence of cultural attitude to change and so on in order to establish whether or not it is worthwhile going ahead with time-consuming and expensive farm interviews.

Paved-road improvements

Generally speaking, improvements are only merited when traffic levels are high and the total operating costs are made up of the product of a large number of vehicle-miles multiplied by a small saving in operating costs for each vehicle. The 'benefits' are determined by the difference between two very large numbers (the operating costs in the 'without' and 'with' project cases). In such circumstances, the small changes in operating costs for the 'with' and 'without' cases are usually critically dependent on assumptions regarding the roughness of the road in both situations. I stress the word 'assumptions' because what the study team has to do is to forecast the roughness in the 'with' and 'without' situations over a 15-20 year period. A horrendous number of assumptions are required regarding the magnitude and efficiency of maintenance (in both situations) in order to make a guess at these, and yet they are extremely critical.

In a recent study I found that in order to justify the construction of a new paved road to replace the existing paved road, all I needed to do was to assume a constant roughness of 3300 mm/km and 1750 mm/km in the 'without' and 'with' cases over the 'project life' and this gave a 12% equivalent rate of return (or internal rate of return) and a Net Present Value (NPV) of £3.5 million on an investment of £18 million. However, if I had assumed that the roughness of the existing road was 3000 mm/km (rather than 3300 mm/km), the benefits reduced from £21.5 million to £12.5 million and the NPV from +£3.5 million to -£5.5 million. Similarly, if the roughness in the 'without' case were assumed to be 4000 mm/km then the NPV increased to +£23 million. This simple example highlights the futility of economic appraisal where benefits are determined from large volumes of traffic and small changes in vehicle operating costs. It is impossible to decide whether the roughness of the existing road over the next 20 years will be 3000 or 3300 mm/km and yet it is this assumption which will result in a good or bad project.

The position is even more polarized when considering the economics of certain maintenance operations where even smaller variations in assumptions regarding roughness will often justify quite large investments.

As an aside I would mention that the measurement of existing roughness is almost irrelevant except to establish an overall appreciation of order of its magnitude. The crucial issue relates to the assumption of the future roughness and this is dependent on the level of future investment in maintenance and its efficiency.

This example suggests that where the magnitude of vehicle-operating cost-savings per vehicle is small, a preliminary analysis be carried out to show how sensitive the results of any detailed appraisal might be to such factors as roughness before embarking on detailed assessment of existing and future traffic levels and vehicle-operating costs. In the majority of cases, such an analysis will indicate the futility of extensive data collection and analysis.

Although it might seem that these comments on economic appraisal are directed at eliminating the role of the transport economist, this would be far from the truth. In fact, I am a very strong advocate of the techniques but I would argue that such techniques are often focused on problems which they are not capable of answering. In such cases, the transport economist runs the serious risk of compromising his professional integrity by peddling plausible pseudo-technical rationalizations to justify investments.

Mr GOODWIN and Mr THURLOW *(Paper 14)*: Mr Bovill is correct to point out that the methodologies used in the economic appraisal of road projects are constantly being developed further. More recent developments have concentrated on the treatment of social and non-economic factors. Clearly wherever these factors can be quantified they should be incorporated into the economic costs and benefits in an appraisal. However, we contest Mr Bovill's belief that this has been done basically to justify the implementation of more projects and because it is fashionable.

Mr Bovill's account of how methodologies have expanded to incorporate more benefits is misleading. The economic theory associated with all his stages of development pre-dates even the earliest of road studies. Furthermore, attempts were made to incorporate most, if not all, the later developments in some of the early feasibility studies. Quite rightly, however, the first essential was to evaluate the more basic factors properly and the early attempts to quantify the peripheral benefits and costs were crude and often unsatisfactory. With more time and more experience to draw upon more satisfactory results have been achieved in evaluating these peripheral factors.

At the stage now reached we believe that there is still no satisfactory method of quantifying the benefits of social factors such as better access to schools, health centres and other facilities. Consequently their incorporation into an appraisal requires the abandoning of the standard economic cost-benefit analysis approach and its replacement by a compound ranking procedure based on weighted economic and social factors. The subjectivity involved in both the evaluation of the factors and the weighting system in our view leads us too far away from a form of measurement which is of real value to the decision-maker.

Far better to quantify, in consistent terms, as much as can be quantified. Then, if the results are not decisive, assess any other factors that have been identified which support or adversely affect the desirability of the project. Central to our argument in favour of the method of approach we use is the need for developing countries to prevent wastage of their own physical and human resources. A ranking procedure based on economic and social factors cannot indicate the existence of net economic benefits and hence assess the value of consuming national resources in making the investment.

An inevitable result of investing in projects without measurable economic net benefits would be a decline in national wealth and greater dependence of Third World countries on international aid - a result which is entirely contrary to the principles underlying international aid schemes.

Paper 14 demonstrates that our preferred method of assessing road projects - by a rigorous economic appraisal - is possible even in extreme situations and there is no need to resort to compound ranking procedures, which have been proposed as simpler alternatives to economic appraisals. In referring to a project in Botswana with which he was involved, Mr Bennett points out that non-economic factors were incorporated in the first phase of the study but but not in the second phase. As a result there were potential problems arising from different rankings of projects using the two methods of assessment. Mr Bennett asks if we are confident that projects rejected by the preliminary ranking of the first phase of our study would not have shown a higher ranking when subjected to the more detailed appraisal of the second phase. We cannot be absolutely confident because detailed appraisal of a particular project can always bring to light information previously unknown which significantly affects the evaluation result. However, a major feature of our study in the Sudan was our consistency in approach throughout, from the selection of areas for study and preliminary ranking of the first phase to the detailed studies of the second phase; the aim being to include all factors from the outset and to use the two phases to increase the precision of the values associated with each factor. The evidence from the detailed evaluation of the eight projects in our study was that the ranking was very similar to that which had been produced by the preliminary evaluation. We consider that introducing social, or other non-economic factors in an early stage of such a study makes it less likely that the projects representing the best investment potential for the country will eventually be selected.

Mr Hartley raises a wider concept of social factors than is usually considered in road projects and at the same time raises the question of social costs, which we admit cannot be adequately incorporated into a purely quantitative economic appraisal. He suggests that improved roads can encourage rural populations to migrate to urban areas, contributing to poverty, unemployment and social problems in these urban areas. If a road project does have major social consequences of this type then we would agree that they should be taken into consideration by the decision-maker, but we believe that issues such as this must be considered on a case-by-case basis and no general approach can be adopted. When the consultants identify potential social effects beyond the area of the road itself, or social costs in the area of the road, then these should be brought to the attention of the decision-maker. They should be considered, but often they cannot be incorporated directly in the appraisal of the road if they cannot be quantified according to the same measurement standard as used elsewhere in the appraisal.

On the particular question of transport projects accelerating the drift of population to urban areas we consider that the direct causal relationship is small compared with the influences of expanding work opportunities and improvements to services in urban areas. In addition improved transport will normally stimulate to some extent the economic activity in a rural area and also enhance the quality of life for the population, which would tend to counteract the drift away from the area.

Mr Mackie draws attention to the questionable

assumptions and methodology employed in the COBA-type schemes to evaluate the value of time for road-users in developed countries. While we agree with much that Mr Mackie states we must point out the vastly different circumstances attending the appraisal of rural roads in developing countries. In our study in the Sudan, for example, we used wage rates as a starting point for valuing passenger time savings, with three categories distinguished, namely: professional, skilled and unskilled. Of these the bulk were, not surprisingly, unskilled. In applying the low values of time that were implied by both the actual and shadow wage rates to these numbers of travellers involved, the aggregate value of time savings was found to be very much lower than the 70% of total user-savings that has been the normal result in the UK. These low wage rates reflect the low general level of productivity in such countries, which surely is the main distinguishing feature between the developing and the developed countries and is the primary reason why travel time has a much lower value in developing countries. In our sensitivity analysis testing extreme values of time savings which could apply in a country such as the Sudan confirmed that the results of the evaluations were not sensitive to the value of time.

Our results in the Sudan study were typical of the results of other studies that we have carried out in other developing countries and emphasize the difference in the significance of the value of time between road evaluation projects in developing countries and those in developed countries.

We did not base the values of time on an analysis of consumers' preferences. The range of choices between alternative modes of transport and road surface standards often prevents such an approach in practice. The consumers' preference approach requires the formulation of demand curves relating user cost differentials between modes to modal demand. Frequently, as in the Sudan, empirical data to construct these curves did not exist nor could be assembled without an allocation of consultant resources totally disproportionate to the significance of the results.

Dr EDMONDS *(Paper 16)*: Paper 16 presents several radical changes in the manner in which we review rural transport planning. I am therefore gratified to note that the comments made are unanimously supportive. One can only hope that this attitude is reflected in future planning exercises.

I am impressed by Mr Kaira's comments and particularly the conclusion from his research study that the capital investment to provide simple transport for a village of 150 people is less than the cost of construction of 1 km of access road. I would agree with Mr Bovill's comments that much of the economic analysis used to justify feeder road investment arrive at conclusions which are stronger than the evidence. I also strongly support his view that the traditional type of economic analysis ignores the transport requirements of the mass of the rural population.

Given the general consensus on the content of my Paper I would hope to see money being invested in a rational assessment of transport demand in the rural areas of developing countries.

17

An appraisal of highway maintenance by contract

DR C. G. HARRAL and E. E. HENRIOD, World Bank, and P. GRAZIANO, Roy Jorgensen Associates Inc.

Recently road authorities in many countries, driven by general government pressures both to improve the quality of road maintenance and hold down costs, have shown increased interest in the use of contractors to perform routine as well as periodic maintenance. This paper evaluates experience in six countries, two of which have had long experience and four of which have recently begun major experiments extending the role of contractors for maintenance. It concludes that prospects for contracting maintenance more generally are promising, that major improvements in efficacy and efficiency may often be gained, and that government establishment can normally be reduced substantially, although government capacity to manage contracts may need to be strengthened.

1. INTRODUCTION

1.01 While many road authorities have contracted for the performance of periodic maintenance--because of special skills or equipment requirements, the short-term nature of the work activity, or the desire to avoid overburdening existing government capacities--the practice in most government highway agencies, at least until recently, has been to perform most routine 1/ maintenance with government forces, using personnel engaged on a full-time basis, and equipment owned and operated by the agencies. This has been a function of government for 100 years or more in most countries.

1.02 However, government highway organizations in many countries are faced with severe difficulties in attracting and retaining well-qualified management and skilled personnel. Moreover, bureaucratic formalities frequently hamper the procurement and effective management of resources and often the full economic costs of government operations, e.g., interest on capital costs of equipment, are not incorporated in management decision making. Most importantly, it is normally not possible to structure strong incentives rewarding superior performance and discouraging poor performance. It should not be surprising that these problems result in ineffective, inefficient and extremely costly maintenance operations in many countries.

1.03 The recent, unusually severe worldwide phenomenon of general budgetary crisis for government, which has hit road authorities particularly hard, is now imposing a new urgency to the solution of these traditional problems. Indeed, in many countries there is need at the same time both for finding of ways to perform maintenance at lower unit costs and for increased levels of funding to meet growing maintenance needs of increasingly aged networks.2/

Contracting: Pro and Con

1.04 One way to deal with this plethora of problems is through the greater use of contractors to perform highway maintenance services, including routine as well as periodic activities. Maintenance contracting has attracted increasing attention in recent years, because it can provide:

(i) strong incentives for improvements in performance and economy;

(ii) a less restrictive operating environment in terms of managing resources, including greater flexibility in scaling resources to suit changing demands, thus facilitating improvements in cost-effectiveness;

(iii) relief to the government from the burden of direct responsibilities for large equipment fleets and work forces;

(iv) contractual commitment of maintenance funds with better control over diversion of resources to other activities;

(v) political support for adequate and more stable levels of funding for road maintenance; and

(vi) a better prospect for developing a lasting institutional capacity, in the form of a pool of local contractors skilled in providing effective and efficient maintenance services. (Development of the potential to expand and undertake more extensive construction activities may be an important byproduct.)

Highway investment in developing countries. Thomas Telford Ltd, London, 1983, 131–140

1.05 Contrary arguments, which are often advanced when proposals for the contracting out of maintenance activities are being considered, include the following:

(i) contracting may not decrease costs where redundant government establishment and work forces cannot be reduced or relocated to the private sector;

(ii) contracting could increase costs because the very process of contracting and contract administration may require additional government resources, e.g., in measuring and certifying work quantities for payment;

(iii) contracting may increase costs to the government where there is lack of effective competition in the procurement process, including abuses such as corruption in contractor selection or price fixing;

(iv) government may not have the capabilities necessary to properly manage contracts;

(v) domestic contractors may not have sufficient capabilities--management abilities, technical skills, equipment, working capital, and other resources--necessary to insure effective execution of maintenance activities; and

(vi) contractors may not be well placed to respond quickly during times of emergency, nor to address small scale maintenance needs in remote areas.

Study Objectives

1.06 In recent years highway authorities in several countries have, for various reasons, turned to contracting of routine as well as periodic maintenance. This study was undertaken to review this experience and evaluate the prospects for contracting maintenance more generally, specifically to provide a basis for determining whether, and to what extent, such contracting is desirable and applicable in developing countries. Since contracting of periodic maintenance is already a well-established, widely accepted practice, the report concentrates on the issue of extending contracting to routine maintenance functions as well.

1.07 This paper is abstracted from a continuing World Bank study, which, at the time of this report, had encompassed a detailed examination of the extensive experience with contracting of maintenance for federal highways in Brazil during the past 11 years, and a briefer review of contract maintenance operations in Yugoslavia, Colombia, Argentina, Nigeria and Kenya.3/ The main characteristics of the systems for contracting maintenance in those countries, as well as the results observed to date, are briefly summarized in the next chapter. Yugoslavia has had even more extensive experience than Brazil with the contracting of routine as well as periodic maintenance, while major experiments are really just beginning in the other cases.

1.08 It is recognized that the countries covered by this study are not typical of many developing countries, whose domestic contracting industries may be much less well developed (or non-existent), although the experience with small contractors in Kenya may be apposite. Moreover, there are some similarities in road maintenance management problems among all countries, and there may be a potential role for contractors in at least some maintenance functions in almost all countries.

II. EXPERIENCE IN SIX COUNTRIES

Brazil

2.01 The Brazilian National Highway Department (DNER) began contracting out maintenance in 1970. Initially, DNER took on contractors to replace their own resources; DNER had been banned from employing new personnel to make up for losses due to attrition. By 1980, use of contractors in highway maintenance had greatly expanded, and of the nearly 60,000 km of federal network, routine maintenance of 25,259 km (42%) was contracted out to the private industry. While the absolute size of the network directly maintained by DNER force account decreased by nearly 75%, the DNER work force was reduced by approximately 60%.

2.02 During those first 10 years of experience with contracting routine maintenance, most of the work was performed under cost-plus-percentage-fee contracts, and DNER retained for the most part direct control of the contractors' resources. Thus, the latter only replaced DNER as a hirer of personnel and holder of equipment. The main advantage gained was a greatly improved flexibility in the hiring and firing of personnel, a considerable help in managing seasonal peaks and troughs of the maintenance activity.

2.03 In 1979 DNER launched three experimental unit-price contracts, seeking to transfer the services of field management to the contractors, and to reduce DNER's responsibility for direct control of day-to-day operations. It also sought greater efficiency and cost reductions, by introducing the incentives of unit-price rates in competitively tendered works.

2.04 The initial results of those experimental unit-price contracts for routine maintenance are encouraging. Table 1 presents a comparison of the observed productivities (in units of production per manday) for the three contracts and also for work performed under force account and cost-plus in various areas. Tables 2 and 3 give a direct comparison of productivity rates and unit costs for a unit-price contract and force account in similar areas of the Ponta Grossa residency in Panama. These figures are directly comparable, as work was carried out under close control under very similar conditions, and the force account costs allow for overheads. Although the Ponta Grossa force account operation

Table 1. Brazil

PRODUCTIVITY* COMPARISONS OF LABOR-ORIENTED ACTIVITIES FOR FORCE ACCOUNT, AND COST-PLUS AND UNIT-PRICE CONTRACTS

			FORCE ACCOUNT		COST PLUS		UNIT PRICE		
	PERCENT TOTAL COST	DNER PERFORMANCE STANDARDS	PONTA GROSSA, PARANA 1979	PONTA GROSSA PARANA 1980	CARUARU PERNAMBUCO 1979	ARCO VERDE PERNAMBUCO 1980	PONTA GROSSA PARANA 1979;80	CARUARU PERNAMBUCO 1980	CURRAIS NOVOS RIO GR NTE 1980
POTHOLE PATCHING (m^3)	36.1	0.71	0.30	0.45	0.43	0.49	0.48	0.29	0.36
CLEANING CURB AND GUTTER (m)	2.7	227.00	170.5	164.6	175.3	202.0	245.3	108.4	403.0
MANUAL DITCH CLEANING (m)	1.4	150.00	93.0	92.0	--	--	166.6	35.1	--
CULVERT CLEANING (units)	0.5	1.00	0.96	1.01	--	--	2.14	0.65	2.02
REPAIR OF BRIDGE RAIL (m)	0.3	3.00	1.75	2.18	--	--	2.67	--	0.41
MANUAL VEGETATION CONTROL (m^2)	20.0	1,363.00	1,355.0	1,351.9	795.3	1,083.9	1,958.9	1,336.5	1,099.8
MANUAL WEEDING (m^2)	11.2	205.00	183.1	172.5	187.3	185.6	96.1	--	
FENCE REPAIR (m)	1.9		31.4	15.5	--	15.5	16.3	--	20.5
SIGN REPAIR (m^2)	2.5	2.00	13.71	7.73	3.95	--	8.31	2.52	--
REPAIR ROADSIDE MARKERS (units)	1.3	20.00	36.13	36.23	12.57	--	10.61	13.85	--
REPAIR METAL GUARDRAIL (m)	.1	3.50	34.28	7.47	--	6.57	4.29	3.80	--
PAINTING CURB AND GUTTER (m)	12.7	182.00	126.8	88.9	144.8	148.2	212.9	118.1	301.1
MANUAL SLOPE REPAIR (m^3)	8.4	1.50	--	1.24	--	1.43	5.22	1.00	2.4
MANUAL SLIDE REMOVAL (m^3)	1.0	7.00	2.68	1.21	--	--	7.04	3.36	--

* PRODUCTIVITY MEASURED IN PRODUCTION UNITS PER MAN-DAY.

is commonly viewed as a very efficient one, the weighted average indicates a 59% greater cost for force account work.

Yugoslavia

2.05 Yugoslavia is a socialist federal republic, but since the mid 1950's government administration throughout much of the economy has increasingly been characterized by the separation of those who invest public monies from those who execute the work. There has also been a general decentralization of decision making to and within the individual republics. This general pattern of devolution has applied also to the highways sector, where the service or work execution agency has been transformed into a type of cooperative enterprise, or contractor. By introducing this form of organization a system of incentives promoting efficiency was created, and employees of the enterprises share in the profits. Virtually all maintenance, including emergency maintenance, has been placed under contract to these enterprises.

2.06 The system has worked well. The roads are generally well maintained, and a high degree of professionalism has evolved, accumulating years of experience into a well codified and well understood system which is administered with a remarkably small government establishment. Indeed, in Slovenia only five inspectors and a director control maintenance for a high standard network of 4,700 km, with only general administrative support from a roads authority which in its entirety consists of 185 people covering all functions (planning, construction, maintenance, operations, safety and administration)--a fraction of the size of most comparable road authorities elsewhere in the world.

2.07 Two key factors in this success are the well qualified inspectors, who are experienced senior engineers, and the use of systematic planning and field control procedures, simplified by the use of sampling. The inspectors spend on average three days per week on the roads recording observations and coordinating directly with the enterprise. The inspector responsible for 500-1,000 km of highways oversees the work of between seven and fifteen crews. He has participated in the work scheduling for the period; he knows the nature of the work to be accomplished at each location, when it is scheduled, what resources are required and the basis of payment. If the activity is paid as cost-plus, he will spot check the crew for resource utilization as it is working; if unit-price, the end result can be checked whenever convenient. A key factor is that the inspector is not responsible for either supervising the

Table 2. Brazil

COSTS AND PRODUCTION FOR FORCE ACCOUNT PROJECT / UNIT PRICE HIGHWAY MAINTENANCE CONTRACT

ACTIVITY	FORCE ACCOUNT[1]			UNIT-PRICE[2]		
	COST [3] (1000 Cr$)	PRODUCTION	COST PER UNIT OF PRODUCTION (Cr$)	COST[4] (1000 Cr$)	PRODUCTION	COST PER UNIT OF PRODUCTION (1000 Cr$)
POTHOLE PATCHING (m^3)	1,726.21	172.52	10.005	1,986.99	235.75	8.43
CLEAN CURB AND GUTTER (m)	92.84	19,420	4.78	128.11	42,546	3.01
MANUAL CLEANING DITCH (m)	906.33	80,123	11.31	216.88	58,490	3.71
CULVERT CLEANING (units)	416.73	488	853.95	177.47	565	314.11
REPAIR SURFACE DRAINS (-)	149.15	--	--	466.70	--	--
REPAIR BRIDGE RAIL (m)	198.62	214	928.13	49.29	50	985.80
MANUAL VEGETATION CONTROL (m^2)	3,487.91	4,307,162	0.8097	2,245.59	8,469.428	0.27
MECHANICAL GRASS CUTTING (m^2)	2,612.14	7,629,990	0.3423	2,522.29	4,721.370	0.53
MANUAL WEEDING (m^2)	611.85	96,435	6.34	76.89	9,900.35	7.77
CHEMICAL WEEDING (m^2)	276.08	123,000	2.24	784.05	338,908	2.31
CLEARING FIRE LANE (m)	--	--	--	381.19	42.260	9.02
GRAVEL PATCHING (m^3)	6.86	17	403.53	--	--	--
BLADING GRAVEL SURFACE (m^2)	1,219.44	1,420,800	0.8583	--	--	--
FENCE REPAIR (m)	387.12	6,105	63.41	--	--	--
SIGN REPAIR (m^2)	1,560.70	1,421.98	1,097.55	38.13	182.91	208.46
REPAIR ROADSIDE MARKERS (units)	1,600.68	10.145	157.78	378.30	1920	197.03
REPAIR METAL GUARDRAIL (m)	45.69	216.5	211.04	379.16	611.35	620.20
PAINTING CURB AND GUTTER (m)	36.92	2400	15.38	158.74	36,862.5	4.31
CLEAN AND PAINT BRIDGES (m)	10.81	180	60.05	40.30	360.70	111.73
OTHER ROUTINE MAINTENANCE (-)	175.76	--	--	--	--	--
RESURFACING (m^3)	3,487.12	567.84	6,141.03	--	--	--
REGRAVELING (m^3)	128.25	502	255.48	--	--	--
OTHER PREVENTIVE MAINTENANCE (-)	2.75	--	--	--	--	--
MANUAL REPAIR OF SLOPES (m^3)	27.73	26	1,066.54	129.28	1,025	126.13
MECHANICAL SLOPE REPAIR (m^3)	1,045.19	19,710	53.03	333.22	2,217	150.30
MANUAL REMOVAL OF SLIDES (m^3)	15.86	17	932.94	317.92	1,994	159.44
MECHANICAL REMOVAL OF SLIDES (m^3)	1,102.90	18,108.3	60.91	1,588.81	11,882.5	133.71
OTHER EMERGENCY MAINTENANCE (-)	90.79	--	--	--	--	--
RESTORATION MAINTENANCE (-)	7,835.53	--	--	--	--	--
CULVERT CONSTRUCTION (units)	24.04	1	24.04	--	--	--
PLANT TREES (-)	5.99	--	--	--	--	--
CONCRETE DITCH CONSTRUCTION (-)	--	--	--	218.73	--	--
EARTH DITCH CONSTRUCTION (m^3)	--	--	--	12.66	108.36	197.67
FENCE CONSTRUCTION (m)	485.25	2,180	222.59	7,255.90	36.708	197.67
OTHER BETTERMENTS COMPLEMENTARY (-)	3,220.04	--	--	--	--	--
LEVELING ROADWAY (m^2)	2,395.34	1,481,890	1.616	--	--	--
OTHER BETTERMENTS MODIFICATIONS (-)	6,239.67	--	--	--	--	--
PREPARATION OF ASPHALT MIX (m^3)	3,049.24	830	3,673.78	894.48	279.16	3,204.18
PLACE CONCRETE CULVERT (m)	38.60	82	470.73	--	--	--
SIGN PRODUCTION (-)	869.71	--	--	--	--	--
PRODUCTION OF ANCHOR POST (units)	--	--	--	333.45	543	614.09
PRODUCTION OF SUPPORT POSTS (units)	--	--	--	5,012.91	12.261	408.85
PRODUCTION OF BRIDGE RAIL (m)	--	--	--	4.15	36	115.28
PRODUCTION OF WOOD ITEMS (-)	469.29	--	--	--	--	--
AID TO FEDERAL POLICE (-)	4,552.73	--	--	--	--	--
AID TO OTHERS (-)	6,182.52	--	--	--	--	--

1/ PONTA GROSSA, PARANA: 185.5km; JANUARY THROUGH NOVEMBER 1980; IN CRUZEIROS OF NOVEMEBER 1980. RATE OF CXCHANGE 1 U.S.$ = 61.3 Cr$.

2/ PONTA GROSSA, PARANA; 75.5km; OCTOBER 1979 THROUGH SEPTEMBER 1980 EXCLUDING DECEMBER 1979; COSTS IN CRUZEIROS OF NOVEMBER 1980. RATE OF EXCHANGE 1 U.S.$ = 61.3 Cr$.

3/ THE COSTS IN THIS COLUMN CONTAIN THE FOLLOWING INDIRECT COSTS DISTRIBUTED IN PROPORTION TO THE DIRECT COST COMPONENT. COSTS IN THOUSANDS OF Cr$: ADMINISTRATION = 22,826.55; PERSONNEL LEAVE = 5,358.86; UNPRODUCTIVE PERSONNEL TIME = 1,044.16; GENERAL TRANSPORTATION = 4,988.35; AND UNPRODUCTIVE EQUIPMENT = 2,300.14.

4/ THE COSTS IN THIS COLUMN CONTAIN THE FOLLOWING INDIRECT COSTS DISTRIBUTED IN PROPORTION TO THE DIRECT COST COMPONENT. COSTS IN THOUSANDS OF Cr$: ADMINISTRATION = 8,676.59; PERSONNEL LEAVE = 165.35; UNPRODUCTIVE PERSONNEL TIME = 202.52; GENERAL TRANSPORTATION = 2,412.74; AND UNPRODUCTIVE EQUIPMENT = 25.46.

NOTE: UNIT COSTS IN THIS TABLE ARE NOT COMPARABLE TO THE UNIT PRICE TABLE BECAUSE THEY CONTAIN MATERAL AND DNER OVERHEAD AND MATERAL COSTS

Table 3. Brazil

COMPARISON OF FORCE ACCOUNT TO UNIT PRICE COSTS PER UNIT OF PRODUCTION

ACTIVITY *	SUM OF FORCE ACCOUNT AND UNIT PRICE COSTS	RATIO** OF COSTS PER UNIT OF PRODUCTION (FORCE ACCOUNT OVER UNIT PRICE)
FENCE CONSTRUCTION	7,741.15	1.13
MANUAL VEGETATION CONTROL	5,733.50	3.05
MECHANICAL GRASS CUTTING	5,134.43	0.64
PREPARING ASPHALT MIX	3,943.72	1.15
POTHOLE PATCHING	3,713.20	1.19
MECHANICAL SLIDE REMOVAL	2,691.71	0.45
REPAIR ROADSIDE MARKERS	1,978.98	0.80
SIGN REPAIR	1,598.83	5.26
MECHANICAL SLOPE REPAIR	1,378.41	0.35
MANUAL DITCH CLEANING	1,123.21	3.05
CHEMICAL WEEDING (SHOULDERS)	1,060.13	0.97
MANUAL WEEDING (SHOULDERS)	688.74	0.82
CULVERT CLEANING	594.20	2.72
REPAIR METAL GUARDRAIL	424.85	0.34
MANUAL SLIDE REMOVAL	333.78	5.85
REPAIR OF BRIDGE RAIL	247.91	0.94
CLEANING CURB AND GUTTER	220.95	1.59
MANUAL SLOPE REPAIR	208.12	8.46
PAINTING CURB AND GUTTER	195.66	3.57
CLEANING AND PAINTING BRIDGES	51.11	0.54
WEIGHTED AVERAGE =		1.59

* ACTIVITIES ARE RANKED IN DESCENDING ORDER ACCORDING TO THE SUM OF THE TOTAL ACTIVITY COSTS FOR UNIT PRICE CONTRACT AND THE FORCE ACCOUNT PROJECT.

$$** \text{ RATIO} = \frac{\text{COST PER UNIT OF PRODUCTION OF FORCE ACCOUNT}}{\text{COST PER UNIT OF PRODUCTION OF UNIT-PRICE CONTRACT}}$$

crews or making the primary measurements of work, which are contractor responsibilities. The inspector performs cross checks against his own field observations, and if discrepancies are noted, payment is withheld until verified in the field and surveillance is subsequently increased. In practice, contractors are said to be careful not to overstate invoices because the costs of being caught are quite substantial, and detailed cost information for comparable activities elsewhere are well known.

2.08 Although practically all maintenance is performed by contract, variations do exist between the types of contracting arrangements used by the various Republics. Generally, contracts for periodic maintenance such as resurfacing are competitively tendered, with larger construction enterprises entering the competition, but routine maintenance contracts are normally negotiated with enterprises which have an assigned area of the road network. The possibility of tendering always exists, if a community is dissatisfied with the performance of a contractor, but in practice, not renewing a contract is a rarely exercised last resort; rather budget and operational control procedures are exercised directly by the community to improve performance.

Colombia

2.09 In contrast to the smoothly functioning systems which have evolved over many years in Brazil and Yugoslavia, it is instructive to examine an unsuccessful initial attempt recently made to employ contract maintenance by the Ministry of Public Works and Transport (MOPT) in Colombia. Four maintenance contracts were let in 1977, and included periodic as well as routine maintenance. The contracts were let for four years, over road lengths which averaged 100 km.

2.10 The following contract features eventually led to serious difficulties in this first experimental attempt at contract maintenance:

(i) Comprehensive well-defined work programs were not developed. The MOPT transferred the responsibilty for specific work identification and scheduling to the contractors. Also lacking were adequate penalties for noncompliance as well as close supervision of work progress and quality.

(ii) The contractors were required to collect tolls, which were their sole source of contract revenues. A formula was provided for escalation of the toll charges, based on daily traffic counts carried out by MOPT personnel. However, some of the original contract toll rates were based on unrealistic initial counts which led to toll collections below the anticipated necessary income. Toll charges were increased, but generally well below true escalation levels.

(iii) Although unit rates were listed in the contract, and a method given for adjusting for price escalation, its application was left to MOPT's discretion. In practice adjustments were not applied on a regular basis, and the adjustments which were effected did not compensate adequately for inflationary rises in basic costs.

(iv) The contractors were required to provide and retain at the work site various items of equipment, including 30 ton-per-hour asphalt hot-mix plants, which significantly exceeded the requirements of the contract work quantities.

(v) Since the contracts included both routine and periodic maintenance activities, and the contractors were free to identify and schedule work without supervision, the tendency was to schedule and perform the more profitable periodic maintenance activities such as seal coating or asphalt overlays, even without prior necessary strengthening or repair of existing pavement structures.

2.11 Other contract features which led to the early cancellation of the four initial contracts included the fact that contractors were permitted to delay the execution of work and thereby increase their holding of toll funds for their own advantage. In one case, the contractor carried out considerably less work than the toll money collected would have covered and, in another, the contractor collected tolls for six months without carrying out any work. All four of these first contracts ended in court cases.

2.12 Based on the experience gained from this initial effort, Colombia's MOPT is proceeding to improve its maintenance planning and contracting practices and intends to extend the application of contract maintenance. Toll collection and maintenance work will be carried out under separate contracts and by different contractors.

Argentina

2.13 In Argentina, contractors have for many years been engaged in periodic maintenance of roads and also, in a few cases, in routine maintenance of private (industrial estate) roads. Late in 1979, the national roads authority, Direcion Nacional de Vialidad (DNV), taking advantage of a law permitting government agencies to reduce their workforce, established a new and bold policy of contracting out 70% of routine maintenance, retaining 30% for its own force account establishment.

2.14 The primary objective was to improve cost-effectiveness by reducing its personnel and improving control over diversion of maintenance resources to non-maintenance functions. To deal with emergencies, contractors were obligated to make personnel and equipment available as required.

2.15 Over a span of about 18 months, maintenance for almost two-thirds of the 47,000 km national roads network was transferred to contractors. At the same time, DNV staff was reduced from approximately 20,000 to 8,300, equipment was sold off and the central workshop closed. Social hardship was avoided by giving pensions to those close to retirement age, by assisting cashiered staff to buy DNV equipment and start up as contractors, and by requiring contractors engaged in maintenance to give DNV personnel first option when hiring new personnel.

2.16 While DNV cadres were reduced in numbers, a new emphasis was placed on procurement, contract management and field supervision skills. The latter function is now being performed by former superintendents, general foremen, gangers and chief clerks, who have become inspectors. Twenty-seven new engineers were engaged to supplement existing district personnel specifically in managing contracts.

2.17 The contracts (on average for 350 km) were initially let for one year, after competitive tendering. The second stage of contracts were let for a 2-year period, with an option to renew for a further year. Successful tenders were, on average, about 7.5% below official estimates, with few extreme deviations from the mean.

2.18 A most important feature of the Argentina system is the retention of 30% of the grid for its own direct action. Work on that segment is managed in each of the districts as if it were a contract, in competition with contracts let in the same district to the private industry. In this way, DNV has introduced a competitive edge to its own work, providing a new source of motivation for DNV staff, and it has retained a valuable gauge for planning and costing maintenance operations under contracts. It has also retained a basic nucleus for direct action, and a capability to expand back, if ever needed or desired, and it will continue its traditional role as a trainer of road maintenance personnel.

Nigeria

2.19 Road maintenance has proven an intractable problem over much of the period spanning the development of the modern trunk road system in Nigeria, and the rapid economic expansion stemming from the oil boom after 1973 brought an explosive rise in traffic. Resources have been concentrated on the construction and improvements program, and development of institutional capacities for maintenance has lagged far behind ever increasing needs. An initial attempt in 1979 to tender maintenance for some 12,334 km of paved federal highways was aborted, when the large variation in bid prices, reflecting uncertainties relative to the amount of work involved, pointed to the need for more specific definition of work activities and quantities. In January 1980 a new program intending ultimately to tender under unit-price contracts the entire federal network (except Lagos State) was initiated. Since the previous situation had been somewhat chaotic, it was still not possible for the Government to present contractors with a very clearly defined work program or performance standards; moreover, the 1-year contract period meant that all mobilization costs and most equipment costs would have to be recouped in one year or risks taken on getting future contracts. Contractors must have perceived a high degree of risk.

2.20 Nevertheless, by the end of 1980 about 30 contracts for routine maintenance (averaging about 250 km) for a total of nearly 8,000 km were awarded, a large number to indigenous and some to multinational contractors. Contract prices, including full mobilization costs, averaged US$8,200 per kilometer. Work progress has varied widely, but is reasonably satisfactory in most cases, and maintenance is now getting done where little was accomplished before. Under these circumstances, this must be appreciated as a significant achievement. The best indicator of performance is that the scheme is being doubled to encompass 16,000 km of trunk roads in 1982, and about 60 percent of last year's contractors will be retained. A crash program for training road maintenance inspectors, oriented to contract management is underway,and these new inspectors will be backed-up initially by four seasoned engineers resident in the states. Presumably, with the accumulation of experience a clearer understanding of the work program, work methods and productivity rates will be achieved, and lower prices should be obtainable as the work becomes better organized and risk premia are reduced.

Kenya

2.21 Kenya has also experienced severe difficulties in establishing an effective road maintenance capacity to meet growing traffic needs in recent years despite large-scale foreign assistance from several sources. Acute shortages of well qualified, experienced management and supervisory staff, fluctuations in available budgets (especially foreign exchange), cumbersome government procedures and a growing burden of redundant staffing have all conspired to undermine the effectiveness and efficiency of government force account maintenance operations.

2.22 For several years the Roads Department has employed small African contractors to provide haulage of gravel in regravelling operations. More recently, these small firms have evolved, with considerable assistance from the Roads Department, into contractors for the complete regravelling operation. By 1980/81 small contractors produced approximately US$5.2 million of regravelling works, which it is estimated, would represent only about half their potential capacity. However, under the circumstances this was a notable accomplishment, accounting for approximately 80% of the entire regravelling operations done in Kenya that year.

2.23 It must be reiterated, that this achievement by small contractors would not have been possible without considerable assistance from the Roads Department. African contractors in Kenya (as, indeed, in many other African countries) are generally not well developed and face a considerable array of problems. Starting from nascent enterprise creation, the owners of these companies have very limited experience and no long-standing tradition of entrepreneurial vocation; management limitations are therefore considerable. There are, of course, a few exceptions but, by and large, the groups which are managing to survive and continue in business are doing so because:

(i) their business span is small (often on an artisan scale) and therefore can be managed with their limited resources;

(ii) they are being paternally guided either by the Roads Department or by larger companies who subcontract their services; and/or

(iii) they are being subsidized by grants or loans from development authorities.

2.24 Thus, any undertaking of road maintenance through such contractors must be cautious, initially small in scale and encompassing only those activities which can easily be supervised (such as mowing, hauling and regravelling), evolving only gradually over time into more ambitious undertakings.

III. CONCLUSIONS AND RECOMMENDATIONS

3.01 This brief review has shown that the use of contractors for routine as well as periodic road maintenance is gaining increasing use in several countries. Motives have varied: some countries have turned to contractors as a last resort when continued efforts to build up government institutions over many years have failed to produce efficacious results; in other cases, the government and road authorities have sought improvements in efficiency so that increasingly hard-pressed highway budgets would stretch further; and in still other cases, road authorities have reluctantly turned to contractors when general restrictions on hiring of government personnel prevented further buildup of force account establishment to meet growing maintenance needs. Different countries, depending on their individual circumstances, have also approached the use of contractors in different ways: some have employed cost-plus contractors as hardly more than suppliers of men and equipment in an indirect extension of force account operations; others have taken maximum advantage of the management capabilities and incentives of unit-price contractors. Some countries have moved with breathtaking speed to thrust the burden of entire networks on contractors in a single year, while others have taken years to gradually expand the role of contractors from minor beginnings to a still moderate role today. What general conclusions can be drawn at this stage?

GENERAL CONCLUSIONS

3.02 The efficacy of contract maintenance. With rare exception, contract maintenance has proven to be a workable undertaking in countries at diverse levels of development and with diverse forms of social and economic organization. With relative freedom from entangling "red tape" and the ability to pay higher salaries to attract, retain and motivate staff, contractors have in some cases, e.g., Nigeria, succeeded in getting maintenance done where all other approaches have failed. Where sufficient profit incentives exist, contractors are normally attracted to maintenance opportunities, even in remote areas; while larger firms tend to prefer larger contracts including periodic maintenance, in many instances small firms have been formed specifically to undertake routine maintenance. Contractors are also commonly obligated to make their men and equipment available to the road authority in the event of an emergency.

3.03 It must be recognized, however, that the introduction of any new system, especially quickly and on a large scale, is risky. The Colombian experience illustrates some of the problems which can be encountered in introducing contract maintenance with inadequate planning and preparation. While the overall burden of responsibilities on the road authority is normally reduced by the introduction of contractors, the nature of the government's responsibilities changes sharply, and there is increased need for contract management skills. Careful planning and the introduction of contracts on a small trial basis initially can therefore reduce risks, permitting the capabilities of government and contractors alike to develop before thrusting too heavy a burden on a new system.

3.04 The efficiency of contract maintenance. While the costs (or more correctly, prices) of contract maintenance can be taken from records of contract administration, and the direct costs of contract supervision by the government could be assessed, quantitative comparisons with the comparable costs of similar operations by government forces are not generally available. Few road authorities in developing countries have management systems which can produce reliable performance and cost information with which to plan and control operations. Systematic recording of the amount of work achieved (as distinct from the amount of resources consumed) is not generally done, and both the amount of diversion to other activities and the amount of idle time are likely to be underreported. Moreover, the economic cost of interest on government investment in plant and equipment is often overlooked, since it is not perceived as a financial cost. Regardless of the prospect of contracting maintenance, there is a great need for improved work planning, performance monitoring and more effective cost accounting in force account maintenance operations,4/ but the prospective increased use of contractors to achieve improvements in cost-effectiveness in the future places particular importance on development of comparable cost information.

3.05 However, certain inferences can reasonably be drawn on basis of the partial information which is already at hand. Unit-price contracts clearly provide contractors with strong incentives for efficiency, and initial cost results from Brazil, Argentina and Kenya (and also the United States)3/ suggest that contractors can perform maintenance at substantially reduced costs. The administrative costs to the government in contract management, quality control, measurement and certification is normally substantially less than in administration of force account works, although the nature of the administrative burden is different and may require enhancement of contract management staff. In every case where contract maintenance has been used on a large scale continuing over time, government has been able to effect a substantial reduction of its own establishment. Yugoslavia, with more than 20 years experience, provides an example of a mature contract maintenance system where as few as five inspectors and one director control (in Slovenia) maintenance for a high standard road network of 4,700 km.

3.06 Those who are familiar with the bloated payrolls and the vast fleets of equipment lying idle, which are characteristic of all too many road maintenance directorates, not only in developing countries, will require little evidence to convince them that contractors can perform maintenance at lower costs. However, many may question whether much of the potential cost savings would revert to the public. The best safeguard of the public interest is lively, open and honest competition; where this is assured, large cost savings are likely to be realized. Since the end of the great road building boom of the post World War II period is at hand in most countries, it is plausible to hope that some of the civil engineering capacities and skills thus developed, which must now seek new outlets, will find their way into road maintenance activities, thus enhancing competition and working in the public interest.

3.07 However, it must be recognized that the twin dangers of monopolization and corruption are ever present. While government force account operations are, of course, not immune to corruption, many may argue that government procedures, while encumbering efficiency, do reduce the amount of corruption--while others might retort that a small element of corruption would be a small price to pay for vastly enhanced efficiency! Retaining at least a nuclear force account operation, as has been done in Argentina, should help greatly to reduce the danger of monopolistic exploitation and provide a useful source of comparative costs.

3.08 Contractors' lobbying of public support for maintenance. Inadequate government budgets for road maintenance are a widespread phenomenon, even in high income countries. Certainly contractors' associations have been active participants in the broad grouping of interests which have formed the lobbies for road construction that have been so potent in recent decades. Where contractors become an established force in maintenance activities, it may be conjectured that their interests will prompt them to lobby for increased budgets for road maintenance, enhancing the much weaker (or nonexistent) lobbies presently promoting such expenditures. This could be an important contribution to the public interest. However, we have no evidence at this time with which to test this supposition.

RECOMMENDATIONS

3.09 The possibility of contracting some part or all of routine as well as periodic maintenance operations should be considered by every road authority. Countries with such diverse circumstances and approaches to social and economic organization as Yugoslavia, Argentina, Brazil, Kenya and Nigeria have all profited from the introduction of contractors (or worker cooperatives) in one facet or another of road maintenance. Retention of a nuclear force account capacity in competition with contractors, where feasible, may provide a valuable spur to efficiency for both.

3.10 Those countries with limited or nonexistent domestic contracting industries also typically have very limited institutional capacities for the execution of works by force account. In these cases the task of developing the necessary institutional capacities will be long and difficult no matter which course is pursued. However, where development of a domestic contracting industry is at all a feasible proposition, the strategy of limiting the burden on government as much as possible, concentrating development of its capacities on the management of contracts, and fostering the development of the domestic contracting industry is an option which merits serious consideration.

3.11 Two principles should guide selection of activities to be contracted: complementarity of existing capacities of government and contractors, and ease of administration. Periodic maintenance--regravelling and resealing--has been contracted so extensively partly because of the ease with which it can be administered. It should be noted, however, that most characteristics of routine road maintenance activities--small scale, technical simplicity, low capital requirements and a relatively stable demand over time--make them particularly well suited as a beginning activity for the nascent contractor. The primary counter-consideration is the dispersed nature of the work, which, as compared with the types of work traditionally contracted, increases the difficulties and costs of contract supervision. However, it is not clear that the need for, and difficulties of, supervision will be less when the same maintenance works are performed by force account, although the costs are usually less obvious.

3.12 An essential ingredient of any cost-effective routine maintenance program is careful planning, scheduling and control. It is an advantage of contracting that it promotes, indeed requires, more careful planning and control of such activities. All of the successful contracting schemes have involved close coordination between the government and contractor in defining the work to be done and planning the work program. Thorough orientation of contractors prior to tendering promotes understanding of the tasks to be done, reduces perceived risks and risk premia included in tender prices.

3.13 The majority of the work should be undertaken under competitively tendered unit-price contracts. A small element, typically not exceeding 10-15% of routine maintenance, would normally be done under cost-plus arrangements to simplify administration of items which are difficult to measure, and also to add management flexibility (as, for example, for emergency operations). Other types of contract which apportion a greater share of risks to the government, e.g., cost-plus-incentive-fee or cost-plus-incentive-or-penalty ("target price") may be considered in some cases, particularly where uncertainty perceived by contractors may be large at the initiation of activities for which they have no prior experience on which to base their productivity and cost estimates.5/

3.14 Simplicity and maximum incentives are desirable features of the contract, and in general unit-price contracts have proved well suited to maintenance activities. However, standard forms based on construction activities are often unsuitable, and contractual instruments should be drawn specifically to fit the needs of routine maintenance. Many governments may wish to consider, at least initially, the use of a consultant or a management services contractor in order to develop management systems and contract instruments.

REFERENCES

1. Road maintenance activities are generally classified as either routine or periodic. Routine maintenance includes operations that normally are repeated one or more times a year (e.g., vegetation control, cleaning ditches and culverts, patching potholes and emergency repairs). Periodic maintenance includes operations which typically need to be repeated over longer than yearly cycles, primarily regravelling for gravel roads, and bituminous surface dressings or seals for paved roads.

2. The overall problems of road maintenance in developing countries are discussed in more detail in the World Bank publication The Road Maintenance Problem and International Assistance (December 1981).

3. Many other countries, industrialized as well as developing, have had varying degrees of experience with contracting maintenance. Some of the more instructive cases are from the United States, the Netherlands, Ghana and Zaire. Recent experience, generally favorable, with contracting of routine maintenance at the city and state level in four American states--Arizona, California, Pennsylvania and Texas--was reported at the US Transport Research Board Annual Meetings (Washington, DC, January 1982). The primary motivation in these cases has been to reduce costs, but the State of Texas has also

sought the additional objectives of reducing government personnel and promoting the development of small contractors.

4. Many road maintenance management information and costing systems have proven to be too complex or too detailed. A simple system addressing the most critical components for an equipment-based maintenance organization is being developed by the Crown Agents in a study for the World Bank.

5. A mixed form of contract involving all three types of payment--unit-price, cost-plus-percentage-fee, and cost-plus-incentive-fee--was introduced in late 1981 in contract maintenance operations in Parana State of Brazil, essentially for the reasons given above. It is not clear whether this type of contract will prove to be easily administered.

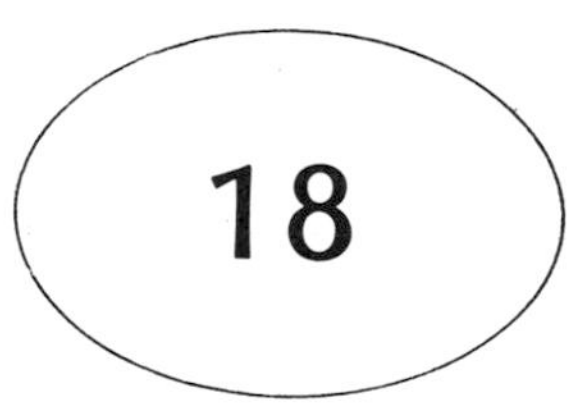

Criteria for assessing rehabilitation and reconstruction techniques in Ivory Coast

B. BAMBA, Public Works Laboratory of Ivory Coast, G. LIAUTAUD, Public Works Laboratory of Ivory Coast and CEBTP (France), and A. DAVIS, CEBTP (France)

A brief review of the pavement survey unit operating in Ivory Coast is given. The various criteria generally used for assessing pavement condition and for defining rehabilitation techniques are described, together with the critical or threshold values normally taken into account and concerning: deflection, roughness, rut depth and cracking.

The main findings resulting from axle-load campaigns carried out over the paved network are briefly discussed and the magnitude of overloading is emphasized. The most aggressive types of vehicles or goods carried and which should be the concern of an axle load control strategy, are described.

The paper finally outlines the various types of materials generally used for strengthening operations in the country, drawing the attention upon two basic documents which have been elaborated by the research department of the National Public Works Laboratory, namely:

. Manual for the structural design of overlays
. an illustrated Guide for routine maintenance purposes.

INTRODUCTION

1. The road network of Ivory Coast extends over a length of approximately 45.000 km, of which approximately 3000 km are presently bitumen-surfaced. The remaining 42.000 km consist of some 11.000 km of gravel roads and 31.000 km of secondary earth roads. Since the beginning of the 1950's, date at which construction of the bituminous surfaced pavements started, two major strengthening or rehabilitation projects have been undertaken: the first one between 1974 and 1977 and the second one initiated in 1981. Both projects included some 500 km of roads to be strengthened.

2. To provide guidance to the local Road Administration for adequately timing and undertaking rehabilitation works, a highway pavement survey division was set up in 1974 within the structures of the Building and Public Works Laboratory of Ivory Coast*. With the equipment and personnel available it has since been possible to survey the entire bitumen-surfaced network, both in terms of deflection and surface conditions. Significant criteria for maintenance operations have been established and maximum permissable values for these various criteria have been developed. Further to defining threshold limits for pavement resistance or riding quality, a manual for strengthening design has been elaborated together with a guide for general maintenance procedures, illustrated by a catalogue of pavement defects.

* Laboratoire du Batiment et des Travaux Publics (L.B.T.P.)

3. This paper briefly outlines the criteria presently used by the Public Works Laboratory for assessing and proposing to the local Road Authorities the rehabilitation and or reconstruction techniques for the paved roads in the country.

THE HIGHWAY SURVEY DIVISION AT THE PUBLIC WORKS LABORATORY OF IVORY COAST

Organization

4. Initially part of the road department of the laboratory, the Survey Division has been attached in 1979 to the Research Department, in order to both profit from the immediate findings of the research works and contribute readily to its activities. Since 1975, the annual budget allowance of that division has doubled, passing from 75 millions CFA to 150 millions CFA (ie 600.000 U.S.$) entirely financed by the Ministry of Transport and Public Works of Ivory Coast. The main purpose of the Survey division is to provide yearly a technical report to the Road Administration, in which priority actions for maintenance or strengthening are defined.

Equipment

5. To carry out the survey, the division has available:

. a Lacroix deflectograph
. a CEBTP Curviameter
. a Benkelmen Beam
. a Transversoprofilograph
. a Viagraph
. a Couple of portable axle-weight devices.

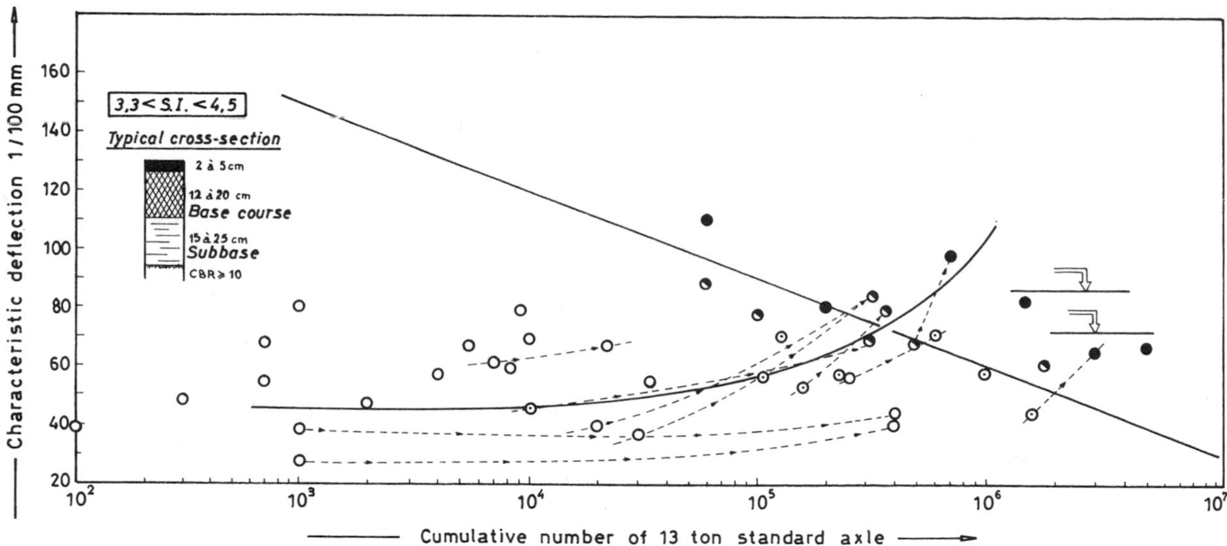

Fig. 1. Trends in the increase of deflection as a function of traffic

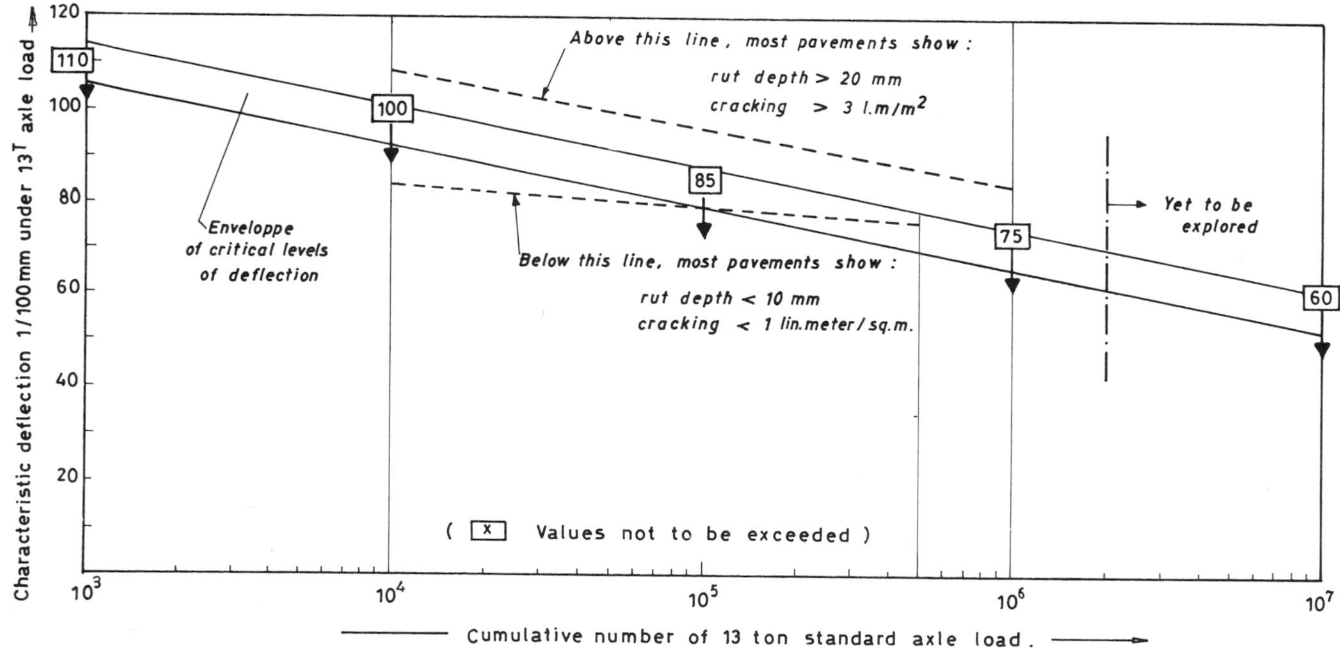

Fig. 2. Allowable deflection levels as a function of traffic

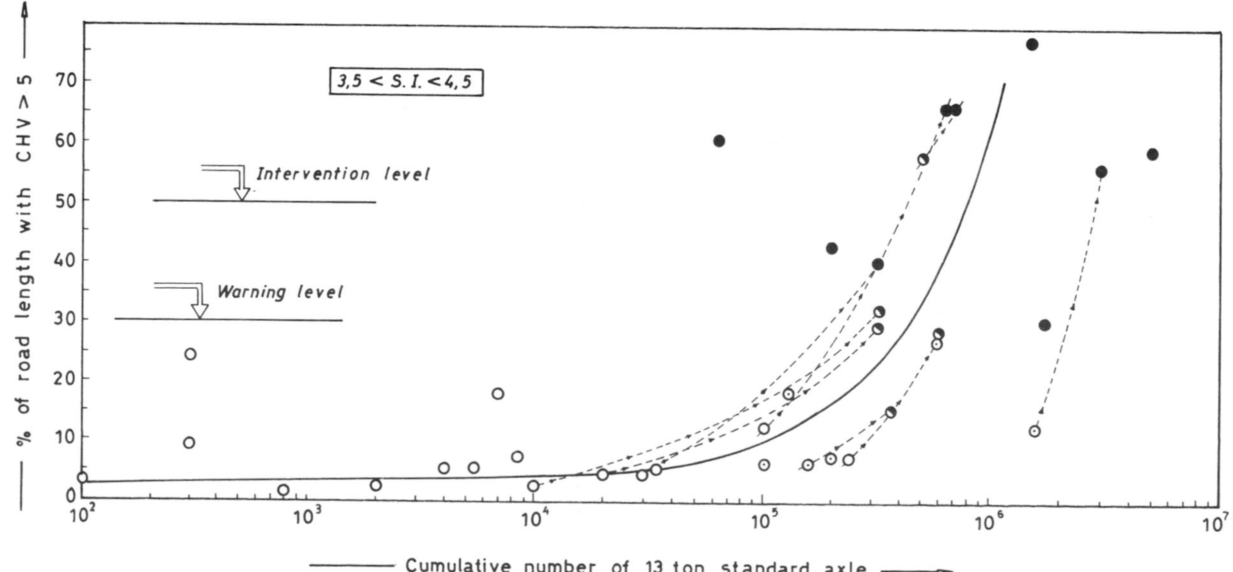

Fig. 3. Trends in the increase of roughness as a function of traffic

This equipment is to be shortly complemented with:

. a fifth-wheel Bump Integrator
. a portable skid-resistance testing device
. an automatic axle-weight unit.

Personnel

6. The personnel includes some 20 people, namely:

. 2 Ivorian Engineers
. 5 Technicians
. 3 office clerks for data treatment, and
. 5 to 8 skilled labourers.

Output

7. During the first 6 years of operation and using the Lacroix Deflectograph the average output of the survey unit was of the order of 5 km per day, i.e. some 1000 km of lanes ausculted per year (or 500 km of two-lane roads in both directions). Today, with the CEBTP Curviameter, the daily output approaches some 30 km. At this rate the entire network may be surveyed within a period of 4 months, thus leaving time for intervention outside Ivory Coast.

CRITERIA FOR ASSESSING REHABILITATION AND RECONSTRUCTION TECHNIQUES

Deflection

8. The majority of pavements in Ivory Coast are of the flexible type, and therefore particularly suitable for deflection measurements. Observations made on the network suggest that deflection is, among other criteria, a valuable index of pavement strength or condition. It is generally found that :

i. Deflection levels often lie within 40/100 to 100/100th mm under a 13 ton axle load.

ii. high values of deflection, i.e. in excess of 1 mm under a 13 ton axle load, are associated either with excessive moisture content at subgrade level or with the presence of schisteous materials or indeed with normal fatigue phenomenom.

iii. the standard type of pavement consisting of 150 mm lateritic gravel sub-base, 150 mm cement-treated lateritic gravel or crushed rock base course with a 20 to 30 mm bitumen surface, start to show irreversible signs of failure whenever some 25% or 30% of the length of the road show deflection higher than 1 mm. This proportion of high deflection levels automatically implies the necessity for strengthening works to be carried out.

iv. an increase in deflection with time or traffic is to be taken into account particularly with crushed stone base courses covered by a double surface-dressing (figure 1).

v. allowable deflection values for various traffic intensities are established and the critical envelope is shown in figure 2.

Roughness

9. Roughness is measured with a viagraph. Permissable and threshold values for maintenance purposes are given in table 1. It is also found that roughness increases with time as shown in figure 3, and moreover is a more sensitive index of the rate of deterioration of a pavement than is deflection alone. Indeed, at some critical period of time, roughness "deteriorates" far more rapidly than does deflection and represents as such a more sensitive indicator as to the suitable time to intervene.

Transverse deformation

10. Rut depth as measured under a 1.20 meter straight edge, becomes critical whenever it exceeds 15 mm. Beyond this threshold value, most types of surfacings show signs of cracking or some sort of deterioration.

11. When measured with the transversoprofilograph, it is evaluated that:

. Transverse pavement deformation is satisfactory whenever the variance of slopes is less than 4×10^{-4}.

. Transverse pavement deformation is critical or unsatisfactory whenever the variance of slopes exceeds 20×10^{-4}.

Cracking

12. Cracking in itself does not appear as a primary criterion for assessing pavement condition. When it is associated with shrinkage stresses it does not provide a good index for assessing fatigue. However, when longitudinal cracking occurs together with rutting, it then becomes a sign of excessive tensile stress at the bottom of the surfacing and as such it may indicate that fatigue has taken place. When measured within a 1 metre square Frame, the threshold value corresponds to approximately a total length of visible cracks exceeding 3 linear meter/sq.meter

Design axle-load

13. As part of its activities, the survey division carries out axle load weighing campaigns. The majority of the paved roads have been subjected to a detailed analysis of the traffic upon them, including data on intensity, axle loads, contact pressure, type of goods transported ... etc. Although the legal limits for single and tandem axle loads are respectively 10 tons and 17 tons, overloading is certainly not uncommon and values of 13 tons on single axle or 26 tons on tandem axles are often encountered.

14. The main findings of the axle load campaigns may be summarized as follows:

i. The daily traffic intensity on the 6 to 7.5 meters two-laned roads ranges from 300 to 5000 vehicles in both directions.

ii. The proportion of Commercial vehicles having more than 3 tons laden weight averages some 27% to 30%.

iii. Most of commercial vehicles, as defined above, i.e. buses, trucks and semi-trailors are overloaded when loaded, particularly on the tandem axles.

iv. The proportions of overloaded axles for a representative population of commercial vehicles generally comprising of 35% empty vehicles are respectively:

on single axles : 9% (max. load registered 22 tons)

on tandem axles : 56% (max. load registered 25 tons).

v. The "agressivity" of a typical commercial vehicle, expressed in terms of the number of equivalent standard axle varies according to the road, between 0.18 and 2.00 "13 tons standard axle" i.e. between 1 and 12 "18 kips standard axle" (figure 4). On average, the equivalence factor of a typical commercial vehicle is 0.85 "13 ton standard axle" or 5 "18 kips standard axle" (using the 4th power law).

vi. The equivalence factors or "agressivity coefficients" of the following various populations of vehicles are respectively:

Population	Equivalence factor	
	Number of 13^T standard axle	Number of 18 kips standard axle
"The" vehicle (including cars)	0.40	2.00
A vehicle having more than 3 tons laden gross weight (i.e. commercial vehicle)	0.85	5.00
Heavy truck having more than 5 tons net weight	1.00	6.00

vii. Among the various types of heavy vehicles the two most agressive are the types "113" and "112" i.e. 1 front single axle + 1 intermediate single axle and one rear triple or double axle. They are equivalent to between 2.2 and 3.5 standard 13 tons axle. As an axle load control measure the import of such vehicles should be strictly limited.

viii. For a given type of vehicle (ex : semi-trailor 112 or 122) the scale of agressivity according to the nature of the goods carried is as follows:

Type of goods carried	Agressivity of fully laden vehicle
Quarry materials	4.92 standard 13^T axle
Vegetable oils	4.31 " " "
Coffee, cocoa	4.19 " " "
Cement	3.95 " " "
Petrol	3.91 " " "
Cotton	3.35 " " "
Timber logs	1.55 " " "

ix. The standard axle load for pavement design is a 13 tons axle having a contact pressure of 6.62 kg/cm^2.

BRIEF REVIEW OF REHABILITATION OR RECONSTRUCTION MATERIALS. GUIDES FOR STRENGTHENING AND MAINTENANCE

15. In 1979, a guide for overlay design has been developed by the Research Department. It provides design thicknesses for overlays as a function of deflection level, traffic intensities and type of existing pavement. Four types of strengthening materials are allowed for in the design:

i. Untreated well-graded crushed rock ii. Cement-stabilized lateritic gravels	for cumulative traffic of up to 4 x 10^6 standard 13^T axle
iii. Bituminous concrete iv. Dense Bituminous macadam.	for cumulative traffic of up to 2 x 10^7 standard 13^T axle.

Table 1. Viagraph values for road maintenance assessment

	% of road length showing a viagraph coef.	
	> 5	> 10
1. Acceptable level for new roads	< 10% (bit. concrete) < 20% (surf. dressing)	< 3% < 5%
2. Allowable level for in-service roads	< 30%	< 10%
3. Critical level for warning	> 30%	> 10%
4. Intervention level for strengthening or improving roughness	> 50%	> 25%

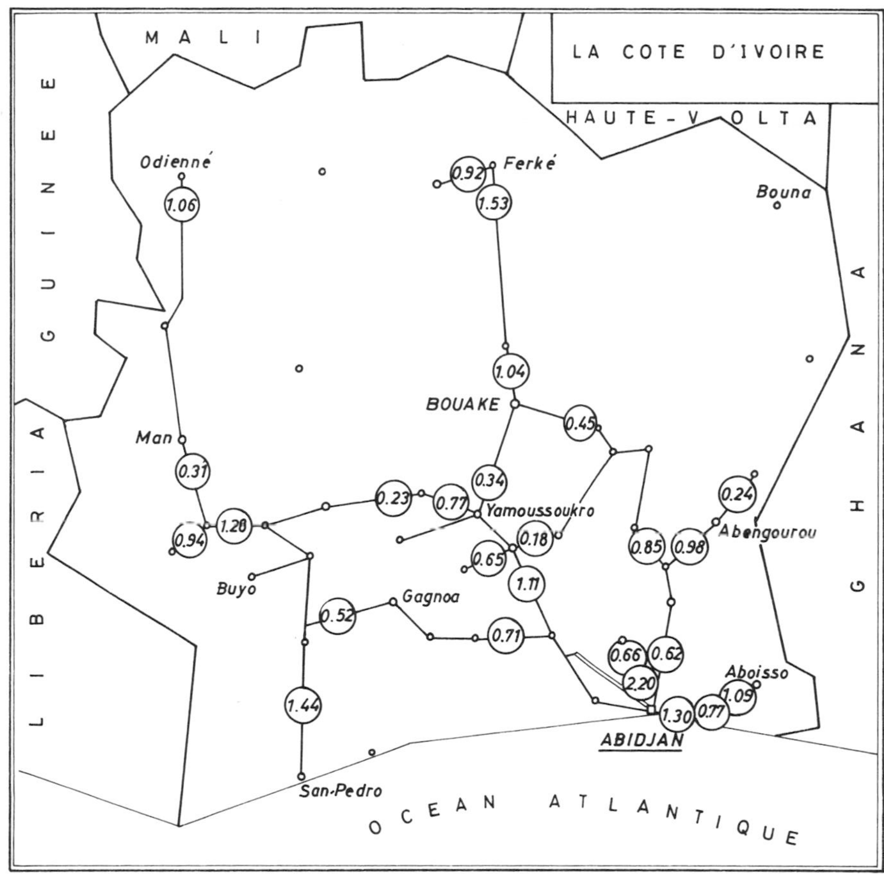

Fig. 4. Aggressivity of the commercial vehicles over the paved network, expressed in terms of the number of 13 tons standard axle

Fig. 5. The CEBTF curviameter for deflection and curvature radius measurements; operating speed 18 km/h, output 30-50 km/day

Fig. 6. The Benkelman beam shows good correlation the CEBTP curviameter; on Ivorian pavements both yield similar results with a correlation coefficient of 0.90

Fig. 7. The viagraph is used to measure roughness; its operating speed is about 5-10 km/h

Fig. 8. The transversoprofilograph measures transverse deformation of the road surface

Fig. 9. Axle weight campaigns are regularly carried out to provide information on the damaging effect of commercial vehicles

16. However, double surface-dressing is also used as a means to restore surface quality (roughness, watertightness ...) for pavements showing less than 75/100th mm characteristic deflection and which are to be subjected to not more than one million standard 13 tons axles.

17. Design criteria are manifold:

- minimum laying thickness of overlay materials
- allowable deflection after overlay
- allowable horizontal strain at the bottom of bitumen-treated courses
- allowable tensile stress at the bottom of cement-treated gravels
- allowable vertical strain on top of existing pavement or subgrade.

CONCLUSIONS

18. The Ivory Coast road administration has been in the process of rehabilitating its bitumen-surfaced network over the last ten years.

19. To provide guidance for a rational system of appraising road strength or condition and devising the best techniques for rehabilitation or reconstruction works, the highway authorities in the country has made available to the Public Works Laboratory a well-equipped unit for pavement survey.

20. This unit, over this same period, has developed a consistent methodology for monitoring the road network and it has defined the significant criteria for decision-making in the field of rehabilitation needs.

21. A manual for overlay design, as an end product to the pavement survey procedure, has been established together with a practical guide for interpreting pavement defects and defining the best routine maintenance actions.

22. These documents are presently being put in application and with the help of further research, the concepts are intended to be progressively reviewed and bettered in the light of present or forthcoming experiments.

REFERENCES

1. "Auscultation of pavements in Ivory Coast. Activity Report for the period 1974-1980" July 1981.

2. "A guide for the strengthening of pavements" Jan. 1981.

3. "Recommendation for the study and assessment of traffic to be taken into account for pavement design" Sept. 1980.

4. "Study on the aggressivity of traffic in Ivory Coast" Research reports Nos. 18, 21, 22, 27, 28, 31, 32. 1980-1981.

5. "Study of the incidence on strengthening costs of various axle-loads policies" Research report No. 20. April 1980.

6. "Research on threshold values of deflection, roughness, rut depth and cracking for paved roads in Ivory Coast" Research report No. 24.

7. Deflection measurement with the CEBTP Curviameter in Ivory Coast" Research report No. 29. 1980-1981.

8. "Stresses and deformations at the interfaces of the structures proposed by LBTP for new pavements and overlays" Research report Jan. 1981.

9. "The overlay solutions of the guide for strengthening: their sensitivity to the variation of some design parameters" Research report No. 41. April 1981.

10. "Graphs of variations of stresses and deformations as a function of overlay thickness and rigidity of the existing pavement" Research report No. 43. May 1981.

19

Highway survey, inventory, and maintenance procedures in Jamaica

D. P. POWELL, BSc, FICE, FIHE, FASCE, T. P. O'Sullivan & Partners, and
G. R. KIRKPATRICK, BSc, MSc, MIEJ, Ministry of Construction (Works), Jamaica

This paper sets out to describe a few of the systems and procedures adopted in Jamaica for use when formulating and implementing road maintenance programmes. Some of the criteria which are applied before works are included in these programmes are also discussed.

Until 1970 the Government had executed very little planned maintenance of its highway network, concentrating more on repairing damage after it had occurred than on pursuing a policy of preventative maintenance. Now, ten years later, the Ministry of Works has a maintenance capability of which it is proud. In that ten years, two major Studies have been undertaken and two major projects, part financed by the international lending agencies, have been embarked upon. The condition of the network has been improved considerably, and despite a shortage of funds is still being improved. But not only is it being improved, it is now being maintained to a consistently high standard in an organised and systematic way. This paper gives an indication of how this has been achieved.

THE NETWORK

1. Jamaica has 10,242 miles of road classified by type as:

Main Roads	.....	2,978 miles
City Streets (Kingston)	.....	886 miles
Parish Council Roads	.....	6,398 miles

2. The most important links are provided by the island's Main Roads which are maintained by the Public Works Department and grouped for administrative purposes into four classes:

"A" or Arterial Roads	.....	474 miles
"B" or Secondary Roads	.....	432 miles
"C" or Tertiary Roads	.....	2,038 miles
Bridle Tracks	.....	34 miles

"A" Roads being defined as roads of national importance, "B" of regional importance and "C" of only local importance.

3. The entire main road network apart from 285 miles has a bituminous surface; most of the "A" roads have asphaltic concrete surfaces, the remainder of the "A", and the "B" and "C" roads are surfaced with double surface treatments.

4. The extent of the network is more than adequate. But because much of it was not designed or built to any specific engineering standards, alignments are very variable, sometimes even tortuous. Many roads have narrow cross sections with no shoulders and poor sight distances. Few roads are provided with adequate drainage structures making maintenance difficult especially as Jamaica's topography is mostly hilly or mountainous and the rainfall experienced is of particularly high intensity (12 inches in 24 hours is recorded at some point on the island almost every year). The problem of having to deal with the high runoffs resulting from these high rainfalls is compounded by so few roads having rights of way in which side drains could be installed. And, landowners are reluctant to allow the discharge of surface water through their properties. Despite these problems though, few pavement failures arise as a result of weaknesses developing in the sub-grade, particularly on the arterial roads. The island is fortunate in having a covering of sub-grade soils possessing high load bearing qualities over much of its area, the major exception being one parish towards the northeast corner of the island where clayey subsoils predominate.

TRAFFIC

1. The most heavily trafficked road on the island is between the present capital Kingston and the old capital Spanish Town, some 15 miles distant. This road has an average daily traffic of about 16,000 vehicles; other short sections of road around Kingston, Montego Bay, and Ocho Rios carry up to 10,000 vehicles per day. But most of the main arterials carry between 500 and 2,000 vehicles per day. Large sections of the network have an average daily traffic of only 50 to 250.

2. Until 1974 traffic was growing at a steady 8% per annum, but following the oil crisis in 1973/74 the rate fell considerably. At one stage in 1978, zero growth rates were being predicted for the five year period 1979 to 1984. However, conditions in Jamaica then worsened and traffic counts made in 1979 showed that there had been a decrease in traffic of up to 16% between 1978 and 1979. During 1980 this decrease probably continued, but there was then an influx of vehicles following the change of Government late in 1980 which resulted in a distinct up-turn. Nevertheless, traffic levels

Highway investment in developing countries. Thomas Telford Ltd, London, 1983, 147–152

are not likely to approach those of the mid 1970's for some time yet.

BACKGROUND

1. Jamaica has spent over US$150,000,000 since 1971 on improving its network and about US$30,000,000 on maintaining it. Both these sums are exclusive of overheads such as salaries and administrative costs.

2. Most Main Roads on the island had remained for many years in the same condition as when they were built until one or two isolated improvements were made to the network in the 1950's. It was not until the early 1960's that the Government of Jamaica begun to realise that much of the network would need improving and upgrading during the late 1960's and into the 1970's if road transport was not going to prove eventually to be a brake on the island's development.

3. The Jamaican Government commissioned a comprehensive islandwide transportation Study in 1967 (ref. 1), one of the objectives of which was to produce a technically sound and economically viable highway improvement programme for implementing through the 1970's. Such a programme was produced; financial assistance was sought and obtained; and work began in the early 1970's on the first few schemes having the highest economic rates of return. It was not long after work got underway on this programme that it became evident that the maintenance capability of the Public Works Department also needed upgrading.

4. Jamaica commissioned the first Study of its road maintenance needs in 1972 (ref. 2). The principal objective of that Study was to "identify an optimum scope and method of execution of road rehabilitation and maintenance to be carried out in the period 1973 to 1977". The Study duly produced a programme of maintenance judged both by engineering and economic criteria and placed in a logical time scale. The programme included 440 miles of overlay work utilising asphaltic concrete and 250 miles of double bituminous surface treatment, as well as other minor repair works.

5. The programme was supported by an analysis of work methods and a budget was prepared, estimated to cost US$27,500,000 adjusted for inflation and taking into account the proportion of contract work to force account work necessary.

6. Work began in 1973 on implementing the programme using finance provided jointly by the Jamaican Government and the World Bank. Work continued until 1978, but the effects of inflation turned out to be far worse than anticipated, e.g. the oil crisis in the mid 1970's resulted in a rise in the price per ton of asphaltic concrete in Jamaica from US$25 in 1973 to US$48 in 1978. So the programme had to be curtailed.

7. A second Study (ref. 3) was then commissioned in 1978 with a view to continuing the programme already started. The programme prepared by this Study team recommended overlaying 76 miles using asphaltic concrete, double surface treating 253 miles and sealing 6 miles of gravel road for the first time. Other minor works such as restoring the profile of some of the roads before resurfacing and carrying out some drainage improvements were also included.

8. This programme is currently being implemented.

SURVEY AND INVENTORY PROCEDURES

1. The procedures adopted for surveying the condition of the network and preparing maintenance inventories during both Studies i.e. in 1972 and in 1978 differed little.

2. On the first Study, road condition surveys were carried out by two teams comprising an engineer and a superintendent travelling in a landrover. The teams covered over 2,600 miles (497 road control sections) in July and August 1972.

3. On the second Study, the field surveys were again carried out by two teams during two months April and May 1978. These teams inspected over 3,000 miles (567 road control sections this time), that is the total mileage of Main Roads plus a number of roads recommended for inclusion by Parish Councils and the Ministry of Agriculture.

4. Their main purpose was to inspect the condition of each section of road and to recommend the appropriate maintenance action required to restore the section to an acceptable condition.

5. To standardise reporting, the principal road conditions and recommended actions were described under 44 codes on the first Study and 60 on the second. Each road control section was reported on in this way. While carrying out the survey, the engineer located the maintenance action in relation to the road control section by HALDA tripmeter reading, noted the main characteristics of the road such as surface width and type, shoulder width, type of terrain, etc., and recorded by code the road condition and the recommended action together with any information considered useful for costing purposes.

6. Survey procedures adopted in both these Studies relied heavily on the engineering judgement of the engineers in recommending the correct maintenance action to remedy the particular condition. However, when developing maintenance programmes from the data collected and the maintenance actions suggested on the surveys, these judgements were always backed by economic justification.

7. At the conclusion of the field surveys studies, Road Maintenance Inventories for each parish were prepared giving:

i. a map showing the parish roads

ii. a schedule of road control sections, naming terminal points

iii. a list of definitions of maintenance work activities

iv. explanatory notes on the entries recorded on the survey sheets

v. the survey sheets complete with recommended maintenance actions

vi. sheets for updating the inventory.

8. Also, in the inventories the recommended maintenance actions were subdivided into five categories: Routine, Periodic, Rehabilitative, Emergency Repairs and Improvements; seventeen actions now being defined, given code numbers, and classed as Routine, ten as Periodic, twenty four as Rehabilitative, and nine as Emergency Repairs; making 60 actions compared with the 44 previously used. Improvements being difficult to codify are described in detail.

9. Thus, the two types of basic information given in the Inventories about each road section are its physical and geometric characteristics which do not change unless major improvements are carried out and; a subjective assessment of the maintenance actions recommended for execution at a particular time. One disadvantage of this is that the second type of information needs updating at regular intervals. Another disadvantage is one mainly concerned with the form of presentation of information.

10. When inspecting a road for the purpose of recommending appropriate actions to restore every section to an acceptable condition, the length of road requiring a particular Routine Maintenance action may not coincide with the length of the section requiring to be resealed (Periodic) during a particular period. Hence the form of reporting becomes complicated. A different type of inventory sheet was therefore devised specifically to cater for the planning and estimating of routine maintenance. This is discussed later under Maintenance Procedures.

11. On the first Study quantitive measurements of surface irregularity (SI) were taken on a representative sample of about 40 per cent of the road network, using a integrating mechanism fitted to a short wheel-base landrover. The average SI of over 400 sections of "A" roads was found to be 76 inches per mile. In terms of a standard towed integrator, this was estimated to be equivalent to about 220 inches per mile. Any road registering this degree of roughness would generally be regarded as needing remedial action to improve its riding qualities. However, the high average on the sections surveyed in this exercise was accounted for by the high proportion of hand-laid penetration macadam and double sealed roads in the sample, and not by high readings recorded over the entire length of road surveyed. The SI values for the machine-laid asphalt concrete sections compared well with those measured in other countries on asphaltic concrete.

DEVELOPING MAINTENANCE PROGRAMMES

1. Having surveyed the network and prepared maintenance inventories, the next stage was to develop the recommended maintenance actions into programmes which can be economically justified. Because of the thousands of alternatives in the patchwork of actions to be considered for inclusion in the programme, the exercise of justification had to be simplified. On the first Study this was done by establishing threshold parameters of traffic and road condition above which the particular action was justified. But the economic benefits of carrying out the particular action were only calculated after first establishing that the particular action was the "best" choice. This was done by answering three questions "Why maintain?", "Why maintain in this way?", and "Why maintain now?"

2. On the second Study, having reported the condition and recommended action for each of the 567 road control sections, it emerged that, because of the heterogenous nature of Jamaican roads, some 7,000 separate recommendations had been made. Four thousand of these were for periodic maintenance actions on 1,900 miles of road.

3. It would have been impossible to carry out full feasibility analyses on such a large number of actions. A method of justifying the programme was therefore devised which involved two procedures being followed:

i. employing a screening procedure to provide a list of recommended actions ranked in order of viability; and

ii. carrying out a detailed economic analysis to compare the viability of patching, resealing and overlay in general terms, and to test the feasibility of the marginal recommended actions by determining their rates of return.

4. In order to develop a simple screening procedure, it was first necessary to decide on which parameters would give the best indicators of road user savings to be derived from maintenance actions. Road user savings are derived generally from improving a road's geometry and/or its pavement surface irregularity. So, because maintenance actions have little or no effect on the geometry of the road, an attempt had to be made to quantify the benefits realised by improving surface irregularity.

5. In recent years, the TRRL has been responsible for a large proportion of the research undertaken to relate the effects on vehicle operating costs of the various external factors acting on a vehicle, work which was undertaken when they were developing the Road Transport Investment Model (RTIM), a model designed to assist planners to decide on optimum road investment programmes (ref. 4, 5 and 6). Their research in Kenya was

further evolved into the Highway Design and Maintenance Standards Model (HDM) developed by I.B.R.D. Neither of these models however provided the relationships we were seeking for use in Jamaica. The deterioration in surface irregularity observed in Jamaica, where pavement structures are strong and traffic relatively light, appears to be more dependent on time than on traffic loading, the parameter which RTIM uses. There is also some evidence in the Kenya data which suggests that surface irregularity varies more with time than with traffic loading, at least in some instances in Kenya. (Figures 12, 13 and 16 of TRRL Laboratory Report 673 ref. 5).

6. We in Jamaica therefore developed a model to relate surface irregularity to time in the form of the following graph (Fig. 1).

7. The deterioration rates represented on the graph were quantified by:

i. comparing the surface irregularity measured during the 1972 Study with the observations made in the 1978 Study;

ii. using basic data abstracts from TRRL LR 673;

iii. observing the performance with time of the periodic maintenance works carried out in Jamaica between 1973 and 1978;

iv. discussing experiences with engineers fully conversant with Jamaican conditions.

8. From a major and separate exercise to relate road users costs with surface irregularity, it was deduced that the average vehicle in use in Jamaica experienced a change in vehicle operating and occupants' time costs of \$0.18¢ Jamaican per inch per mile of surface roughness.

9. Before screening, the Viability Index (V.I.) for each recommended action was calculated using the following expression:

$$V.I. = \frac{365 \times T \times V \times R}{C}$$

where T was the average daily traffic using the section of road on which the action was recommended

R was the change in surface irregularity likely to result from the proposed works (in inches per mile)

V was the change in operating cost of an average vehicle associated with the change in surface irregularity (in \$ per inch per mile)

and C was the estimated cost (in \$) of the recommended action.

10. The Index thus represents a single year rate of return to the traffic using the section of road to which the recommended action refers.

11. Having completed the screening procedure, the other recommended actions such as patching, resealing and overlay were subjected to economic

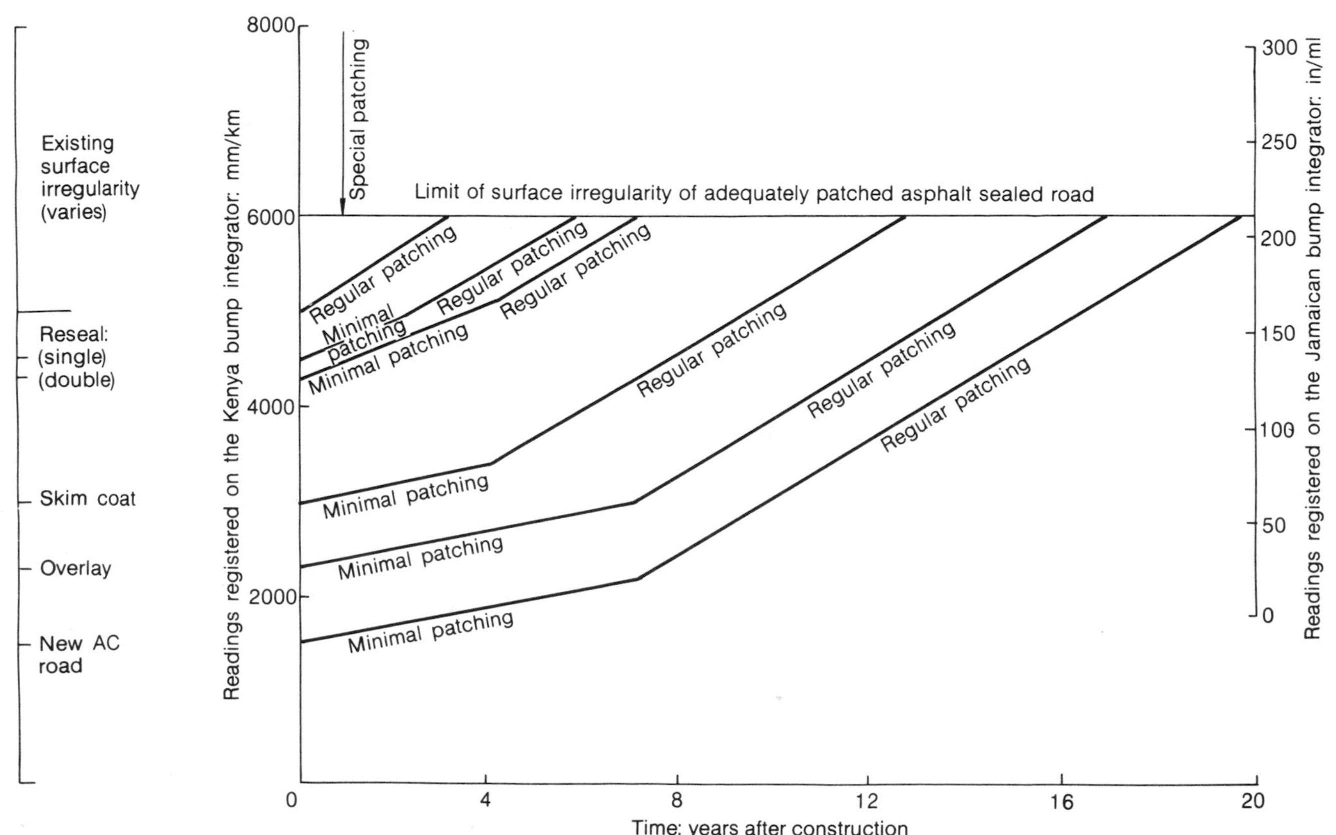

Fig. 1. Relationship of surface irregularity with time

analysis. Some of the interesting conclusions to come out of this exercise were that:

routine patching of all paved roads, other than those in need of rehabilitation, should take precedence over the periodic maintenance programme whatever levels of traffic the road carries; and

priorities for patching should be proportional to traffic levels and inversely proportional to the average area of each patch. In other words, roads carrying the heaviest traffic and needing the least amount of patching should be patched first.

MAINTENANCE PROCEDURES

1. The Public Works Department in Jamaica has, over a period of many years, established a reputation for keeping the island's roads open to traffic against formidable odds presented mainly by the weather, and for carrying out its maintenance tasks to consistently high standards. However, there was a danger that some of the time honoured practices and procedures used by the Department would be lost unless they were recorded. A decision was made therefore in the early 1970's to prepare a single comprehensive document containing a record of the basic systems and procedures to be adopted for highway maintenance by the Department. The document called "The Highway Maintenance Manual" was duly published on the 19th October, 1977, a day which marked an important stage in the introduction of systematic and uniform maintenance procedures in Jamaica.

2. The manual sets out to inform the entire staff of the Department how the complete system of maintenance management adopted by the Department works. In the manual, maintenance work items are defined, common defects that staff are likely to observe during the course of their work are described, and methods by which these defects should be rectified are suggested. The manual also includes short sections on Highway Law and contract procedures and a chapter giving a brief introduction to road design and construction.

3. Whereas prior to 1973 most maintenance used to be carried out on an ad hoc basis as and when the need arose, the Ministry's field staff management, by using the manual, is now able to develop maintenance programmes ROMGRAM (derived from Routine Maintenance Programme), to budget and allocate resources, to authorise and schedule work and to report on and evaluate performance. A computerised system of maintenance costing (RMCS) has also been installed.

4. An exercise known as ROMGRAM surveys is currently being conducted to assess the regular routine maintenance needs of each road control section in terms of costed work items, using the type of survey sheet mentioned earlier in this paper. It is taking some time to quantify and cost the amount of work to be performed on each maintenance action in each control section. However, the exercise is proving ideal as a means of training staff, and it has become evident from the information collected to date that it should be possible to quantify accurately the shortfall between funds actually provided and those necessary to maintain the network to its desired standard. The ratios of funds needed to funds available calculated so far with the limited amount of available information vary between 2.5 and 6.

5. To facilitate the effective utilisation of resources in Jamaica, maintenance works are classed by type as either Routine or Periodic, Rehabilitative or Emergency Repair and work items are related to the five elements of the road, that is, Pavement, Roadside, Drainage, Traffic and Structures. Work items are also defined and given a code number so that most maintenance works can be authorised, implemented, costed, and reported on, with little chance of misinterpretation.

6. The following routine maintenance actions are all carried out almost entirely by labour intensive methods in Jamaica:

Patching
Regulation and edge ravelling
Reinstatement of public utility trenches
Bush and trim banks
Overhanging limbs and stumping
Bushing (drainage)
Cleaning, shaping and grading drains
Cleaning culverts and culvert basins
Maintenance warning and directional signs
White line routine maintenance
Cleaning and painting structures
Repairs to structures
Painting of bridges

Other routine actions such as:

Sweeping, cleaning, clearing minor slips and debris
Local reconstruction
General regulating and patching

are also mainly labour intensive operations although some machine assistance is required, whereas most periodic maintenance actions (apart from drainage works) such as:

Resealing of surfaced road - one coat
Resealing of surfaced road - two coats
Resurfacing of surfaced road - overlay course
Resealing shoulders
White lining - thermoplastic
White lining - road marking paint

require extensive machine inputs. The exceptional actions which are classified as routine in Jamaica and which require machine inputs are:

Regrading and regravelling
Grading shoulders and verges.

7. When labour is employed it is almost always employed on a task work basis. Supervision is provided by Works Overseers who are usually

responsible for the maintenance of about 40 miles of the network. And work is authorised and performance judged by standards laid down in the Maintenance Manual. The Works Overseer normally sets tasks which can be conveniently completed within a fortnight as payment procedures are set up on a fortnightly cycle.

8. Periodic Maintenance works, on the other hand, are almost all executed by contract, tenders being invited from prequalified contractors.

ACHIEVEMENTS

1. Whereas Jamaica, at the start of the 1970's was pursuing a policy of maintenance based mainly on the repair of damage after it had occurred, now in the early 1980's its capability is such that not only are comprehensive programmes of preventative maintenance being implemented effectively, but also its maintenance needs for five years in advance are being prepared for. All the systems and procedures required to facilitate this have been introduced within the last ten years and are now operating effectively.

THE FUTURE

1. The one overriding certainty with which Jamaica and most other developing countries will have to contend in the forseeable future is that funds for maintenance are going to become increasingly limited. If roads are not going to be allowed to deteriorate, new materials for roadmaking and maintenance, which can be produced much more cheaply, will have to be found. The alternative of course is for large deposits of the materials we now use (eg. bitumen) to be sought, found, and exploited at cheaper cost than they are at present.

2. One other course of action open to us which could achieve the results we desire much more easily than by seeking new raw materials is for each and everyone of us to try to use the materials we now have at our disposal more efficiently. This can be done in two ways, one by having as our goal the elimination of all waste, and two by improving communications, and management information systems. Everyone should be able to contribute in some small way to the achievement of both of these goals. This conference is providing one of the means by which we can learn of better ways to communicate with each other. This paper - "Highway Survey, Inventory, and Maintenance Procedures in Jamaica" - we hope has gone some way towards informing you of the ways in which we manage our maintenance in Jamaica. We trust that people will go on developing management systems and procedures so as to allow the most effective use to be made of the funds made available for maintenance in future even though they may be limited.

REFERENCES

1. JAMAICA, MINISTRY OF COMMUNICATIONS AND WORKS. Jamaica Transportation Survey. Lamarre Valois International Ltd. June 1970.
2. GOVERNMENT OF JAMAICA. Road Maintenance Study. T.P. O'Sullivan & Partners and Peat Marwick Mitchell and Co. April 1973.
3. GOVERNMENT OF JAMAICA. Road Improvement and Maintenance Study. Stage II. T.P. O'Sullivan & Partners. October 1978.
4. HIDE H. The Kenya road transport cost study: research on vehicle operating costs. Transport and Road Research Laboratory. 1975. Report LR 672.
5. HODGES J.W. The Kenya road transport cost study: research on road deterioration. Transport and Road Research Laboratory. 1975. Report LR 673.
6. ROBINSON R. A road transport investment model for developing countries. Transport and Road Research Laboratory. 1975. Report LR 674.

20

Evaluation of flexible road pavements in Kenya

F. J. GICHAGA, BSc, MSc, PhD, FIEK, MICE, University of Nairobi

Experience from some of the recently completed roads shows that road pavements have at times failed prematurely thereby leading to unplanned expenditure in the exercise of rehabilitating them. This paper outlines results of studies carried out to establish long-term behaviour of road pavements under tropical climatic conditions. The studies involved measurements of elastic deflections, pavement distortion and rutting, cracking as well as establishing traffic loading patterns for typical high standard trunk roads of varying design in Kenya. The results of the studies show that while pavements are weakened by repeated wheel load applications pavements also tend to develop strength with age. The results further showed that for a pavement approaching failure elastic deflections are a function of cracking and rutting; and that higher elastic deflections are obtained during the months of high rainfall and high temperatures. The paper recommends that there is need for Road Authorities to regularly monitor factors that relate to road pavement performance such as traffic loading, pavement condition etc. in order to help in the financial planning for pavement strengthening and maintenance works and that the necessary funds should be set aside in the budget.

INTRODUCTION

1. The rate of development of developing countries is a direct function of the efficiency of the transportation systems in these countries. The road transport system remains the primary means of transport as can be evidenced by large financial investment that goes into the road building programmes in developing countries. It is of cardinal importance that expected benefits should accrue from the construction of roads in terms of accommodating traffic thereby leading to accelerated economic growth of the country. It is generally recognized that road building stimulates development in developing countries especially in the rural agricultural areas.

Several road pavements of high design and construction standards have failed prematurely leading to unexpected expenditure in reconstruction works. It is not uncommon to see extensive deformations, cracks and potholes along some of the trunk roads especially during the rainy seasons. In addition to the finance involved in the construction and reinstating such failed roads, poor road surface conditions also lead to increased vehicle operating costs and consequently high transportation costs. In order to establish the factors that lead to pavement failures it is desirable to carry out studies on road construction materials and on full-scale pavements, in addition to assessing the pavement design procedures. The exercise can also be extrapolated to incorporate the strengthening required for a failed pavement.

2. Design of Flexible Road Pavements in Kenya

Almost all roads in Kenya are of the flexible pavement standard. This has been mainly due to tradition (and the inbuilt skill) and partly due to the high degree of quality control required for concrete roads. The Ministry of Transport and Communications is responsible for planning, designing, construction and maintenance of roads in Kenya and has continued to carry out commendable research to establish design and maintenance procedures suited to the country's climate.

Most roads of the 1940's and 1950's in Kenya were generally designed on the basis of the rule-of-thumb with double seal (two layers of surface dressing) as a form of surfacing. During the 1960's most major road pavements were designed on the basis of the Road Note 31 (10) and Road Note 29 (9). The design involved selection of pavement structure from design charts for which the thickness of road base was generally fixed at 150mm to 200mm and surfacing was in form of surface dressing in the case of Road Note 31 and 50mm to 100mm bituminous mixtures in the case of Road Note 29. The subbase thickness was selected on the basis of subgrade strength determined in terms of CBR.

A road design manual (11) for Kenya was eventually developed in 1970 and this required determination of traffic loading on the basis of expected heavy vehicles per day (24 hours) five years after the road was opened to traffic. Thicknesses and type of subbase, road base and surfacing were selected from design charts. A more comprehensive materials and pavement design manual (12) was adopted in 1980. This design manual requires the designer to determine the

Highway investment in developing countries. Thomas Telford Ltd, London, 1983, 153–160

KENYA
Isiolo
Nakuru
Nairobi
1
2
3
4
5
6
Voi
Mombasa
INDIAN OCEAN
ES1 Thika road site
ES2 Langata road site
ES3 Airport road—Belle Vue
ES4 Airport road—GM factory
ES5 Mombasa road—Kenchic
ES6 Mombasa road—KMC

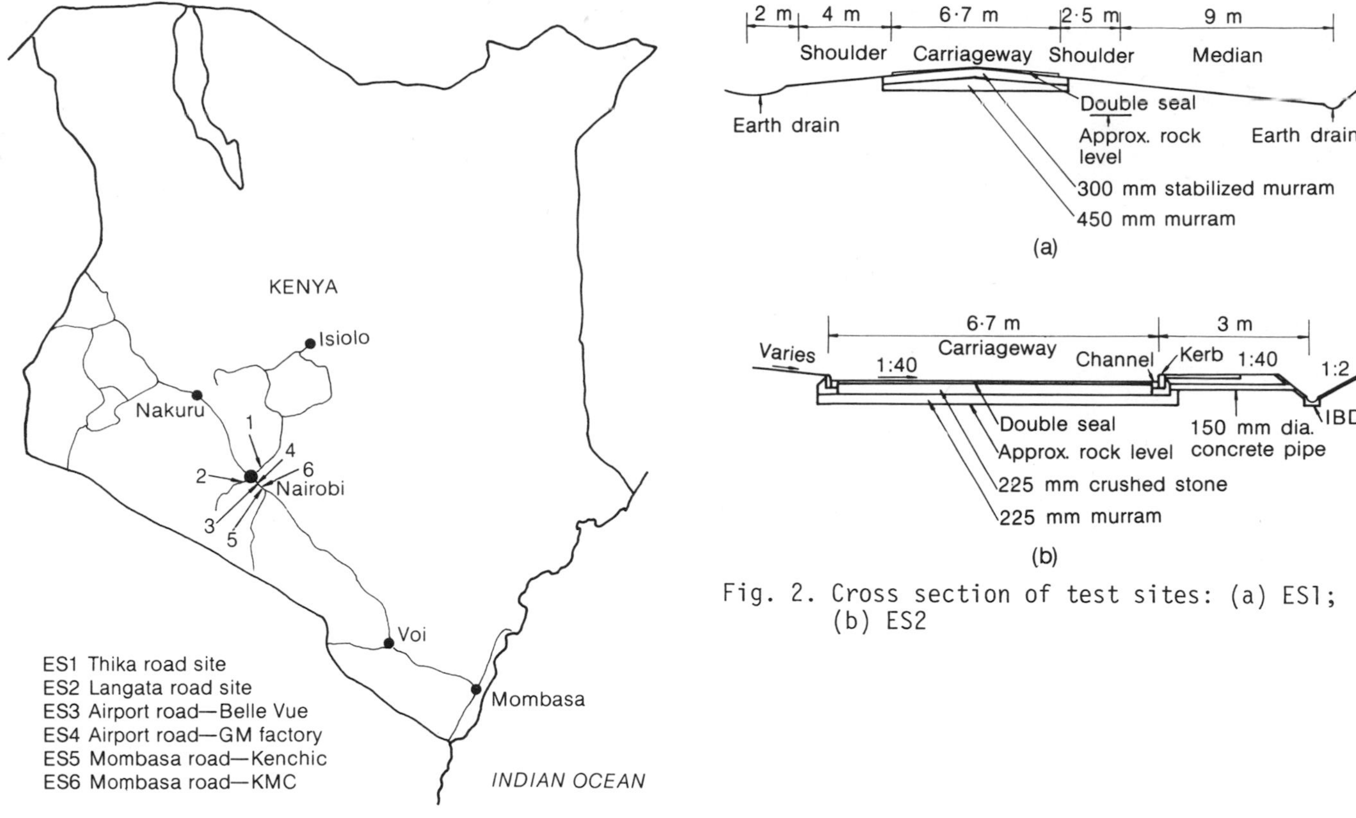

Fig. 1. Location of road test sites

2 m
4 m
6·7 m
2·5 m
9 m
Shoulder
Carriageway
Shoulder
Median
Earth drain
Double seal
Approx. rock level
Earth drain
300 mm stabilized murram
450 mm murram
(a)
6·7 m
3 m
Varies
Carriageway
1:40
Channel
Kerb
1:40
1:2
Double seal
150 mm dia. concrete pipe
IBD
Approx. rock level
225 mm crushed stone
225 mm murram
(b)

Fig. 2. Cross section of test sites: (a) ES1; (b) ES2

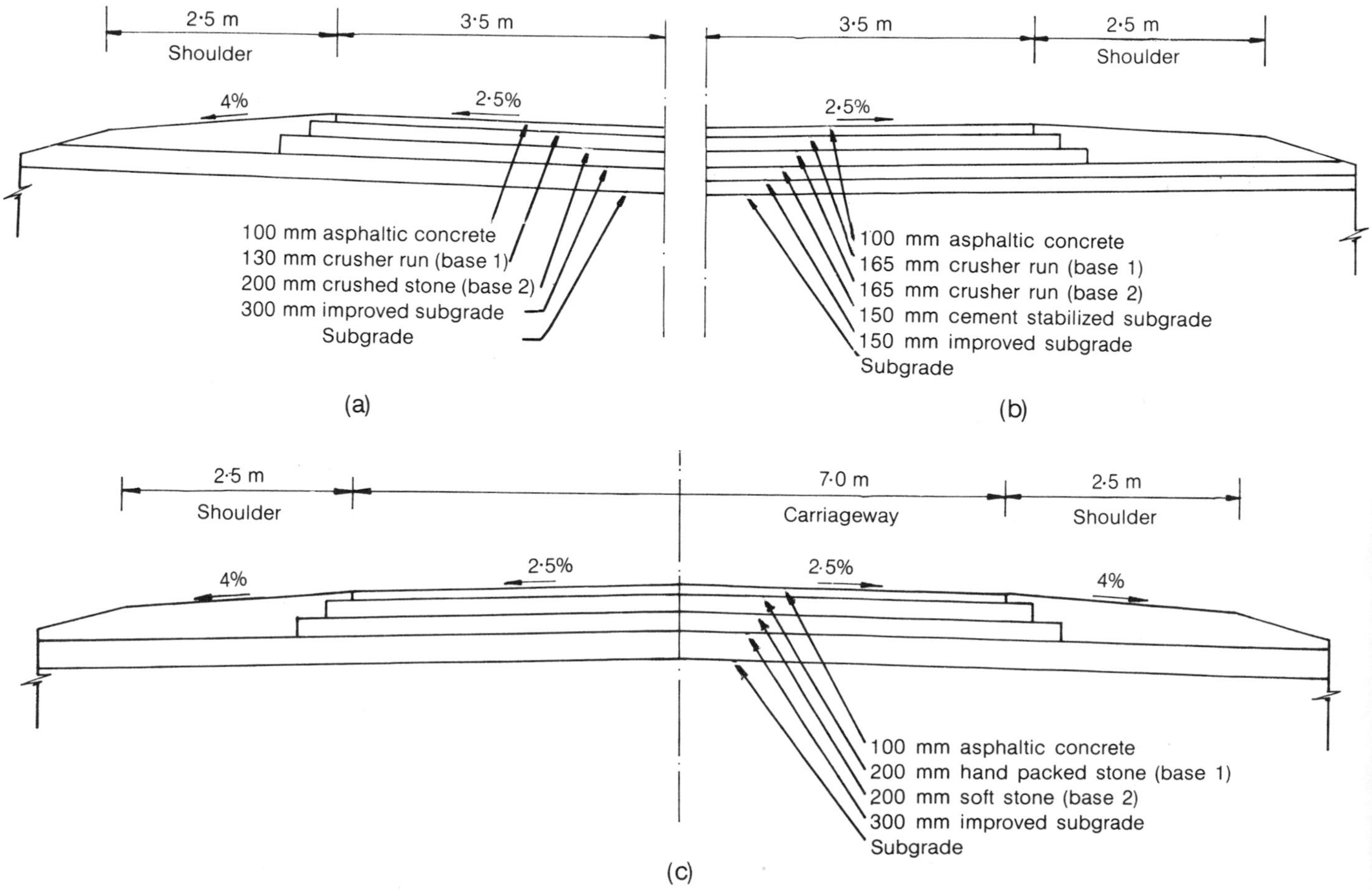

2·5 m
3·5 m
Shoulder
4%
2·5%
100 mm asphaltic concrete
130 mm crusher run (base 1)
200 mm crushed stone (base 2)
300 mm improved subgrade
Subgrade
(a)
3·5 m
2·5 m
Shoulder
2·5%
100 mm asphaltic concrete
165 mm crusher run (base 1)
165 mm crusher run (base 2)
150 mm cement stabilized subgrade
150 mm improved subgrade
Subgrade
(b)
2·5 m
Shoulder
7·0 m
Carriageway
2·5 m
Shoulder
4%
2·5%
2·5%
4%
100 mm asphaltic concrete
200 mm hand packed stone (base 1)
200 mm soft stone (base 2)
300 mm improved subgrade
Subgrade
(c)

Fig. 3. Cross-section of test sites: (a) ES3; (b) ES4; (c) ES5 and ES6

subgrade quality (in terms of CBR) and the traffic loading during the design life of the road (in terms of cumulative standard axles) and then select the pavement structure from a catalogue of structures depending on the materials available for road construction.

3. Objectives

This paper emphasizes the need to facilitate proper planning of expenditure on road building, maintenance and reconstruction by carrying out field studies to establish long-term structural behaviour of road pavements. It further stresses the need to carry out regular evaluation of the state of roads in developing countries.

METHODS OF EVALUATING ROAD PAVEMENT CONDITIONS

4. There are many methods of assessing pavement condition and they generally include:-

(a) Pavement Condition Survey

Pavement condition survey involves visual inspection of the condition of the pavement. The rating of the pavement condition is based on the pavement distress features such as patched up areas, cracks, potholes, fretting and ravelling, pavement distortions, pavement edge failure etc.

It is desirable that data on the above distress features is documented regularly so that planning of road maintenance works can be made more effective.

It is important that highway rating team is properly instructed on how data is to be documented and used in monitoring pavement condition so that road sections requiring attention are attended to before deterioration occurs. The data could also be used to specify appropriate maintenance measures.

(b) Use of Mechanical Devices

Several mechanical devices are available for assessing the structural behaviour of pavement structures and they include:- straight edge, Benkleman Deflection Beam, Deflectometer, field CBR, Plate Bearing test etc.

The straight edge is probably the cheapest and simplest and is used to measure rut depth on a pavement surface. It measures the pavement distortion due to repetitions of wheel loads and other environmental factors.

The highway inspection team would find the straight edge easy to use but would require guidance on documenting and monitoring the data obtained to enable maintenance measures to be effected before deterioration of pavement occurs. It would for example be possible to specify the rut depth which would require overlay as a form of maintenance.

Measurements of deflections using Benkleman Beam or Deflectometer are useful in evaluating the structural strength of pavements. The Benkleman Beam is widely used in Kenya for non-destructive test of pavements.

The field CBR test and the Plate Bearing test are rarely used in evaluating pavements in Kenya.

It is desirable to carry out pavement condition surveys and at the same time carry out tests based on mechanical devices. The data so obtained would be helpful in specifying when and how a failing pavement can be reinstated.

EVALUATION OF FLEXIBLE ROAD PAVEMENTS IN KENYA

5. Experimentation

This chapter outlines the field tests that were carried out on six selected test sections to establish structural behaviour of typical high standard road pavements in Kenya.

The selected test sections were in sound physical surface condition at the time of selection. Each test section had five cross-sections and twenty test points arranged in such a manner that the test points coincided with typical outer and inner wheel paths at 0.91m (3ft) and at 2.58m (8.5ft) from the pavement edge. Figure 1 shows the general location of the test sites and figures 2 and 3 show the cross sections of the six test sections. Other data relating to the test sections is shown in table 1.

The following field tests were carried out on the selected test sections:-

(i) Traffic Studies

Traffic studies were undertaken to establish ADT (Average Daily Traffic), vehicle classification, axle-loadings (using portable weigh bridges) along the selected test sites. Results of this test series are shown in figures 4 and 5 for test sites ES1 and ES2.

(ii) Benkleman Deflection Measurements

Benkleman deflection beams were used to determine elastic deflections using rebound procedure (5). Tests were carried out on the twenty marked test points and these points were used every time the test series was run. The test truck was loaded to give rear axle load of 14,000lb (6350kg) and rear tyre pressures were set at 85 lb/in^2 (5.98 kg/cm^2) (tyre size 900 x 20D).

Results of the test series are shown in figures 6 and 7 and table 2.

(iii) Pavement Distortion Measurements

Pavement settlement and rutting were determined using precise levelling (for ES1 and ES2) and 1.8m straight edge for all test sites. Figure 8 shows typical results of rutting study and table 3 gives a summary of rut measurements for the six test sections.

(iv) Pavement Cracking Measurements

Pavement cracking index was determined on the basis of the length of visible crack on the road surface in metres over an area of one square metre surrounding a test point. Summary of results of this test series is shown in table 4.

TABLE 1 - DATA ON SELECTED TEST SECTIONS

TEST SITE	YEAR WHEN OPENED TO TRAFFIC	COMMERCIAL VEHICLES CARRIED PER DAY
ES1 - THIKA ROAD	1961	1500 in 1976 2000 in 1980
ES2 - LANGATA ROAD	1965	1100 in 1970 2400 in 1980
ES3 - BELLE VUE, (Nairobi-bound carriageway of Airport Road)	1977	600 in 1978
ES4 - GM FACTORY, (Mombasa - bound carriageway of Airport Road)	1977	600 in 1978
ES5 - KENCHIC, (Mombasa Road)	1977	1200 in 1978 (both directions)
ES6 - KMC, (Mombasa Road)	1977	1200 in 1978 (both directions)

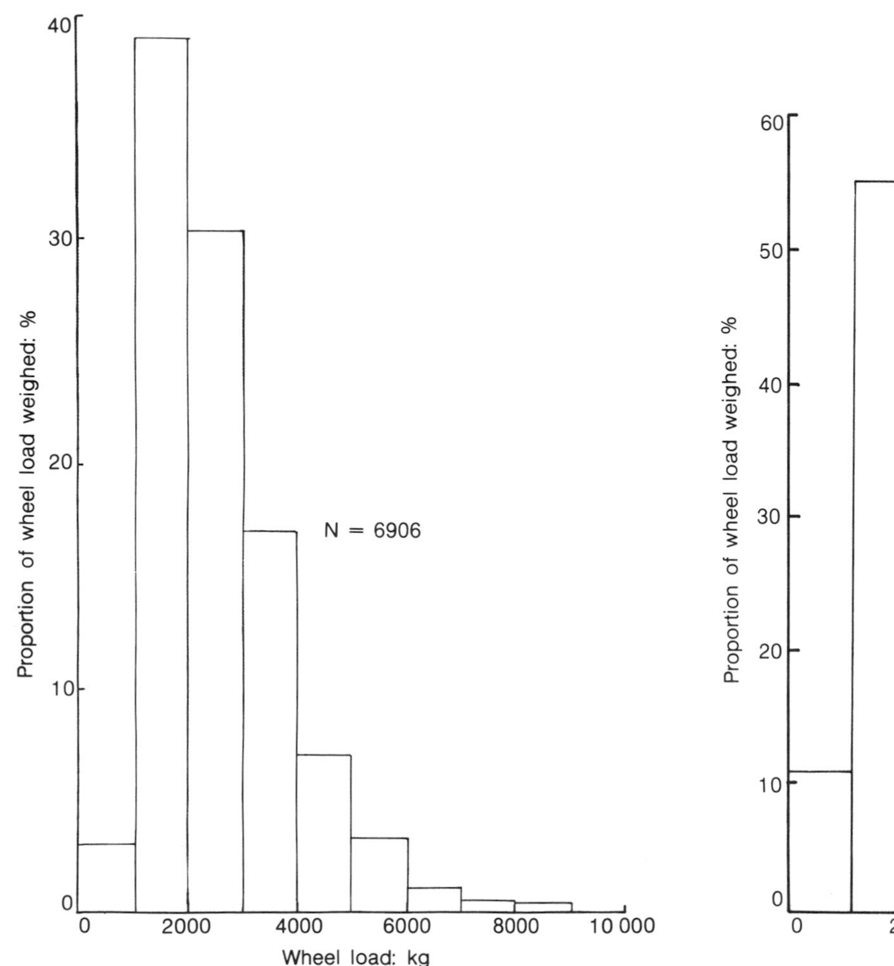

Fig. 4. Wheel load distribution for trucks along Thika road site (ES1), 1976

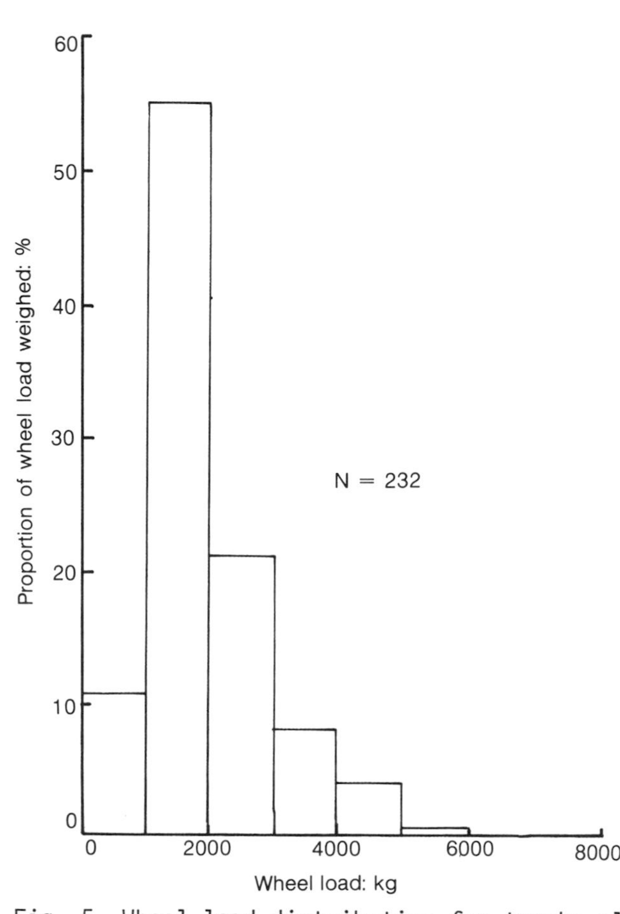

Fig. 5. Wheel load distribution for trucks along Langata road site (ES2), 1970

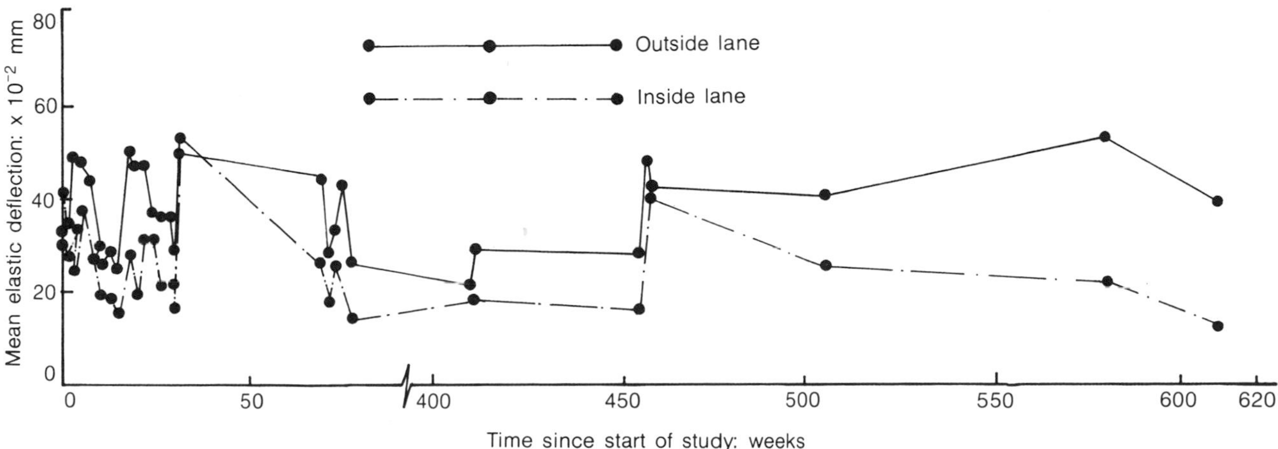

Fig. 6. Variation of elastic deflection with time for ES1

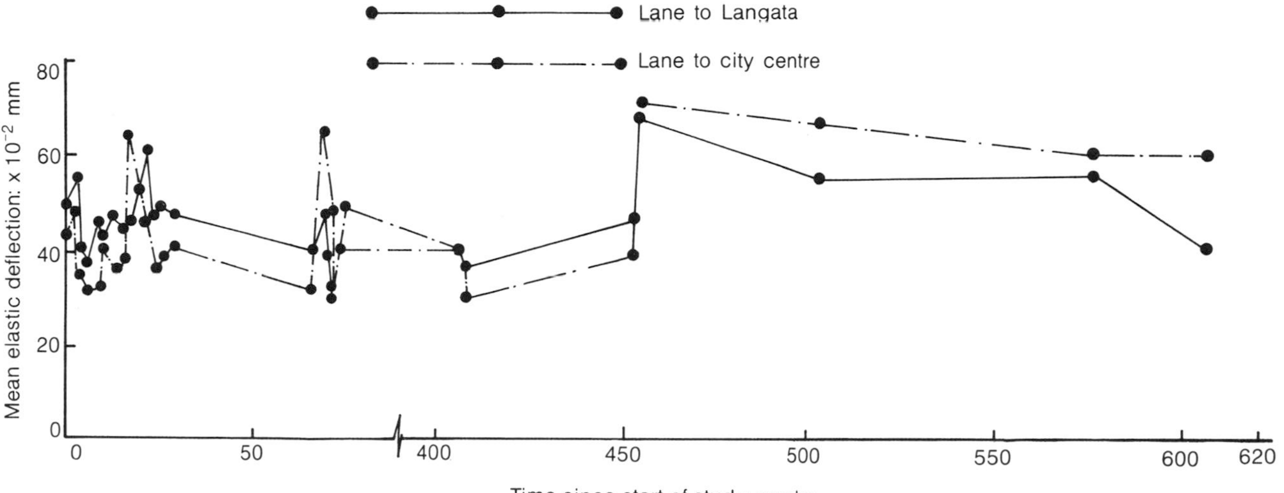

Fig. 7. Variation of elastic deflection with time for ES2

TABLE 2 - ELASTIC DEFLECTIONS OBTAINED IN 1981 ($x10^{-2}mm$)

TEST SITE	DATE OF TEST	LANE	MEAN FOR LANE*	STANDARD DEVIATION FOR LANE	MEAN FOR TEST SITE**	STANDARD DEVIATION FOR TEST SITE
ES1 - THIKA	17/1/81	OUTSIDE	39	27	26	24
		INSIDE	12	10		
ES2 - LANGATA	24/1/81	TO LANGATA	40	12	50	21
		TO CITY	60	23		
ES3 - BELLE VUE, AIRPORT ROAD	17/1/81	OUTSIDE	60	14	49	18
		INSIDE	37	15		
ES4 - G.M. FACTORY, AIRPORT ROAD	24/1/81	OUTSIDE	22	9	22	9
		INSIDE	21	9		
ES5 - KENCHIC MOMBASA ROAD	7/2/81	TO MOMBASA	42	14	38	15
		TO CITY	34	16		
ES6 - K.M.C. MOMBASA ROAD	14/2/81	TO MOMBASA	22	9	22	13
		TO CITY	22	18		

* Mean of ten

** Mean of twenty

Test Axle load was 14,000lb (6,350kg).

TABLE 3 - SUMMARY RESULTS OF RUT DEPTH MEASUREMENTS IN 1981

TEST SITE	DATE OF TEST	LANE	MEAN RUT DEPTH OF LANE* (mm)	MEAN RUT DEPTH OF TEST SITE** (mm)	RANGE OF RUT DEPTH OF TEST SITE (mm)
ES1 - THIKA ROAD	23/1/81	OUTSIDE	5.7	5.2	1.2 - 13
		INSIDE	4.7		
ES2 - LANGATA	23/1/81	TO LANGATA	16.8	15.1	1.2 - 33.8
		TO CITY	13.4		
ES3 - BELLE VUE, AIRPORT ROAD	13/2/81	OUTSIDE	2.9	2.1	0 - 5
		INSIDE	1.3		
ES4 - G.M. FACTORY AIRPORT ROAD	13/2/81	OUTSIDE	1.8	2.4	0.5 - 5.3
		INSIDE	3.1		
ES5 - KENCHIC, MOMBASA ROAD	13/2/81	TO MOMBASA	1.5	1.6	0 - 3.5
		TO CITY	1.7		
ES6 - K.M.C. MOMBASA ROAD	30/1/81	TO MOMBASA	2.2	2.1	0.1 - 3.4
		TO CITY	1.9		

* Mean of ten

** Mean of twenty
1.8m straight edge was used.

TABLE 4 - SUMMARY RESULTS OF SURFACE CRACKING TEST IN 1981

TEST SITE	DATE OF TEST	LANE	MEAN CRACKING INDEX OF LANE* (m/m^2)	MEAN CRACKING INDEX OF TEST SITE ** (m/m^2)	RANGE OF CRACKING INDEX OF TEST SITE (m/m^2)
ES1 - THIKA ROAD	23/1/81	OUTSIDE	0.342	0.184	0 - 1.85
		INSIDE	0.027		
ES2 - LANGATA ROAD	23/1/81	TO LANGATA	2.396	1.564	0 - 10.78
		TO CITY	0.732		
ES3 - BELLE VUE, AIRPORT ROAD	13/2/81	OUTSIDE	0.182	0.091	0 - 0.90
		INSIDE	0		
ES4 - G.M. FACTORY, AIRPORT ROAD	13/2/81	OUTSIDE	0.073	0.036	0 - 0.38
		INSIDE	0		
ES5 - KENCHIC MOMBASA ROAD	13/2/81	TO MOMBASA	0	0.057	0 - 0.91
		TO CITY	0.114		
ES6 - KMC, MOMBASA ROAD	30/1/81	TO MOMBASA	0.043	0.021	0 - 0.43
		TO CITY	0		

* Mean of ten
** Mean of twenty.

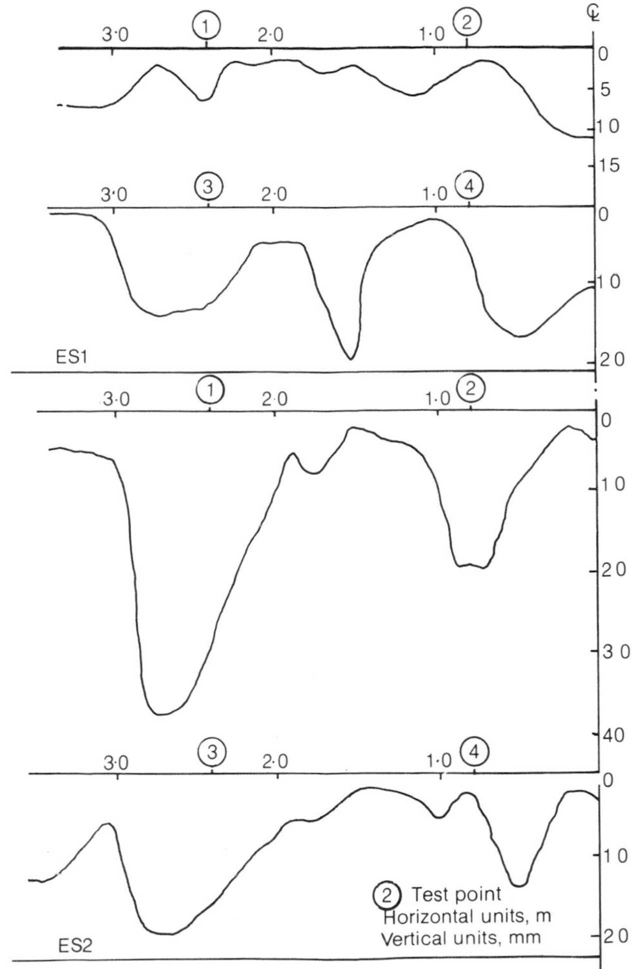

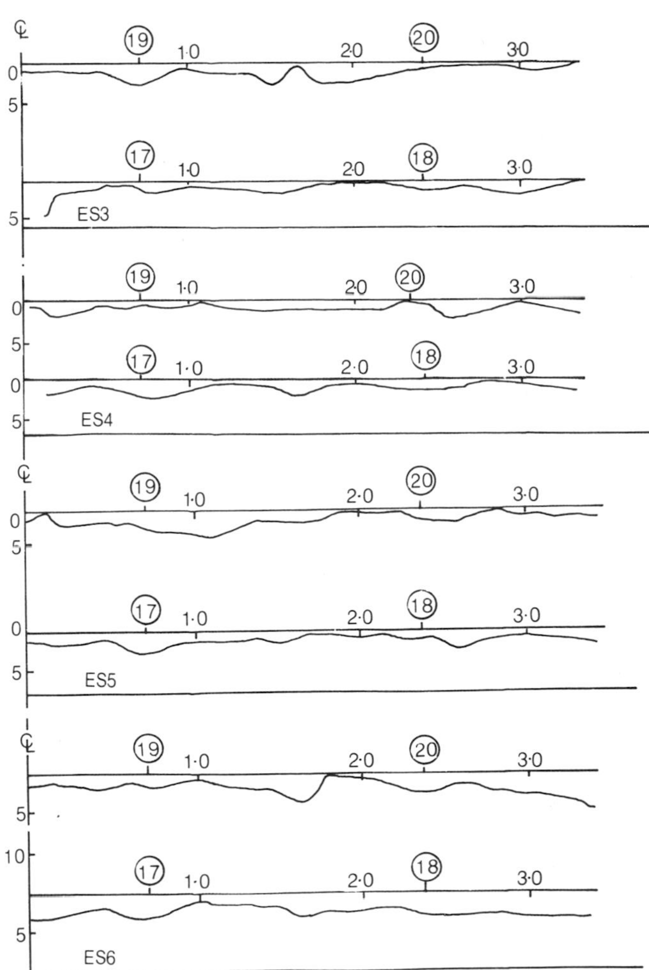

Fig. 8. Rutting profiles of the test sections

6. Discussion of Results

(i) Traffic Studies

The wheel load spectra shown in figures 4 and 5 are fairly typical of the type of wheel loads found on trunk roads in Kenya.

(ii) Elastic Deflections

The results of this test series show that the lane carrying slow and heavier traffic generally gives higher elastic deflections than the lane carrying faster traffic.

(iii) Pavement Distortion

As seen in figure 8 test sites ES1 and ES2 had considerably higher rut depths than the newer test sections in 1981. Test site ES2 shows rut depths which are high enough to justify application of overlay as a form of strengthening the pavement.

(iv) Pavement Cracking

This test series shows that test sites ES1 and ES2 had generally higher rate of cracking than the newer test sections in 1981. Test site ES2 displayed widespread cracking thereby suggesting that there is need to strengthen the pavement structure.

(v) Effect of Environmental Factors

The data collected was analysed to find out the influence of the following factors on elastic deflections:-

Age of pavement (weeks)
Traffic loading (in standard axles)
Pavement temperature (^{o}C)
Air temperature (^{o}C)

The multiple regressions obtained had weak correlation coefficients. The expressions indicated that elastic deflection tends to fall with age, which suggests development of pavement strength with age, whereas increased repetitions of wheel loads lead to higher deflections.

An analysis, relating three point moving averages of monthly mean elastic deflection and monthly mean rainfall and monthly mean air temperatures during the period 1969 to 1981 for test site ES1, was carried out and gave the following regressions:-

Outside Lane (Slow Lane)

$d_3 = 32 + 0.16r_3$;correlation coefficient = 0.66

$d_3 = -11 + 2.58t_3$; " " = 0.61

$d_3 = 5 + 0.11r_3 + 1.5t_3$ " " = 0.73

Inside Lane (fast lane)

$d_3 = 21 + 0.11r_3$; correlation coefficient = 0.54

$d_3 = -3 + 1.48t_3$; " " = 0.42

$d_3 - 10 + 0.09r_3 + 0.65t_3$; " " = 0.56

Where,

d_3 = 3 point moving average of monthly mean deflection ($x10^{-2}$mm)

r_3 = 3 point moving average of monthly mean rainfall (mm)

t_3 = 3 point moving average of monthly mean air temperature (°C)

These expressions, although showing weak correlation coefficients, indicate that higher deflections are obtained during the months of high rainfall and high temperatures.

Further regression equations relating mean deflections and cracking index and rut depths were derived for site ES2 giving the following expressions:-

d = 56 + 31C, Correlation Coefficient = 0.38
d = 55 + 0.21r; " " = 0.13
c = -1.25 + 0.22r; " " = 0.69
d = 56 + 32C - 0.05r; " " = 0.37

Where,
C = cracking index (m/m^2)
r = rut depth (mm)
d = elastic deflection ($x10^{-2}$mm)

Although the correlation coefficients obtained are very weak, the expressions suggest that for a pavement approacing failure elastic deflections are a function of cracking and rutting.

7. This case study on the six test sections shows that if data on road pavement conditions is collected and analysed regularly it can give important leads on the change in behaviour of pavement and can be useful in deciding on the priority of where money could be expended, and can also be used in planning maintenance works.

RECOMMENDATIONS

(1) In order to ensure that road maintenance works are planned properly and pavement strengthening policies effectively formulated it is recommended that Road Authorities should monitor factors that relate road performance to such factors as traffic loading, pavement condition etc on a regular basis.

(2) In order to achieve the above recommendation is necessary that Road Authorities should institute systems of monitoring road pavement conditions. Such systems should be accompanied by catalogues specifying corrective measures corresponding to various distress features so that acceptable standards of road serviceability are maintained.

(3) In order to ensure effective implementation of the above recommendations it is necessary for a Road Authority to set aside necessary funds in the budget.

ACKNOWLEDGEMENTS

This paper is based on long-term studies carried out at the Department of Civil Engineering, University of Nairobi. The contributions by the technical staff of the Department in data collection and the students who were involved in the various phases of the studies are recorded with gratitude. The contribution of the secretarial staff in the Department in typing the paper is also gratefully acknowledged.

REFERENCES

1. GICHAGA F.J. - Structural Behaviour of Flexible Pavements in Kenya. PhD Thesis. University of Nairobi. 1979.

2. OVERSEAS UNIT - TRRL - Pavement Engineering in Developing countries. Development in Highway Pavement Engineering -2, P.S. Pell (Ed.). Applied Science Publishers. London 1978.

3. LISTER N.W. - Deflection Criteria for Flexible Pavements. TRRL. Report LR 375. 1972.

4. GICHAGA F.J. - Report on Highway Maintenance in the city of Nairobi. Nairobi 1972.

5. GICHAGA F.J. - A study of Deflection characteristics and Elastic response of Flexible Pavements in Kenya. M.Sc. Thesis. University of Nairobi 1971.

6. HALL G.A. - A study into the Behaviour and Use of Asphalt concrete surfacings over unbound bases in Kenya. PhD Thesis. University of Nairobi. 1979.

7. CRONEY D. and BULMAN J.N. - The influence of Climatic Factors on the Structural Design of Flexible Pavements. Proceedings. Third Int Conference on Structural Design of Asphalt Pavements. London 1972.

8. LISTER N.W. - The transient and long-term Performance of Pavements in relation to Temperature. Proceedings. Third Int. Conference on Structural Design of Asphalt Pavements. London. 1972.

9. ROAD RESEARCH LABORATORY - A guide to the Structural Design of Flexible and Rigid Pavements for New Roads. Road Note 29. Second Edition. H.M.S.O. London. 1965.

10. ROAD RESEARCH LABORATORY - A guide to the Structural Design of bituminous-surfaced roads in tropical and sub-tropical countries. Road Note 31. Second Edition. H.M.S.O. London. 1966.

11. ROADS DEPARTMENT - Roads Design Manual. Ministry of Works. Kenya. 1970.

12. ROADS DEPARTMENT - Materials and Pavement Design Manual for New Roads. Ministry of Transport and Communications. Kenya. 1980.

Discussion on Papers 17–20

Mr J. REICHERT *(Belgian Road Research Centre)*: The development of road systems and growth in traffic, particularly heavy vehicles, have resulted in greater awareness of the economic importance of maintenance. Users are more and more exacting with regard to safety, travel comfort and continuity of service and maintenance techniques are being improved. Many countries have opted for a programmed preventive maintenance which makes it possible not only to protect the structures but also to adapt them to the change in traffic. When seeking the optimum balance between budgetary constraints and user requirements, the medium- and long-term profits and costs for the whole of society must be taken into consideration.

A maintenance policy must be based on: the assembly of data on the condition of the surfaces; aids to decision-making consisting of standards, and also recourse to data bank and analytical accountancy; the adoption of decisions based on a simple estimate of requirements and an economic analysis; the utilization of appropriate techniques; and adequate organization. A powerful tool for such a comprehensive analysis is modelling, that is, setting up an overall algorithm that incorporates all the relevant parameters and permits prediction, by calculation, of the medium- and long-term consequences of a proposed construction and maintenance policy.

A report by the Organization for Economic Co-operation and Development suggests that, when setting standards, it is essential to specify various thresholds at which each deficiency or pavement characteristic should be evaluated. Basically, intervention or restoration should be initiated when one of these thresholds is passed. There are six levels of threshold: warning threshold; optimum intervention level; public acceptability threshold; legal acceptability threshold; level after restoration; and the actual intervention level, which is a floating level, determined in practice by the available funds. Ideally, this actual intervention level should coincide with the optimum intervention level, and it should remain above the public acceptability threshold.

Paper 17 reports on a detailed study into the possibilities of using contractors to perform periodic and routine highway maintenance. A critical analysis of the experience gained in six highly different countries has made it possible to draw relevant and differentiated conclusions. The study is well presented and could be useful to road Authorities both in developing and in industrialized countries. It also defines the new task to be fulfilled by those responsible for the management of roads.

It seems to me, though, that before making any recommendations, it would be interesting to have more practical examples. I would like the Authors to specify the numerical criterion which can serve as a basis for deciding that maintenance is inadequate.

Paper 18 discusses the concrete results of a long-term effort made by the Building and Public Works Laboratory and the Road Administration of Ivory Coast. For the network of paved roads of Ivory Coast, the Authors have been able to quantify permissible values for the various maintenance criteria and to define limit threshold values. The methodology which has been developed and the organization which has been set up have turned out to pay off well; it would be worthwhile to follow the example in other developing countries.

The Authors quite rightly draw attention to the frequent overloading which has been recorded during axle-load weighting campaigns for commercial vehicles, and on baleful influence of this on the life-expectancy for pavements. This information complements the studies made by C.I. Ellis of the Transport and Road Research Laboratory (TRRL) in 1967 and by J.P. Serfass (France) in 1973. I recommend two basic documents from Ivory Coast, which are of great practical value: the Manual for the Structural Design of Overlays and the Illustrated Guide for Maintenance Purposes.

Paper 19 is undoubtedly very interesting because it presents a clear and critical report on the two studies made in 1972 and 1978, and because it gives a description of the original methodology which has been developed in Jamaica. I would emphasize the following points.

(a) The influence of execution methods on surface irregularity.
(b) In contradiction with the road transport investment model, RTIM (TRRL) and the highway development model, HDM (World Bank), the deterioration in surface irregularity observed in Jamaica appears to be more dependent on climatic conditions than on traffic loading. I would like to know more about the possibilities of each of the three models.
(c) The proposition to look for new materials which can be produced more cheaply, as well as the proposition to improve community and management information systems.

Would the Authors specify the parameter used in the design calculations?

Paper 20 describes the results of studies made to establish long-term behaviour of road pavements in tropical regions. The methods of evaluating road pavement condition are described and the results obtained on six test sections are given. It would be worthwhile to compare the experimental correlations found in Kenya between certain conditions of deterioration and climatic factors with correlations obtained on similar structures in other tropical countries.

Mr R.L. MITCHELL *(University of Zimbabwe)*: I would firstly like to refer to Paper 17 by Harral *et al.*. I was long associated with the Rhodesian Ministry of Roads which had an outstanding departmental maintenance organization, which was under the control of engineers and superintendents reared on construction. Pot-holes did not exist and reseals were carried out after routine panel inspections. However, such experience and dedicated staff no longer exist in most of Africa - a spin-off from inflation and fixed salaries - and maintenance by contract is a necessary evil. However, does Dr Harral consider that increased use of contractors is consistent with increasing applications of socialism? Surely the concept rather than the technique of contract maintenance represents the major hurdle.

Messrs Bamba *et al.* (Paper 18) and Professor Gichaga (Paper 20) have produced some most interesting technical data. However, I would like to query Fig. 1 in Mr Bamba's Paper. How was the straight (limit) line of deflexion derived? Even assuming its validity, concave upward traces have been drawn for measured deflexions with repetition of load - but in my experience, deflexions do not increase with repetition of load. It is suggested that the scatter of data is too great to define the tendencies, but even if such tendencies are true, the intersection with the limit line could be erroneous to an order of magnitude of 10, i.e a pavement could last for, say, 10^5 or 10^6 repetitions!

I must criticize both Papers for the use of different axle loads - of 13 tons by Bamba and of 6.35 tonnes by Gichaga - as opposed to the international reference axle of 80 kN.

Bamba *et al.* in paragraph 15, have under-stressed an important statement: they recommend granular overlays with surface dressing (as I have done in Paper 12) with up to 4×10^6 loadings of a 13 t axle - some 3×10^6 E80 axles in conventional parlance - and state, in paragraph 16, that surface dressing is sometimes permissible for up to 6 or 7 million E80 axles. This means that surface treatment on granular overlays, which is far cheaper than bituminous concrete, is adequate for overlays on most rural roads in Africa. I fully agree. There are no rural roads in Zimbabwe, with axle-load control, which carry as much traffic.

Yet conversely Professor Gichaga illustrates test roads in Kenya with 100 mm of asphaltic concrete - which must surely be economically unjustified? However, in his Figs 6 and 7 he shows no consistent pattern with deflexions against time (with 6.35 t axles), which tends to confirm my previous suppositions that regressions are not statistically significant. I agree with him that 15 mm of rutting is as good a criteria of failure as any - particularly as this appears to be the critical depth beyond which ruts hold surface water.

From these comments two significant conclusions appear to be valid: firstly that granular overlays with surface treatment should be the norm, and secondly that deflexion data should be restricted to research, and not used as a tool for quantitative overlay design.

Mr J.J. GANDY *(Scott Wilson Kirkpatrick & Partners)*: Dr Harral (Paper 17) has been pressing the case for maintenance with great clarity and vigour for a long time. I agree entirely with what is said in this Paper and hope that it has the success it deserves. In paragraph 3.14, the Authors say that contractual instruments should be drawn specifically to fit the needs of routine maintenance. As this is an important recommendation, would they care to comment more fully on the required form of contract? Also, does this comment imply that periodic maintenance is best dealt with by a standard form of construction contract?

Dr J.B. METCALF *(Australian Road Research Board)*: Paper 17 seems to contain a number of internal inconsistencies which I would seek to clarify. For example, the Authors say (paragraph 1.04) that contract maintenance can produce a less restrictive operating environment and reduce the burden of direct responsibility and at the same time provide better control of diversion of resources and ensure adequate and stable funding. In a later section (paragraph 2.01) they refer to the use of contract because direct labour recruitment was limited. Is this not a way of eroding a constraint rather than any demonstration of the effectiveness of contract maintenance? Could the Authors discuss these points and also respond to two questions. First, what measures of and data on maintenance efficiency do they propose? Experience in Australia has shown that unit rates, even under closely comparable conditions, do not necessarily reflect the quality of the work and thus performance of the maintained route. Second, in an area where many minor tasks must be done correctly and quickly for such quality, is there not likely to be a different personnel training problem in the contract approach to routine maintenance without which the massive efforts described become necessary?

Paper 18 attacks the central problem of linking maintenance to road performance and I would like to contrast recent Australian results with the data in the Paper and seek the Authors' comments on the differences. First (referring to Fig.2), the tolerable deflexion levels seem to be 20% less than comparable roads in Australia and yet show higher rut depths. What is the Authors' explanation of this? Second, have they discovered any robust relation between roughness and traffic (Fig.3)? Australian data (ref. 1) has indicated relationships can vary from linear to (high) power function and initially we suspect this to be the result of differences in maintenance policy.

Mr P.W.D.H. ROBERTS *(Overseas Unit, TRRL)*: Many countries are not at present in a position to allocate the foreign exchange and other resources required to establish the detailed monitoring systems described in Paper 18. Moreover, the experience reported by Mr Mitchell (in Paper 12) throwing doubt on the value of deflexion measurements for predicting the life of pavements with thin bituminous surfacing, has been duplicated elsewhere. At present, simpler direct techniques provide more useful measures of pavement condition.

The Building and Road Research Institute of Ghana (assisted by TRRL) began a national study of road pavement conditions in 1978. The study was carried out on behalf of the Ghana Highway Authority with support from the World Bank. Among the indications of this study it is interesting to note that

(a) the quality of surface treatment work is of overriding importance in dictating the amount of maintenance required
(b) with very little maintenance, most of the unsealed running surfaces - carrying up to 150 vehicles a day - tended to remain in a moderate-to-poor condition without further rapid deterioration
(c) through careful planning and sampling, the relatively simple techniques used were well able to provide the information necessary for the Highway Authority to determine rational priorities for investment in maintenance and improvement.

On the basis of the above and other studies and the experience of many engineers, TRRL has recently published Overseas Road Notes 1 and 2 giving guidelines on road maintenance management and techniques for District Engineers in developing countries.

Our studies confirm the very poor performance of maintainance organizations. Moreover, we find that many projects to improve this performance have also been disappointing. The efforts of many countries are primarily constrained by continuing shortages of foreign exchange and, even more important, of experienced manpower at middle management and other levels. It is a matter of great concern that increasing efforts to raise the capabilities of road maintenance organizations will fail if the existing constraints and resources available are not carefully taken into account. On the other hand, organizations will be seriously over-committed (as many already are) with the result that the efficiency of their operation is even further reduced. On the other hand, there is a risk of underestimating and therefore underutilizing those resources which are available locally (such as particular labour skills or a well-established administration within the community).

Capacity cannot be rapidly expanded to meet the full needs of road maintenance in most developing countries. Some short-term improvement may be achieved by supplementing middle management and supervisory manpower through recruiting technical assistance from other countries. It is more important to ensure that the available resources are concentrated so as to be employed effectively on those sections of the network deserving highest priority. This will also establish a suitable working context for staff to gain the sound practical experience which is so important in complementing formal training. Thus, in the longer term, it is this manpower development which will permit an expansion of the organization's capacity for recurrent support of the infrastructure.

Mr J.A. TURNBULL *(Mott Hay and Anderson)*: Papers 18, 19 and 20 outline the types of failure measured during inspection i.e., roughness, deformation, cracking and so on. The solutions proposed range from localized patching through resealing to overlay.

No mention is made in the Papers of drainage of the whole pavement structure. It has always been my experience that a free-draining layer at either base or sub-base level is essential. A saturated sub-base subjected to heavy or excessive axle loads leads to failure of the surface layers above and any amount of overlay will not overcome surface failure in these conditions. What measures do the Authors take to assess this problem during their inspection and, if the problem is shown to exist, what measures are taken to overcome this problem?

Mr C. FREDERIC *(Belgian Road Research Centre)*: Within the framework of the activities of the Road Research Programme of the Organization for Economic Co-operation and Development, a working group was set up at the end of 1979, with the task of writing a report on the maintenance of unpaved roads in developing countries. This working group is chaired by Mr J. Reichert. The working group includes representatives from some ten countries and from the World Bank. An inquiry questionnaire was prepared by the working group and sent to 107 countries, including 85 developing countries. Replies have been received from 47 countries, including 32 developing countries. These replies make it possible to draw some conclusions including the following.

Unpaved roads

Unpaved roads represent about half of the world's roads, but make up two thirds of the total network of roads in developing countries. Traffic plays a major role in the decision on whether or not to upgrade a road. On average, a traffic density of 50 vehicles a day is required to upgrade a road from an earth- to a gravel-road, 200 vehicles a day to upgrade a gravel road to a surface-treated road and 1 500 vehicles a day to upgrade a surface-treated road to a bituminous-premix surface road.

Road network

A road network inventory is an indispensable tool for sound management. However, such an inventory may be of no use if it is not regularly updated. This updating is done by more than 60% of the countries that replied to the inquiry; traffic censuses are very widespread, but the enforcement of permitted axle weights raises serious problems in developing countries.

Despite all the efforts made to prevent them, deficiencies in roads appear nearly everywhere. Pot-holes, rutting, corrugation and losses of material are problems which can be encountered everywhere - as they were mentioned in more than 85% of the replies. Unevenness and loss of camber were also mentioned in 70% of the replies.

These deficiencies affect the vast majority of unpaved roads. Traffic is the major cause of deficiencies and is by far the most important element in determining the maintenance requirements of this category of road, although other elements should also be taken into consideration (such as the materials used, the quality of maintenance, climate).

Maintenance

The most common maintenance activities are grading, regravelling, and the filling of potholes. Some techniques, such as dust-reducing treatments, are limited to a small number of countries. The grain size distribution curves of the materials used in wearing courses are fairly similar, despite major climatic differences.

The decision on whether to contract work out or to allow directly employed labour to carry it out or to use a combination of the two is made in various ways. In half the countries, direct labour preponderates; in 15% contracting out is the most usual method; while in the rest a combination of the two is used. One method of dividing work encountered in about a fifth of the replies is to allow direct labour to carry out routine maintenance and to contract periodic maintenance out. In two thirds of cases, maintenance is carried out by a combination of hand labour and machines - exclusively by hand labour in one sixth of cases, and exclusively by machines in one sixth of the replies.

Budgeting

There is a lack of information with respect to the way in which the required budget is provided. Previous practice plays an important role in the allocation of maintenance budgets, the assessment of requirements and the setting of priorities. The assessment of requirements is often based on visual inspections; not many countries have developed a system of measuring deficiencies based on the seriousness and the extent of defects. A catalogue of deficiencies, which is necessary for consistent and accurate data-collecting, exists only in a few developing countries (Algeria, Senegal, Tunisia). In other countries (Morocco, Nigeria, Togo) a catalogue of this type is under preparation. As to the choice of priorities, rating systems for assessing road-surface condition are infrequent and are largely based on the visual assessment of defects, these inspections being made mainly at the district level. Budgets are most often allocated at the national level, in response to known requirements and in view of an annual works programme.

Mr W.J. WYLEY (*Howard Humphreys & Sons, Leatherhead*): I should like to emphasize the importance of collecting actual road-pavement life-history data, for as wide a range of conditions as possible, in order to develop greater understanding of pavement-deterioration mechanisms. Howard Humphreys has recently completed a study in the UK of available data on maintenance undertaken on individual sections of road. This study has shown that pavement performance is extremely variable (even after taking account of the analysis of actual standard-axle accumulation). This means that elaborate models based on laboratory work - and even full-scale (but closely controlled) trials - must be treated with caution.

Most importantly, this has real implication in the allocation of resources for the development of models for design purposes on the one hand, and for surveys of road condition and the determination of maintenance priorities, on the other.

Mr A.V. LIONJANGA (*Ministry of Works and Communications Botswana*): I share the view that maintenance should get much more attention than it has previously received from the authorities concerned, and therefore warrants a share of the budget which will enable it to cope reasonably with the work-load. However, the practical situation is often very different, in that there is always a shortfall of funds in the maintenance budget to look after the capital invested in new roads. In such situations, the maintenance engineer is faced with the problem of deciding where to concentrate his activities in the road network given the financial restraints. Should maintenance funds and work be concentrated on the worst sections of the network (usually a smaller proportion of the road network)? Leaving the less worse-off sections (usually a greater proportion of the road network) invites the risk of further deterioration with no guarantee that funds will be available to maintain the whole network in future. Or should the limited funds available be used to maintain a greater proportion of the network, i.e. the less worse-off sections leaving the worst sections to further deterioration in the hope that funds will be available in future for major rehabilitation/reconstruction? I should be most grateful to have comments from the Authors of Papers 17 and 18 on this issue.

Mr H. VAN SMAALEN (*Agricultural University, The Netherlands*): Deterioration of the pavement structure is often caused by drainage problems. We should not wait until increasing deflexions ask our attention and then take remedial measures. These problems can and should be foreseen in many cases: we should pay more attention to the design of a good drainage system. 'Good' means a complete system, which does not end at the right of way.

Very often the constraints on a good drainage system are a result of the construction of the road - not only has the water falling on the road itself have to be carried, but there will also be an improved drainage of the areas on both sides of the road. Therefore, the total discharge will increase and very often the existing system is not suited for this. This can make it necessary to design some additional features far from the road, in close co-operation with the responsible agencies.

Dr HARRAL, Mr HENRIOD and Mr GRAZIANO (*Paper 17*): Mr Gandy has asked for advice on specific types of contract for maintenance. Cost-plus percentage-fee contracts should be avoided wherever possible. This is not just because under a cost-plus maintenance contract, the contractor's incentive is to utilize more resources in order to increase his profit, it is also because under the cost-plus contract, the government's administration retains the supervisory burden of organizing, mobilizing and directing the use of resources purchased from the contractors (at least if it desires to ensure efficient production and quality output). Under unit-price contracts, the contrary is true: the contractor must assume the burden of management and the responsibility to produce an output, and the contractor's incentive is to utilize fewer resources in achieving his work objective. The burden on the government's administration to organize, mobilize and direct resources to accomplish maintenance work is greatly reduced, and the government's efforts can be focused on the higher level functions of planning, scheduling and control. The government needs 'only' to identify specific maintenance work needs for the contractor and verify the quality and quantity of the completed work.

This is a significantly reduced role in terms of the government's supervisory manpower requirement. However, it is also a rather different and critically important management role, so that it requires somewhat different and some higher level skills and experience. Also, the contractor clearly must assume a much greater responsibility and direct supervisory burden. Thus, touching on one of the points raised also by Dr Metcalf, some training and experiential development of both government administrators and contractors will be required for each to successfully assume its role under a contracting scheme.

On the last aspect of Mr Gandy's question, concerning the specific type of contract for maintenance, it need only be observed that periodic maintenance is very easily contractually specified and administered on a unit-price basis, and that no doubt explains in part why periodic tasks are so commonly, and successfully, contracted. The traditional question arises with respect to how one contractually specifies routine maintenance tasks. Recent experience has shown that this is not the insuperable task that it has been thought to be - at least not where there is a well-defined routine maintenance programme, encompassing an inventory of the network, work programming and schedule. Since such improved planning is essential to achieve more cost-effective execution of the work - whether by force account or by contractor - contracting of this work poses no substantial additional burden. In fact, some consider it an advantage of contracting that it is more likely to evoke such improvements in maintenance programming than is continuing force account execution, where very large inefficiencies have gown to be tolerated. We should note, however, that it may be sensible to contract a few small items of routine maintenance on a cost-plus basis to simplify administration.

Dr Metcalf saw 'a number of internal inconsistencies' in our Paper. Yet the one instance he cites constitutes no inconsistency. The fact that certain (but by no means all) road authorities were first motivated to use contractors because of hiring restraints on government personnel does in no way conflict with the finding that such contracting has also proved to be both more effective and more efficient than that by the remaining government forces. On another point, Dr Metcalf has queried the relationship of efficiency and effectiveness. A number of measures of efficiency (assuming a fixed quality standard of the work achieved) can be employed. The most important of which are the intermediate indicators of equipment availability and utilization and ideally, the full economic costs per unit of output achieved. Unfortunately, reliable data are rarely available in developing countries on these or any other indicators of efficiency, a matter that we are seeking to address in current work of the World Bank, including a study with the Crown Agents on improving management of equipment in highway authorities in developing countries.

Of course, the quality of the work and subsequent performance of the road also vary widely, and constitute another major problem, a problem which we all must struggle with constantly. However, we would abjure if Dr Metcalf's implication is that the problem of quality control would be worse with contractors than with direct government execution. The problems we have observed in quality control in developing countries are at least as great for government executed work as for contracted work. An important consideration is that under contracting the quality control function is separated from those who are responsible for execution of the work, and it is no loss to the government to require a contractor to go over unsatisfactory work, whereas the opposite situation prevails under force account.

Mr Mitchell questioned whether the use of contractors is consistent with socialistic forms of government. The answer is that it certainly can be. Indeed, it is a socialist federal government in Yugoslavia which has gone further than any other government in contracting out maintenance. A contracting firm does not have to be privately held. The Workers Co-operative of Yugoslavia may constitute a unique form of ownership and management of capital, but para-statal contracting companies are common. A para-statal company may not have as strong incentives for efficiency as a private contractor (since it is usually the government's capital which is at risk and the prospect of bankruptcy is more remote) but, compared with a civil service bureaucracy, important gains in flexibility, salaries, incentives and staff retention may be possible.

Mr Lionjanga asked our views on a separate topic - the issue of maintenance priorities under budget constraints and, specifically, whether maintenance funds should be concentrated on the worst sections or the less-worse sections. This is a very important question which every road authority should ask itself whenever budget funds are too limited to do everything that is needed. Either strategy may be optimal - the answer for a given case will depend very much on circumstances, particularly, traffic volume, the existing pavement condition and most likely future pavement condition under alternative maintenance strategies for each road segment. This problem has been addressed by Messrs Harral, Watanatada and Fossberg of the World Bank, using the highway design and maintenance standards model (HDM). These papers are available on request to Dr Harral at the World Bank, Washington D.C. **Mr Frederic quotes the OECD study findings on the threshold traffic volumes required to justify upgradings of pavement stan-**

dards the level of traffic at which a higher standard pavement (be it gravel, bituminous surface dressing asphaltic concrete, or other) can be economically justified will vary very widely with local circumstances, particularly the unit costs of construction and maintenance and the opportunity cost of capital (which may be extremely high, even over 100% per annum, where budgets are constrained and road paving would pre-empt funds for much needed maintenance). The values quoted from the OECD study (or indeed any fixed values) could be seriously misleading in many cases, particularly where there are serious budget constraints. This issue requires specific study case by case, and we should avoid any fixed rule-of-thumb values.

Mr BAMBA, Mr LIAUTAUD and Mr DAVIS *(Paper 18)*: Mr Mitchell asks how the line of deflexion in Fig. 1 of Paper 18 was derived. The straight line of the critical or permissible deflexions is drawn, as it is done conventionally, to separate the plots representing pavements in good condition from pavements showing signs of deterioration (as shown in Fig. 2 of Paper 18). The absolute precision of that line can of course be questioned and it must be used only as a guide. The increase of deflexion with repetition of load is observed on some roads in the Ivory Coast, although there again, the error in assessing the point of intersection with the critical deflexion line may be, it is true, quite large. However, trying to assess the tendency for deflexion to increase with time can be very valuable in deciding maintenance priority action. When a rapid increase is observed, it may suggest a need to increase observation or indeed to carry out preventive action.

A 13 ton axle load is used for deflexion measurement, as it corresponds to the reference load used in pavement design in the Ivory Coast. Results can be easily converted in terms of 80 KN axle, as research has shown that deflexion is directly proportional to applied load, within the range of axle loads generally used.

As Dr Metcalf points out, tolerable deflexion levels in the Ivory Coast are indeed quite low compared with some other countries' results. In our view, this is due to generally favorable climatic conditions and to the fact that current practice in the Ivory Coast systematically uses at sub-grade level a layer of 30-40 cm selected lateritic gravel with a high modulus and low strain at failure. As the bulb of pressure from the applied load lies mostly within the sub-grade layer, relatively low deflexions are generally measured.

Rut depth exceeding 20 mm is fairly rare in the Ivory Coast, and when it occurs most of the rut is usually the result of the consolidation or plastic flow of the surfacing, the base course or the sub-base (and not necessarily of the subgrade). That may explain the apparent discrepancy between the low level of deflexion and these exceptionally high rut depths, whenever they occur.

Figure 3 shows some relationship obtained on one particular road between roughness and traffic. Research is continuing to assess the validity of that relationship for other roads. But, it is suspected that other shapes of relationships may be observed, depending essentially on the quality of the construction works, drainage efficiency, type or adequateness of surfacings.

In reply to Mr Roberts, it is true that the very sophisticated type of monitoring system described in Paper 18 requires a high initial investment from highway authorities, and that simpler direct techniques may provide useful measures of pavement condition. However, when the extent of the road network is important or when information is required very rapidly and, in some cases, more than once a year over a substantial part of that network, we think that the use of high-rate measuring equipment becomes compulsory. Reliability and repetitivity of results are improved, and in the long run, the unit cost of the test itself (including treatment and interpretation) is substantially reduced, particularly when large volumes of information are to be obtained.

Mr Turnbull is right to emphasize the problem of drainage. To assess this problem, we proceed in the following way.

(1) High values of deflexions are often associated with drainage problems. Peaks of deflexions are therefore pin-pointed on the deflexion diagram.
(2) Visual observation on site of the existence or deficiency at these points of superficial drainage systems will, in some cases, confirm a fault in design or in maintenance.
(3) Should visual observation fail to do so, test pits are dug in order to establish the presence of water or excessive moisture content within the pavement (including subgrade). This should generally provide the answer as to whether or not drainage is concerned.

Measures taken to overcome the problems vary according to the diagnosis made. They may include

(a) the construction or rehabilitation of surface drainage systems (culverts, side drains and so on) or of deep drainage
(b) the removal of unsuitable or wet pavement layers, followed by their reconstruction
(c) the sealing or waterproofing of a permeable running carpet. (Permeability of the pavement surface is previously tested with conventional means.)
(d) or indeed, earthworks operations aimed at upgrading the vertical profile of the road, above some critical level.

Mr Lionjanga refers to the problem of distribution of funds. The limited funds generally available for maintenance imply that they cannot be used to maintain the greater proportion of the less worse-off sections of the road and that they should preferably be used (although they often barely suffice) to the proper repair of the smaller proportion represented by the worst sections.

The type of relationships obtained, under the impulse of the World Bank, between roughness and vehicle operating costs or indeed between surface deterioration rate and pavement standards, is certainly the best tool available for answering this problem and for assisting in defining the type and budget allocation of maintenance.

In the Ivory Coast, the present practice suggests that maintenance work is concentrated on the worst sections of the network whenever the area concerned is less than 20% of the total area, per lane, of the road. Should deterioration concern a greater proportion, general strengthening or overlay works are envisaged over the whole length of the road. It is hoped that this purely empirical practice will, in the future, be refined or modified on the basis of research work aimed at establishing in the country a proper set of relationships similar to those obtained in Kenya, Brazil or India under the auspices of the International Bank for Reconstruction and Development.

Mr POWELL and Mr KIRKPATRICK *(Paper 19)*: Before trying to develop our model for pavement deterioration in Jamaica, we used the RTIM and the HDM but found that the conclusions we reached by doing so were so sensitive to the values we assumed for sub-grade CBRs (the values at equilibrium moisture content) and the resulting structural numbers, that we decided to discard the theoretical approach. We then resorted to basing our predictions for pavement performance on assessments made by Ministry staff who had been responsible for the maintenance of these pavements over many years.

Although traffic loading must have an effect on the rate of deterioration of pavements, and certainly a devastating effect on the rate of deterioration of weak pavements, it does not appear to have the dominating effects in Jamaica that it has elsewhere. In Jamaica somehow where pavements are mostly founded on very strong sub-grades even though the pavements themselves are old and not particularly strong, pavement deterioration due to weakening does not appear to increase with age at anything like the same rate as the surface roughness which can be attributed to the effects of weather. It is for this reason therefore, that we chose to represent pavement roughness with time to enable us to predict pavement performance, and to develop resealing and overlay maintenance programmes for Jamaica. Incidentally, we have also found that many more overlays are needed in Jamaica to improve riding quality than to improve pavement strength.

Mr Turnbull raises the problem of drainage. Because Jamaica is fortunate enough to have much of its highway network founded on subgrade soils with very good bearing capacities, fewer failures of the surface layers occur than is usual elsewhere. About 80% of the primary roads are built on strong sub-grades. The other 20% do present problems, particularly where drainage is poor.

The measures taken to assess the problems arising from poor drainage during inspection are confined simply to identifying and recording those lengths of pavement exhibiting any form of distress, from those just beginning to show uneveness to those which have already failed. These lengths are then included in the programme of rehabilitative work to be implemented in the period under consideration.

At the time of implementation the programme more thorough investigations into the cause of such failures are made.

The measures taken to overcome the problem are many and varied. They range from carrying out a major reconstruction of the lengths involved while at the same time raising the pavement, to deepening side drains, to sometimes installing a free draining layer at sub-base level to lead away sub-surface water. One of the added difficulties which have to be overcome in Jamaica though is that often long lengths of pavement which are poorly drained, are lined by private properties. This renders it necessary to often indulge in heavy expenditure to obtain easements over adjoining properties for disposing of the surplus water.

Professor GICHAGA *(Paper 20)*: I agree with Mr Reichert that it would be desirable to obtain comparable correlations for similar pavement structures in other tropical countries.

In reply to Mr Mitchell, regarding the use of 6.35 t (14,000 lb) axle load, I would mention that the research project was initiated in the 1960s at a time when the standard axle for deflexion measurements was 14,000 lb (6.35 t). In order to maintain continuity and consistency in the analysis of the data, the axle load has been maintained at a constant level over the years. However, the standard axle load in Kenya is 80 kN (18,000 lb).

Concerning the thickness of asphaltic concrete surfacing in four of the test sections, I would like to clarify that pavement structures with 100 mm asphaltic concrete surfacing in Kenya are the most heavily trafficked and are designed to cater for over 10 x 10^6 standard axles (80 kN). It is also essential that road projects in Kenya are economically evaluated to establish viability.

On deflexion patterns, it is true that deflexions alone are inadequate to determine the structural integrity of a pavement structure due to the high degree of scatter of that deflexion data (this is mainly due to the numerous factors that influence the distortion of the pavement structure), but when deflexion measurements are accompanied with measurements of other distress features a more comprehensive picture of the structural status of the pavement structure emerges.

On the significance of regressions, it is desirable to carry out statistical analysis to obtain regressions as this helps in detecting trends that may not be obvious. It is also important to examine regressions critically in order to assure oneself that the pattern of behaviour indicated by the regressions is realistic. One way of doing so is by instituting significance tests to obtain correlation coefficients corresponding to the regressions. Although the regression equations shown in paper 20 have weak correlation coefficients the trends indicated by these equations are considered realistic.

On the subject of drainage of the whole pavement structure, I would state that the test sections used in the research project were selected in such a way as to avoid obvious features that lead to early pavement failure and the test sections were selected from portions of roads with good cross-section elements. I agree that there is need for a free-draining layer at sub-base level and if this is not provided, subsequent overlay would not overcome surface failure. It is therefore important in overlay design to evaluate all the structural components of the existing pavement structure.

REFERENCES

1. POTTER D.W. The development of Road Roughness with Time - an investigation. Australian Road Research Board, 1982, Internal Report AIR 346-3.

21

Urban transport investment criteria for developing countries

N. D. LEA, BASc, SM, and J. A. C. ANDREWS, BSc, N. D. Lea & Associates Ltd

INTRODUCTION

1. Adequate urban transport is very important in achieving vitality for large urban areas. The quality of the urban transport contributes greatly to the quality of life in the urban area, and thereby impacts the true output of the city and of the country. The per-capita demand for such transportation increases as the standard of living increases. This creates very severe problems for developing countries where the rate of population growth in urban areas tends to be very high. The severity of the problem is increased because most developing countries already have a balance of payments problem and a shortage of financing capability. Capital investment in high-cost urban transport facilities creates a severe drain upon financial resources involving mortgaging of the future, whereas not to invest in such facilities may involve greater consumption of scarce and expensive petroleum fuels which also deplete the available foreign currency. This creates a very great dilemma.

2. Investment in transit improvements is sometimes promoted as a simple answer to this dilemma, but, on close examination in each particular set of circumstances, the simple solution is found to be clouded by many uncertainties. The decision between the various possible courses of action and between the various investment options, including improved roadway and transit possibilities, becomes exceedingly difficult. The objective of this paper is to suggest some guidelines for evaluating and comparing urban transport investments for developing countries, including specifically roadway and transit alternatives.

GENERAL PROBLEM APPRECIATION

3. Before coping with a particular set of investment options, it is important to gain an adequate appreciation of the overall problem of the particular city and country in regard to the important determinants and policies for both the short term and long term options. Government regulations, licensing and subsidies may have an important influence on urban transport operations in either encouraging or discouraging the part played by the private sector. Generally the private sector adjusts its prices to establish some sort of market balance and it is important to have some understanding of this balance.

4. There is one fundamental urban transport condition, however, which is not optimized by free market forces. This is congestion. In the case specifically of motor vehicle congestion on roadways, the marginal total societal cost of one more vehicle is much greater than the cost perceived by that one user because he increases the congestion and imposes costs upon all other users. The price charged the marginal user under congested conditions is less than the economic efficient price. Various measures are currently being devised to increase the price with a view to increasing the system efficiency during congestion. These measures may be called auto disincentives and include increased parking charges, enforcement of parking regulations, and special licensing such as the Singapore scheme. Some estimates have been made for the U.S.A. of the magnitude of the most efficient highway user charges (ref.1). The results indicate that current rural highway user charges are already about 50 percent above the most efficient level, whereas the urban auto user charges would need to be increased by over 300 percent to reach the most efficient level. Overall the total highway user charges in the U.S.A. are estimated to be only about half the optimal amount. Many developing countries face just as serious a disparity between the actual and the optimal level of the road user charges, and this seriously exaggerates the problem of providing adequate transport infrastructure.

5. The pricing of transit services also needs to face the issue of congestion pricing. When facilities and equipment are provided only for use in a short peak period, the marginal cost of serving the marginal peak period passenger is very high. Some developed countries are starting to use higher rush hour prices to reflect these higher peak period costs and to achieve a more efficient pricing system. The Washington Metro in U.S.A. is one example. Higher peak period pricing has been common for a long time with privately operated developing country transit systems, and the departure from this sound practice by public systems has probably come about by following developed country practice without adequate questioning. The cost of providing service in remote districts is also very much higher than the average, and where adequate higher revenues cannot be obtained because of uniform fares, the service tends to be withdrawn. Similarly, most privately operated

Highway investment in developing countries. Thomas Telford Ltd, London, 1983, 169–175

systems do not provide service on routes where the patronage is so low as to be non-compensatory under the allowable pricing structure.

6. Another appreciation which is important when approaching an assessment of urban transport investment is the strong propensity of consumers to buy personal transportation. This is shown by the direct relationship between per-capita income and motor vehicle ownership, as illustrated on Figure 1. Several studies have indicated that the elasticity of auto per-capita ownership to per-capita income is about 1.0 (ref.2). This probably assumes that the motor vehicle pricing policy remains constant. There is no doubt also an elasticity of auto ownership to pricing policy such that if a country, through auto purchase taxes or through regulation, makes it more costly to own a motor vehicle, then the ownership rate declines. This is suggested on Figure 1 by the deviation from the trend of China, Russia and Korea, where, because it is much more difficult or costly to own a motor vehicle, the motor vehicle ownership rate is much lower. Possibly the policies in western countries make it too easy to own an auto, whereas those in Russia and China may make it too difficult. It is important for a developing country to strike a reasonable balance on this issue which can affect greatly the demand for urban transportation.

USER PAY PRINCIPLE

7. A good starting point for any transportation investment assessment is the full cost recovery, or user pay, principle. This is not necessarily the only or the final criterion, but, for any project or comparison between alternatives, it is helpful to have an appreciation of the value which the users place upon the project, which is demonstrated by their willingness to pay the full cost. The full cost to be considered is the full marginal cost including the cost of externalities. This full marginal cost must include the cost of congestion and the cost of the investment which is under consideration.

8. If the user is charged less than the full cost, then he is being subsidized, and the amount of the subsidy requires justification. One of the shortcomings of transportation programs in many developed countries is that there have evolved over time practices of paying large direct and indirect subsidies to various forms of transportation and the reasons for the subsidies may have been lost sight of, if indeed they were ever explicit. For example, in North America rapidly increasing subsidies were paid to transit operators from all levels of government during the 1970's. Some of these were allocated on as vague an objective as to increase transit riding as though this in itself were a public good. Now, as the federal transit subsidy program in the U.S.A. is being cut back, better justification is being required to differentiate between the various transit subsidy options. This problem has become more acute in the western world, where in some instances, with the increasing subsidy, there has come less attention to the user's desires and more attention to the funding agencies' financial problems. In a number of respects, transit systems in developing countries are more responsive to the user's desires. For example, some users are willing to pay more for a faster, higher quality transit ride, whereas others prefer lower cost and lower quality; so in many developing countries several levels of service are offered with several levels of price.

9. One of the major differences between roadway and transit investments is that the user's response for the transit investment is immediately apparent through the farebox; whereas for roadway investments, there is no ready means for the user to respond unless it is a toll facility. It is possible, however, to predict with reasonable accuracy how users would respond to a toll charge, and this analysis procedure may be used to estimate what the road user would be willing to pay. There are many instances, particularly with urban roadway systems, where the users of the roadways would be willing to pay the full cost of a greatly improved roadway system, but because of institutional and financial constraints, he is not being given the option of purchasing such improved urban transport. This is somewhat analagous to the situation in transit in many European and North American cities, where only one standard of service is being offered and the user or potential user who would be willing to pay the full cost of a higher standard of service is not allowed to do so because the private sector is not allowed to provide such a service.

10. The user pay principle is, therefore, perceived as a basically sound one which should not be departed from without good reason. There are, however, a number of good reasons for departing from this principle and these are considered in the balance of this paper.

VALUE CAPTURE

11. 'Value Capture' has emerged in North America as a term describing the process whereby the users of property abutting upon a new transportation facility are induced to pay a portion of the cost. In principle, therefore, it is consistant with the user-pay principle with the concept of 'users' expanded to embrace some who benefit indirectly. This indirect benefit to land abutting rapid rail transit stations, for example, is reflected in higher land prices and thus transit agencies attempt to capture the incremental land value to cover a share of the cost of the transit facility. One of the most extensive and effective uses of this practice is in Hong Kong where a substantial part of the cost of the Metro expansion is being financed through development of properties which are adjacent to stations and which have been made available to the Mass Transit Railway Corporation.

12. Although a valid fund raising expedient, there is some question concerning the validity of using the value capture amount as a benefit

to deduct from the total cost of the facility before calculation of the appropriate user cost.

Influence Upon Urban Form

13. One of the reasons given for heavy investments in exclusive right-of-way transit facilities in U.S.A. and in Canada is to encourage a more compact urban form and, in particular, to reinforce the Central Business District. In some transit studies, in the City of Toronto for example (ref.3), various transit options have been evaluated in terms of their comparative impact upon the urban form. It seems well established that rail transit does encourage a more compact form with concentrations at the transit stations and in the city centre where the rail lines converge. The principal reason for the Downtown People Mover programme in the U.S.A. was to encourage the rejuvenation of city centres. It cannot be assumed, however, that rail transit will always reinforce the centre. The streetcar lines which were built extensively during the 1920's in North America were usually associated with city dispersal at lower density. The rail lines built from Manhattan across the rivers led to a rapid exodus of population from Manhattan.

14. Much judgment is required to assess the urban form impact of transit systems and also to decide which urban form is more desirable. As yet, there has not been enough experience with urban busways to provide a sound basis for judgment concerning their urban form impact in comparison with that of rail transit. Developing country cities often have already a very compact core because of dependence upon walking and further encouragement of centralization may be undesirable.

INCOME DISTRIBUTION

15. One of the reasons given for transit subsidies is to help the poor - to service the auto disadvantaged - to improve the equity of income distribution. In developing transit systems and setting fare structures, the impact upon the urban poor is important and somewhat complicated. Some transit systems, particularly C.B.D. oriented rail transit systems, have been found to serve largely the more affluent; thus, a subsidy to such systems is a subsidy to the rich. It is clear from Exhibit 2 that in developing countries, the transit price varies greatly in relationship to the income of the poor. It is a serious problem if the transit fare is so high that it requires 20 percent of a household's income for the breadwinner to travel to and from work. Subsidizing the transit system in total, however, is a very inefficient way of getting money to those who have the need. Less than 10 percent of the subsidy might be used in the end by those who need it. Financial support directly to those who have the special need may be more efficient and effective and may also encourage transit efficiency by forcing it to compete in the marketplace.

SPECIAL RIGHT-OF-WAY CONSIDERATIONS

16. In an economic evaluation, right-of-way cost may be considered as a transfer payment, or a lower time preference factor may be applied, or right-of-way may be considered as a resource in limited supply which will increase in real value, achieving a salvage value cancelling the initial cost. For these reasons, right-of-way costs deserve special treatment.

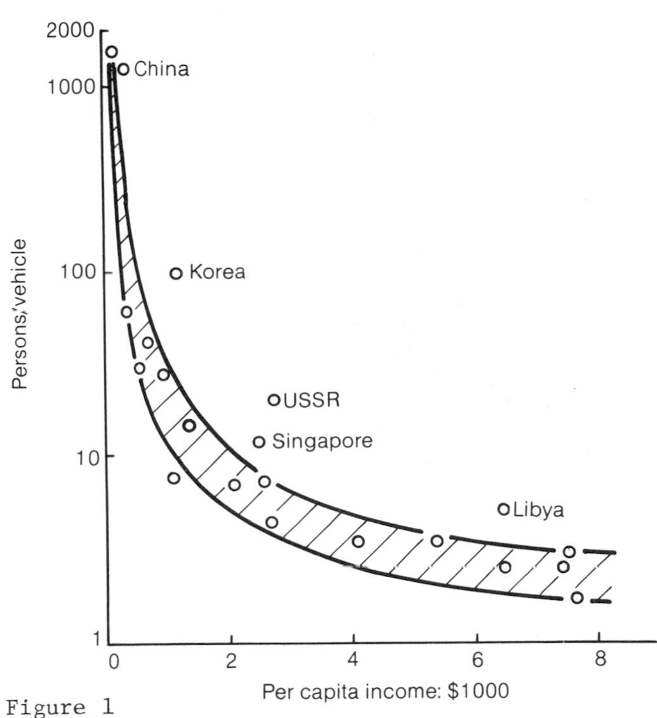

Figure 1

17. Furthermore, cities with the most serious urban transportation problems are those without good transportation corridors or rights-of-way. If wide transportation reserves have been provided during the early stages of a city's development, it becomes possible to provide a good urban transport system at much lower cost than if the rights-of-way are narrow and inadequate.

18. Therefore, serious attention should be given to the provision of generous right-of-way reserves or transportation corridors, before subdivision development takes place. Also, when a project involves the obtaining of right-of-way which improves the corridor network, the cost of this right-of-way should be given special treatment in economic evaluations.

CONGESTION RELIEF

19. If it is not possible to relieve congestion through an efficient user charge and it is costly to increase roadway capacity to relieve the street congestion, then some congestion relief benefits may be achieved through the provision of improved transit facilities using an exclusive right-of-way and either rail or bus technology.

20. The literature contains many fallacious ways of quantifying the benefits of congestion relief. An appropriate approach is to use as the neutral path a programme of maximum on-street traffic engineering improvements. Any rail transit, busway, or freeway option will then be compared with the neutral path in terms of net present value with any reduction in operating, maintenance or time costs because of congestion relief entering as a benefit.

EXTERNALITIES

21. Externalities such as air pollution, noise and environmental degredation are costs which need to be taken into account when evaluating any urban transportation proposal. The benefit of improvement in such impacts may sometimes be considered as a public good to be enjoyed and properly paid for by the population at large. It may be very difficult, however, to estimate the relative merits of schemes with differing environmental impact. For example, the scheme with lower noise levels may have higher air pollution. One widely used procedure is a simplistic set of absolute, and somewhat arbitrary, standards which classify all proposals into one of two categories: 'acceptable' or 'not acceptable.' A better method of differentiating between varying amounts of externalities is desirable.

ENERGY

22. With the increase in international petroleum prices, energy efficiency is often used as an argument to support investment in transit systems. Under some circumstances, it is a good argument, but it must be used with care.

23. Exhibit 3 gives the result of one analysis of the energy intensity of several person transport modes including the indirect energy required for vehicle manufacture and for construction of the roadway. The comparison on the basis of seat-kilometre is only a rough general guide. For any particular application, the comparison should be based upon incremental consumption per passenger kilometre. The efficiencies per seat-km are sufficiently close that for a particular application the preference could be switched by variations in vehicle occupancy.

EXHIBIT 2

TRANSIT SYSTEM COMPARISONS

City	Population in Millions	Income/ Capita (US $) (1970)	Percent Motorized Households	Motorized Person Trips/ Capita/ Day	Transit Trips/ Capita/ Day	% Income Spent on Transit by 20th Percentile	Transit Seat-Kms Per Capita Per Day (e)
Metro Kuala-Lumpur (1973)[a]	0.91	660	53.0	1.76	0.71	3.0	2.7
Singapore (1973/74)[a]	2.30	1100	41.0	1.68	0.88	4.5	4.7
Jakarta (1974)[a]	5.73	325	25.0	1.10	0.72	20.0	0.7
Bangkok (1977)[a]	4.60	525	25.0	1.22	0.81	3.0	3.2
Metro Manila (1971)[a]	5.33		18.0	1.27	1.02	2.9	1.6
Hong Kong (1974)[a]	4.10		14.0	1.30	1.10	2.0 to 5.7	5.9
(1973/74)[b]		850	20.0	1.45	1.27	5.0	11.7[c]
(1980)[b]					1.39	10.0[d]	
Greater Bombay (1975)[a]	6.90	390	8.5	1.11	1.0	1.7	2.3
Madras (1970)[a]	4.40	180	4.5	0.44	0.4	.5.9	1.2
Calcutta (1965)[a]	9.70	270	8.5	0.98	0.92	7.9	1.1
Karachi (1971)[a]	3.44	360	6.3	0.62	0.36		1.6
San Jose (1973)[a]	0.46	430	25.0	0.59	1.22	2.2 to 2.7	1.6
Porto Alegre (1979)[b]	2.38			0.82	0.55		
Washington DC (1971)[a]	0.76	5399	80.0	3.39	0.09	3.9	1.4
Greater London (1971/72)[2]	7.1	2550	50.0	1.83	0.59		

(a) Source: Lim's 1979 World Bank paper (ref.6).
(b) Source: World Bank Transit Pre-Feasibility Guidelines, by N. D. Lea & Associates Ltd, 1982 (ref.7).
(c) Breakdown - 5.7 by regular bus, 5.5 by PLB, 0.5 by train, excluding 4.5 by taxi and other variable route.
(d) Using only Metro in 1981. PLB's are more costly and regular bus less costly.
(e) Taxis excluded.

For example, on low density feeder routes, bus occupancies are low, and a taxi with a high load factor might be more fuel efficient. For rail cars, the crush load may be three to four times the number of seats whereas the off-peak load may be quite small. Therefore, each particular situation must be explored on its own.

24. In some situations, rail transit and trolley bus installations have the advantage that they make use of electrical power which may be generated from some source other than petroleum.

INSTITUTIONAL ARRANGEMENTS

25. In North America, transit system productivity has remained relatively constant for the past 30 years, but tends to decline as system size increases. Private operators tend to be a little less efficient in producing vehicle-hours of service, but more efficient in delivering the type of service which the users are willing to pay for (ref.4).

26. In most developing countries, however, the productivity and efficiency of publicly-owned transit systems is dramatically poorer than that of private systems, particularly when compared with owner-operated systems and systems paying the driver an incentive per passenger. Although a comprehensive and reliable statistical analysis of the differences between the public and private systems is not available, there are enough documented specific examples to call for extreme caution in estimating operating costs for publicly-owned transit systems in developing countries. For example:

- In Jakarta, as a result of government regulation and eventual nationalization of the bus system in 1980, the number of buses in operation has decreased from 4,000 under the private system in 1969 to 2,700 under the public system in 1980. The public operation requires a subsidy equal to the farebox revenue, and is not able to provide for vehicle replacement (ref.5).
- In Bogota, part of the regular bus service is provided by a publicly-owned operator and part of the service by privately-owned operations. With essentially the same fares, the private operation receives a subsidy of 80 percent of the fares and makes a profit, whereas the publicly-owned operation requires a subsidy of 225 percent of the fare. A competing and higher quality mini-bus operation (called buseta) operates privately and profitably at three times the fare of the publicly-owned regular bus operation, and with costs per passenger for the higher quality service slightly less than the public operator's costs for the regular service (ref.5).
- In Port of Spain, Trinidad, privately-owned shared-ride taxi operations compete with the publicly-owned bus operation. In one of the main urban corridors, the taxi operators have been taking a steadily increasing share of the market until they now have about 75 percent. The taxi fares are about ten times the bus fares and the taxis cover their full cost from fares, whereas the bus farebox covers only about 10 percent of the bus cost per passenger-km.
- A study by Feibel and Walters of private versus public bus costs in Calcutta, Bangkok and Istanbul, concludes that where such operators compete at the same fare, the private operators provide equal or better service without subsidy, while the public operators require massive subsidies (ref.11).

COMPARISON OF BUSWAY AND RAIL TRANSIT

27. One issue which may bring into focus all of the above considerations is the assessment of exclusive right-of-way transit and, in particular, a comparison between bus and rail options.

28. Exhibit 4 suggests passenger flow thresholds at which exclusive right-of-way transit may be expected to become viable for several right-of-way conditions in developing countries. Exhibit 5 lists most of the exclusive right-of-

EXHIBIT 3

ENERGY INTENSITY

Mode	Number of Seats	Energy Intensity (MJ per seat-km)[1] Direct	Total
Automobile on Arterial	5	0.62	0.93
Van on Arterial	12	0.63	0.75
Diesel Urban Bus	40	0.50	0.54
Trolley Coach	40	0.23	0.26
Subway[3]	76	0.13	0.42
Streetcar (PCC)	52	0.22	0.43
Commuter Rail	160	0.57	0.66

1) Megajoules per vehicle kilometre.
2) Assumes 1985 CAFE standard 27.5 miles/US gallon = 3.00 MJ/veh-km (i.e.0.87 MJ/seat-km total).
3) Excluding cooling which in hot climate may equal direct transport entity.
4) Source: Metropolitan Toronto Area Transportation Energy Study, Background Report II, December 1980 (ref.8)

EXHIBIT 4

PRELIMINARY THRESHOLD PASSENGER FLOWS
ABOVE WHICH EXCLUSIVE WAY TRANSIT SYSTEMS MAY BE FINANCIALLY VIABLE

Transit Systems and Right-of-Way Conditions	Minimum Threshold Flow: Daily Volume In Passenger Kms/ Line Km/Day (a)	Minimum Threshold Flow: Peak Volume In Passengers/Hour (one-direction) (b)
Low cost because right-of-way is available for pre-metro (LRT) or for busway. Little if any grade separation is needed.	50,000	4,000
Considerable grade separation and other right-of-way expense is required for pre-metro (LRT) or busway or full metro extension.	150,000	12,000
Much of the length must be in high-cost tunnel, suggesting full metre, i.e. Rail Rapid.	300,000	24,000

Source: Reference 7.

EXHIBIT 5

SYSTEM AVERAGE RAIL RAPID AND LIGHT RAIL TRANSIT USE IN THIRD WORLD COUNTRIES [c]

City	Kms of Transit Line [d]: 1975 Built	Kms of Transit Line [d]: 1980 Built	Kms of Transit Line [d]: 1980 Under Construction	Annual Passenger Kms (Millions)	Annual Passenger Kms/Line Km (Millions)	Average Weekday Kms/Line Km [b]
Moscow	164.5	196.3	9.8	18,334	111.5	372,000
Tokyo	163.2	186.7	26.6	12,306	75.4	251,000
Mexico City	42.0	44.0	40.0	4,256	101.3	338,000
Beijing	23.0	43.0	27.0	1,224	53.2	177,000
Buenos Aires	31.6	34.7	10.2	1,210	38.3	128,000
Sao Paulo	17.2	21.4	21.1	1,176	68.3	228,000
Seoul	10.3	10.3	29.8	720	70.0	233,000
Kharkov	9.8	9.8	7.6	390	36.8	123,000
Hong Kong	Nil	15.6	10.7	245	75.0	249,000
U.S.S.R.	276.0	347.5	94.4	25,033	90.4	301,000
Asia	322.6	418.4	151.3	19,137	59.3	198,000
Latin America	105.9	144.3	97.0	6,812	48.0	160,000
North America [a]	885.4	1000.1	159.5	18,356	21.0	70,000
Europe [a]	1266.7	1504.7	320.3	22,783	18.0	60,000

(a) Without Light Rail.
(b) Average Weekday taken as 1/300 of annual.
(c) Source: Pushkarev 1980 Table A2 (except Hong Kong) (ref.12).
(d) 'Line' is commonly measured C to C of stations and may contain two or more tracks.

way rail transit systems presently operating in the developing world.

29. Observe that passenger flows on rail rapid transit lines in North America and Europe are, on average, only a fraction of those in the rest of the world. Additions to the system in U.S.A. have tended to have only about 10 percent of the minimum threshold flows suggested in Exhibit 4 (ref.12). Even in the developing world, exclusive right-of-way transit systems are usually put in place before they are financially viable. There is hardly any rail rapid system in existence which recovers its full cost through the farebox. Thus, most exclusive right-of-way transit systems justify their existence by reasons other than user-pay, even though the reasons may not be made explicit.

30. One of the interesting circumstances in which it has become necessary to make these reasons explicit is in comparisons between busway and rail rapid proposals. Numerous such comparisons have been done for North American conditions and they usually show little difference in total cost, with the busway scheme having lower capital cost and the rail rapid system having lower operating cost. In North America, the decisions have been swung in favour of rail rapid using non-financial arguments. For developing countries, similar decisions are being faced, with the non-financial arguments carrying less weight for a variety of reasons. In-depth comparisons between bus and rail have been in progress for many months in Singapore, for example, and also in Bogota. Serious questions have been raised concerning the usefulness of the Caracas Metro, which is now under consideration (ref.9). Even in North America, it is being suggested that recent rail rapid advocacy rests upon a set of illusions (ref.10). Certainly there has been a significant change in U.S.A. federal government policy concerning urban rail investment.

CONCLUSION

31. Developing countries are already, because of increasing vehicle ownership, facing severe traffic congestion in their urban areas. Means of improving their urban transportation may include provision of more highway facilities, restrictions on motor vehicles and provision of transit facilities. Generally, all means need to be provided to provide an improvement. This paper deals mainly with investment criteria for transit improvement.

32. Care must be taken not to transfer to developing countries, without careful examination of applicability, the procedures and systems used in the developed countries. One way to evaluate such urban transport systems is to rigorously apply the user-pay principle. Departure from the user-pay principle should only be made for special reasons, such as right-of-way costs.

ACKNOWLEDGMENT

33. Part of the information presented in this paper has come to the authors' attention in the process of developing transit pre-feasibility guidelines for the World Bank, but, of course, the World Bank is in no way implicated in the opinions expressed.

REFERENCES

1. LEE D. B. Efficient Highway User Charges, Appendix D to the Final Report of the Federal Highway Cost Allocation Study. United States Department of Transportation, FHWA, January 1982.
2. LEA N. D. & ASSOCIATES LTD. Impacts of Petroleum Supply and Pricing upon the Demand for Transportation Services. Report to the Ministry of Transportation and Communications, Government of Ontario, 1981.
3. Metropolitan Toronto Transportation Plan Review. Choices for the Future, Report No. 64. January, 1975.
4. BARNUM DAROLD T. From Private to Public: Labour Relations in Urban Mass Transit. College of Business Administration, Texas Tech University, 1977.
5. URRUTIA MIGUEL. Buses Y Busetas. Fedesarrollo, 1981.
6. WEI-YUE LIM. A Comparative Study of Public Transport in Principal Asian Cities. A draft report of the Urban Projects Department of the World Bank, March, 1979.
7. LEA N. D. & ASSOCIATES LTD. in collaboration with World Bank staff. Transit Pre-feasibility Guidelines. Expected to be published in 1982 by the World Bank, Urban Projects Department.
8. Metropolitan Toronto Area Transportation Energy Study, Background Report II. Energy Consumption and Intensity of Transportation Modes. December 1980.
9. SPERLING DANIEL. Caracas Metro: A Luxury? Transportation Research Board Record 797.
10. HAMER ANDREW MARCHALL. The Selling of Rail Rapid Transit. A critical look at transportation planning. Lexington Books.
11. FEIBEL CHARLES and WALTERS A. A. Ownership and Efficiency in Urban Buses. World Bank staff working paper No. 371. February 1980.
12. PUSHKAREV BORIS and ZUPAN JEFFREY. Urban Rail in America. An exploration of criteria for fixed guideway transit for UMTA, U.S. Department of Transportation. 1980.

Urban roads and traffic in Cairo

DR A. K. F. LASHINE, Vice Chairman, Ministry of Transport, Arab Republic of Egypt, and DR J. C. R. LATCHFORD, Jamieson Mackay and Partners

Cairo, with a population of some 9 million people, has an annual rate of growth of 3% and a car ownership level of 22 vehicles per 1,000 people. The annual increase in GDP is 6% and there is a 15% increase in traffic per annum. Traffic congestion is increasing particularly in the CBD and parking and road safety problems are also increasing. Long and short term plans have been produced to accommodate the future travel demands and these include construction of an underground metro system, the building of new satellite cities and the improvement of all forms of public transport and also of pedestrian facilities. Highways and traffic management schemes are also being implemented. These include new arterial routes, junction flyovers, highway maintenance and transport management.

A plan for highway improvements and traffic management over the next five years is described in this paper which requires an investment in highways of approximately LE 30 million per annum. The highway strategy concentrates on making the most efficient use of existing highway facilities, eliminating major bottlenecks and improving the maintenance and paving of the distributor road system. Improvements to traffic management are a fundamental part of the short term plan and include a new circulatory system for the CBD, police training and equipment, accident recording and enforcement measures.

1. INTRODUCTION

1.1 The City of Cairo, the capital of Egypt, is the largest city in Africa and the Middle East with a population of some 8.85 million which represents about 22% of Egypt's total population. With the current annual growth rate of 3% it is estimated that the population of Greater Cairo could reach 15 million by the year 2000. To counteract the inevitable urban growth the Egyptian Government have embarked on an ambitious programme and are constructing new cities in the desert at Sadat City, Tenth of Ramadan and King Khalid City.

1.2 Greater Cairo is administered by; Cairo Governorate to the east of the River Nile with a population of some 5.5 million all in Greater Cairo, Giza Governorate to the west of the River Nile with an estimated population of some 2.5 million, of which 80% is in Greater Cairo, and Kalubeia a rural Governorate to the north of Cairo which contains the remainder of the urban population.

1.3 The extremely rapid growth in population of urban Cairo over recent years has resulted in very high population densities, for example in excess of 100,000 persons/square kilometre in the older districts of Rod El Farag and Bab El Sharia. The newer residential areas of Nasr City and Heliopolis have been developed on modern lines and the population density is of the order of 9,000 persons/square kilometre.

1.4 Commercial activities are generally located along the major axes. The main commercial centres are the Central Business District (CBD) and to the east of this area in Mouski for all traditional trades. Banking and services activities are concentrated in the Central Business District. Some 210,000 job opportunities exist in the CBD and the adjacent areas of the Ministries, Tahrir Square and Ramses Street. The major manufacturing areas are located in Shubra El Kheima in the north of the city and at Helwan some 30 km south of Cairo. Helwan is the centre of heavy industries; steel works; car industries; chemical industries, etc.

1.5 In 1979 it was estimated that the average household income in Egypt (excluding agricultural workers) was LE 473 per annum, although in Greater Cairo the average was estimated at nearer LE 600 per annum. The GNP in Egypt for 1979 was LE 7,651 million which represents a figure of LE 188 per capita (LE 1 = £0.70).

1.6 Over the last 10 years a series of transport studies have been undertaken for Cairo. The most comprehensive of which was produced by SOFRETU (ref. 1) and comprised a plan and feasibility study for a new metro line for the whole of Cairo with the first phase being under the CBD from Ramses Square to Bab-el-Luk and an overall highway plan for the whole city and a traffic management scheme for the CBD. Construction of the first phase of the new metro system began in 1980 and a number of the highway schemes on the SOFRETU plan have now been constructed. The Entrances to Cairo Study (ref. 2) recommended inter-urban routes connecting to the urban highway plan. The Government has over the last decade invested heavily in a full range of transport facilities in Cairo and has engaged the

Highway investment in developing countries. Thomas Telford Ltd, London, 1983, 177–188

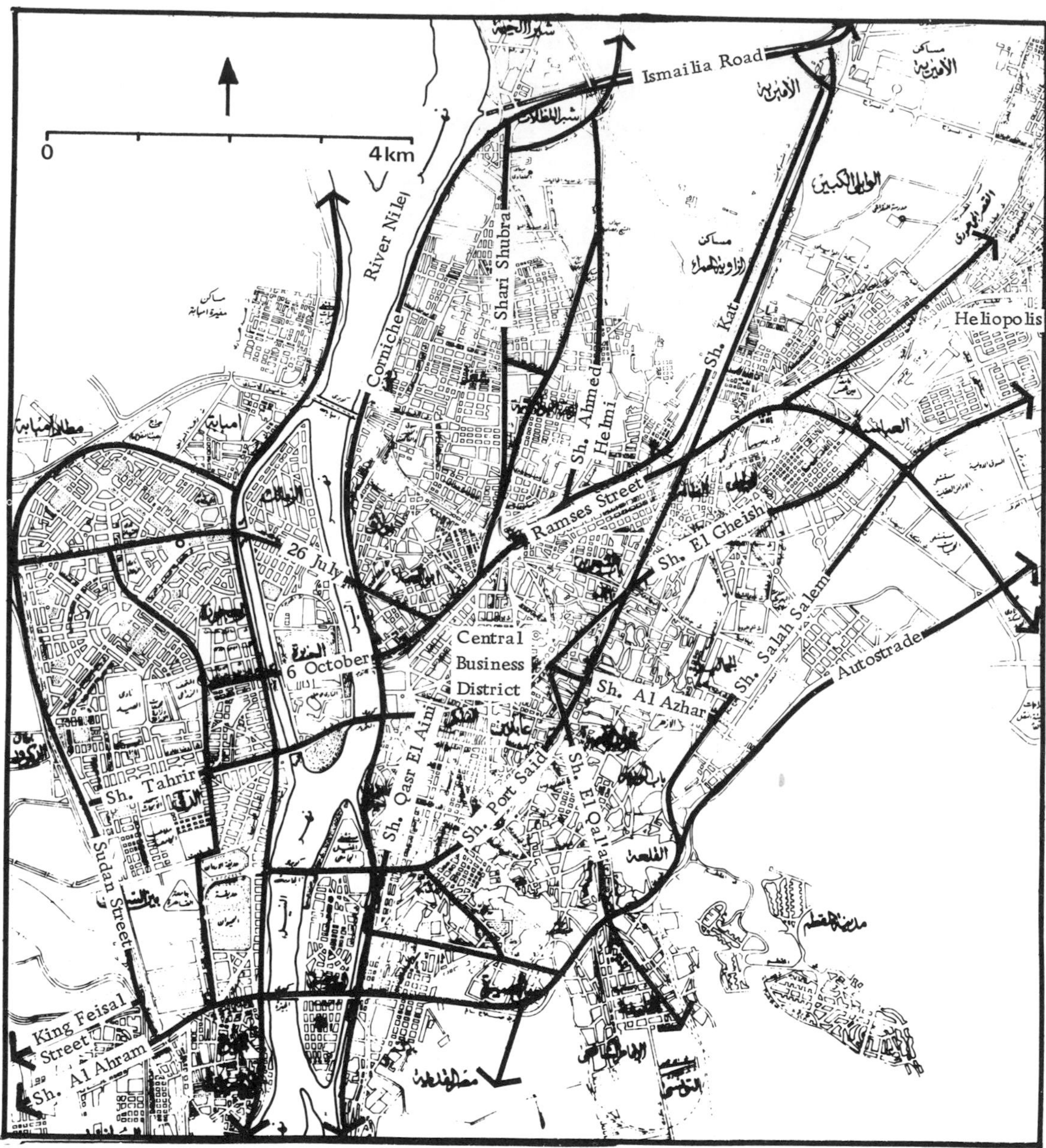

Fig. 1. Greater Cairo existing highway network

Cairo University, in conjunction with the Massachusets Institute of Technology, to collect statistical information, (ref. 3, 4) and undertake analyses (ref. 5) to enable them to monitor the situation (ref. 6).

1.7 In 1979, the Transport Planning Authority of the Ministry of Transport, in conjunction with the Governorates and other transport agencies, initiated a Cairo transport management programme to be implemented by the mid-1980's and parts of which were to be suitable for financing from the World Bank. The Transport Planning Authority was assisted in its technical work by Jamieson Mackay and Partners under assignment by the Overseas Development Administration U.K. The recommendations and conclusions of the preparation for this project (ref. 7) are described in this paper.

2. EXISTING TRAFFIC CONDITIONS

2.1 The average person trips recorded in Greater Cairo in 1979 was one trip per person per day of which 85% are made by motorized vehicles and 15% by foot. A major factor in the transport usage is the income levels of the population of Cairo and the inability of the large majority of the travelling public to pay economic fares or to buy private cars or motorbikes. The current vehicle ownership of 22 vehicles/1,000 persons is low, but with a forecast annual increase in GPD of 6% and a situation where the demand for cars exceeds supply, it is expected that the total vehicle fleet will continue to increase by approximately 15% per annum for the foreseeable future. Local manufacture and assembly can produce 24,000 units/annum which could rise to 40,000 units per annum during the next 5 - 7 years.

2.2 The growth in numbers of commercial vehicles in Greater Cairo has also been rapid in recent years. In 1978 some 28,000 commercial vehicles were registered, representing approximately 13% of all vehicle registrations. The annual growth rate has been approximately 20% since 1978.

2.3 Although the vehicle ownership is comparatively low, traffic congestion is severe in many parts of Greater Cairo, and in particular, in the Central Business District (CBD). This congestion is mainly due to the concentration of the traffic on the primary distributor roads. In addition, it is estimated that the operating capacity of these roads is approximately 25% less than in developed countries where maintenance standards are higher and driver and vehicle performances are better.

2.4 The CBD covers an area of approximately 88 hectares and generates over 50,000 person trips per hour in the peak periods. Although the central area of the city has a substantial network of wide streets, the conflict between the various road users combined with this very high trip generation rate leads to considerable traffic congestion and accidents. Within the CDB the congestion and disorder caused by parking are of particular concern and parking represents 10-15% of the total area.

2.5 Accident rates in Cairo are currently high at 79 fatalaties and 601 injuries per 10,000 vehicles. These can be attributed to poor vehicle performance, driver capabilities and awareness, poor pedestrian behaviour, failure of drivers and pedestrians to obey traffic controls and poor road surface maintenance.

2.6 The growth of traffic and the poor road safety record is particularly evident on the major routes and in the CBD where there are large pedestrian movements. The highway network of the City has been improved over the recent years but the funds needed to totally redevelop the highways would be very large and the existing level of funding cannot keep pace with the level of traffic demands.

3. EXISTING HIGHWAY NETWORK

3.1 The existing highway network for Greater Cairo is illustrated in Figure 1, and comprises peripheral routes, radial routes, north-south and east-west routes. The eastern peripheral route, the Autostrade, a dual three lane carriageway, runs from Cairo International Airport in the north-east of Cairo, through Heliopolis and then in a south-westerly direction through old Cairo to connect to Giza Bridge, at the southern end of Manial Island, and the Pyramids Road. Port Said Street is a peripheral route to the CBD from the Ismailia Canal Road in the north to the southern end of Qasr El Aini in the south.

3.2 One of the main north-south routes is the Corniche El Nile which runs along the east bank of the River Nile and provides the long distrance routes from north Cairo through to Maadi and Helwan in the south. The route is a mixture of dual carriageway, single carriageway and one-way systems in the Garden City area. In the east-west direction to the east of the Nile, the northerly east-west route is along the south bank of the Ismailia Canal and provides a main collector and distributor for the national routes to Tanta, Alexandria, Benha and Port Said.

3.3 In the northern half of the city the only complete east-west major route is the continuation of the 6th October Bridge along Galaa Street to Ramses Street, Abassia Square and Oruba Square. The route is a dual carriageway in total and between Ghamra Bridge and Abbassia Square a one-way system operates which has a road width of up to 6 lanes in each direction. In addition to acting as an east-west facility, the eastern section from Ramses Square is a main commuting facility for traffic from the large residential districts in north-east Cairo and from Giza.

3.4 In Giza, the network is basically structured in the north-south, east-west axes with 3 main north-south routes, the central routes passing through the Cairo University campus at the southern end, through Giza, Mohandissin and Aguza to the north. The second main north-south route is the Shari Nile which runs along the west bank of the River Nile for some 90% of its length. On the western boundary of the present

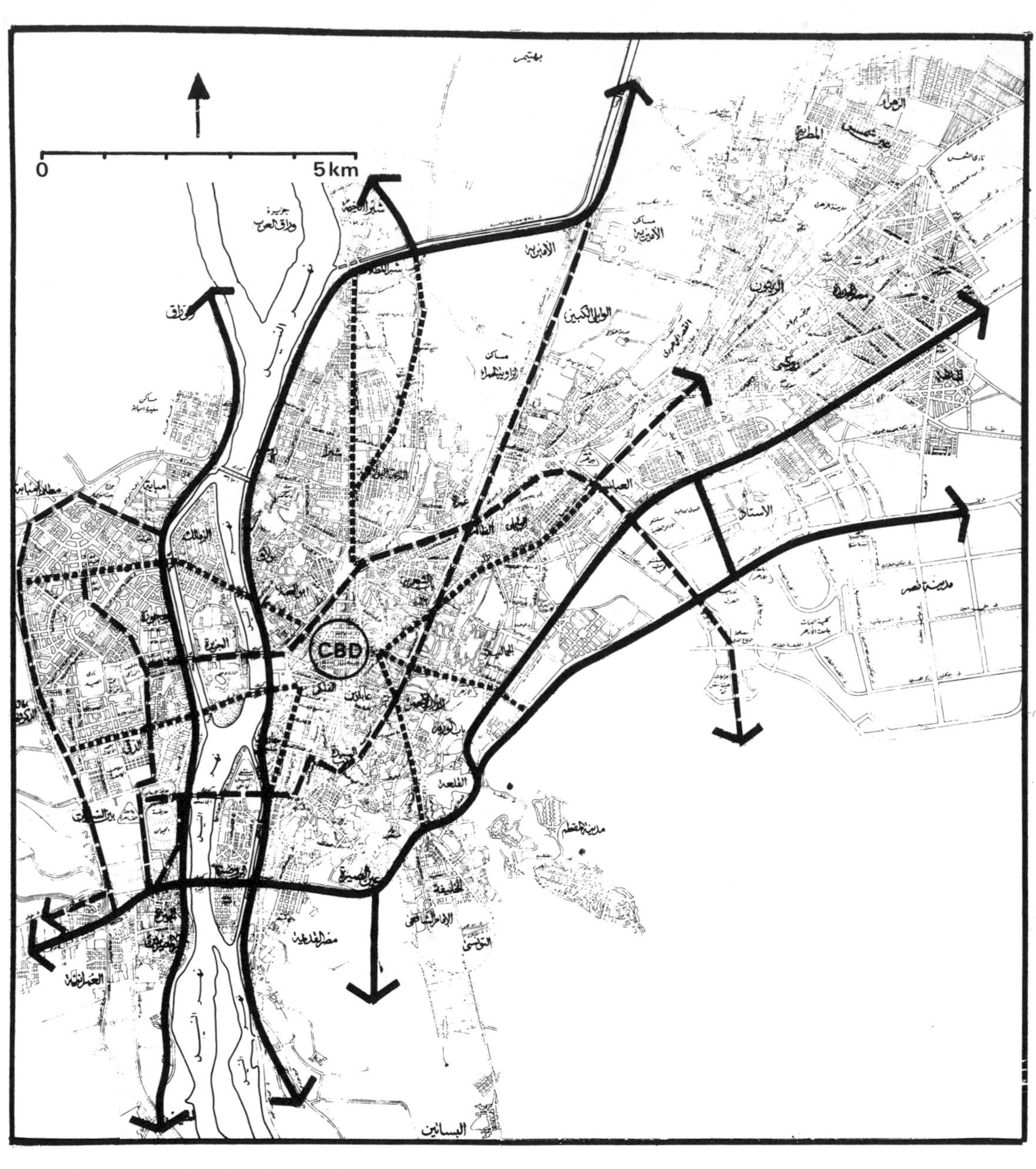

Peripheral routes for long distance traffic.

Inner peripheral routes for intra-urban traffic.

Radial routes for CBD traffic.

Fig. 2. Greater Cairo short-term highway strategy

urban area the third north-south route has been partially completed. The east-west routes basically act as collectors for the cross river facilities.

3.5 An important limiting feature of the highway network is the lack of effective distributor roads and this forces traffic, which normally could be spread over a wide network, to use only the main roads that are reasonably surfaced and this has a significant impact on the routeing of buses and commercial traffic.

3.6 The expenditure on highways in Greater Cairo is some LE 30 million per annum including roads, bridges, tunnels and flyovers which contribute the major proportion of the expenditure. Recently major highway expenditure has concentrated on the improvements to the Galaa Elevated Road and its approaches, its extension (the 6th of October axis) to Abbassia, the 26th of July elevated road across Zamalek and its new 2 Nile bridges and a number of junction flyovers. In addition, traffic managements improvements have been made at Ramsis Square and Tahrir Square and also a traffic signal control system is operating in the central area of Cairo and another contract has been signed for major intersections at Giza.

3.7 Until recently, highway planning and maintenance was the responsibility of both the Government and the Governorates but since September, 1979, as a result of legislation, the Governorates have the primary role on policy decisions on traffic management and highway planning within their boundaries as well as executive responsibility for implementing major projects. The Governorate of Cairo has additional powers on measures affecting more than one Governorate within Greater Cairo.

4. HIGHWAY STRATEGY

4.1 The immediate transport aims for Greater Cairo of the transport planning agencies are to improve the infrastructure of the capital city by increasing the level of investment, and in particular:-

i) To make more efficient use of existing transport facilities particulary roads and public transport.

ii) To enhance the use of high occupancy vehicles such as buses and to discourage the use of low occupancy vehicles such as private cars.

iii) To eliminate major bottlenecks in the road network by traffic management measures and low cost physical works.

iv) To give special attention to the transport requirements of lower income groups.

4.2 The longer term highway context for the short term strategy, Figure 2, are the SOFRETU and the Entrances to Cairo plans. The immediate highway implementation plan is aligned to:-

i) An improvement of the primary road network to expressway standard.

ii) An increase in the effective capacity of major junctions by management or physical means.

iii) An early completion of the construction of bridges and intersections partly constructed.

iv) An acceleration of the paving of the distributor road system.

v) An initiation of a programme for area improvements to provide better accessibility and create improved local environment.

vi) The development of the existing road maintenance programme.

4.3 These improvements basically relate to the CBD and the older areas of Greater Cairo as the newer development areas of Heliopolis and Nasr City are being built to modern develop ment standards and the same pressures for improvement do not exist.

5. THE SHORT-TERM HIGHWAY PLAN

Ramses Street

5.1 The main east-west route serving and bypassing the Central Business District is the Galaa Street/Ramses Street corridor. This links to the Corniche in the west and to Giza by way of the 6 October Bridge, and by way of Ramses Street, Abbassia Square, Oruba Square and beyond to the east. Major improvement has taken place by the construction of the Galaa Elevated Road, the Abbassia Square flyover and the north-south overpass at Oruba Square. Ten alternative options ranging from LE 45 million to LE 7.6 million were considered for improving the next critical section of this route from Ramses Square to Abbassia Square. Of these, two alternatives were evaluated in detail as they provided the best economic return and satisfied other transport operational and planning criteria.

5.2 The first stage of the recommended scheme is now under construction where an elevated road of 4 lanes is being built between Ghamra and Ramsis Square, overpassing Heliopolis Metro and Egyptian Railways lines at Ghamra and will be joined to the elevated road on Galaa Street at its present end in Ramsis Square. The estimated costs of this scheme range between LE 15 and LE 20 million. Proposals of extending the new elevated road to Abbassia Square are now under study.

5.3 The scheme was considered as a priority because of the imminence of the new metro construction and therefore a start was made in 1981 and the scheme should be completed by 1983. The existing westbound carriageway of Ramses Street will be utilised by buses two-way and the existing eastbound carriageway for other eastbound traffic.

5.4 Traffic management measures at Ramses square, Figure 3, will improve the operations of the existing traffic movements and create improved pedestrian access and safety. The scheme allows for the re-connection of the

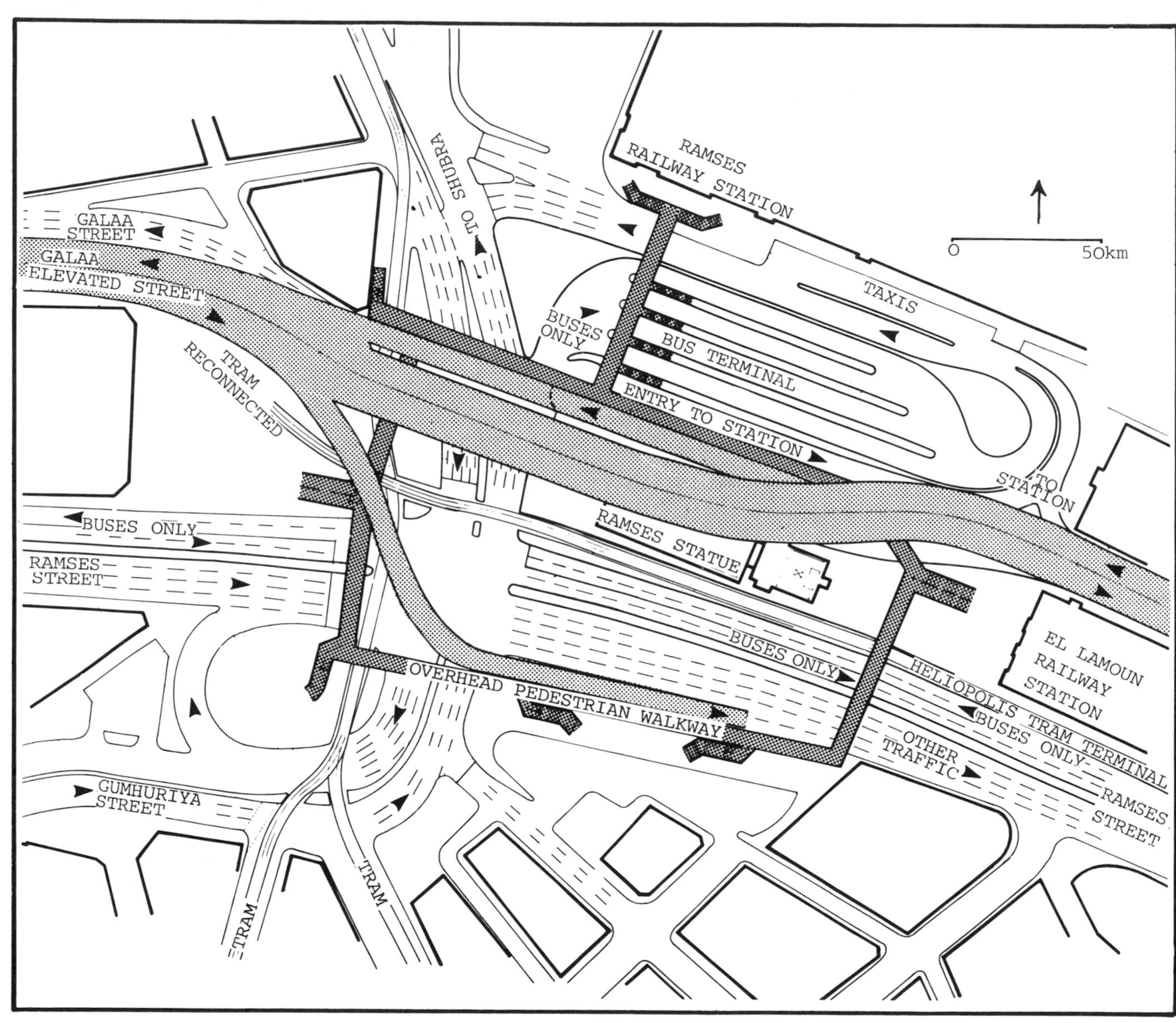

Fig. 3. Ramses Street extension: Ramses Square

Helipolis Metro through Ramses Square to the tram tracks in Galaa Street. A new access and stance arrangement is planned for the bus terminal and the adjacent access to the railway station.

5.5 Alternative junction improvements were also considered at Ramses Street/Misr Sudan. A three lane one-way overpass along Ramses Street was compared to a two lane two-way overpass along Misr Sudan. It was concluded that the latter alternative was not feasible in traffic operations and road safety terms. Therefore, at Ramses Street/Misr Sudan junction a three lane overpass on Ramses Street is proposed.

Corniche El Nile Corridor

5.6 The Corniche is an important north-south route serving long distance through traffic, inter urban traffic and city traffic. Between the 26th July Bridge and the Malek El-Salen, the Corniche operates as a mixture of one-way systems, dual carriageways and single carriageways and the objective was to upgrade it to expressway standard.

5.7 A flyover with slip roads over the Corniche was proposed at Abu El Ella to form part of the comprehensive scheme for the construction of a new 26th July Bridge and a new Zamalek Bridge, both as dual 4 lane facilities, and subsequently as an elevated road through Zamalek connecting the new high level bridges.

5.8 There is a long term plan to provide a second carriageway to the Corniche from opposite the Nile Hilton Hotel south to the junction with Salah Salem. Alternative traffic management systems were considered, diverting south-bound traffic to the Corniche and all northbound traffic to Shar Qasr El Aini, but the existing management system was retained. In view of the longer term need to improve the Corniche over this middle section a design feasibility study was proposed to develop a scheme to maintain, protect and enhance the environment of the river frontage and meet the increasing traffic demands.

Salah Salem - El Rodah Corridor

5.9 This corridor is an outer peripheral route, running in parallel to the Autostrade, connecting from Cairo International Airport through Helipolis and then in a south-westerly direction through Old Cairo to Giza Bridge, at the southern end of Manial Island and Al Ahram beyond. It is a dual three lane carriageway throughout and improvement works costing some LE 1 million are required to provide a consistent standard of facility. The works comprise channelisation, automatic traffic signals, carriageway patching and resurfacing and two crawler lanes to improve operating conditions on the steep gradients north of the Citadel.

Shari Al Ahram - King Feisal Street Corridor

5.10 Sahri Al Ahram is a dual carriageway road 40 m wide of some 8 kms in length with frequent accesses. It is a multi-purpose facility carrying commuter traffic, long distance traffic from Alexandria and Fayum, heavy commercial traffic, tourist traffic, public transport and horse-drawn-agricultural traffic. Minor junction improvements, wearing course overlays to specified road sections and a pedestrian over-bridge at the Zomar Canal are proposed.

5.11 Giza Square is located at the east end of the corridor and is the focal point of the area with a local shopping centre which has very large pedestrian movements, three out-of-town shared taxi terminals, a bus terminal and four main roads entering the square. A detailed study is recommended for the design and implementation of improvements at Giza Square and at the junctions of Al Ahram and Shari El Rabi El Giza having regard for the opening of King Feisal Bridge.

5.12 The King Feisal Bridge is under construction and when completed King Feisal Street will provide an additional main road for long distance traffic to Cairo and for traffic generated by new developments in Giza. The northern carriageway of King Feisal Street has been partially constructed for a distance of 8 km east to the Alexandria Desert Road. On completion of the westbound carriageway of the bridge from Sudan Street to King Feisal street, it is planned that the other carriageway of King Feisal Street should be completed.

Other Routes

5.13 The main north-south route in Giza is a dual 2 or 3 lane carriageway and improvements are planned to junctions on the southern section of the route at Dokki Square, at the junction with Shari Sharwat and at the main entrance to Cairo University Campus.

5.14 Port Said Street runs from the Ismailia Canal Road in a south-westerly direction to connect with Qasr El Eini and the Corniche El Nile at the Sore El-Eyone (Aquaduct) Area. The proposed improvements, costing some LE 1 million, include junction layout modifications and traffic signals at major junctions, extensive road and footway repair and, where street activities are intense, the application and enforcement of parking and other restrictions.

5.15 Shubra Street is the main radial from the northern side of the CBD serving the large residential areas in the north of the city and connecting to the national routes at the north of the city. The areas are served by both tram cars and buses, and two-way tram lines run down the centre of Shubra Street. Corridor improvements are proposed of surfacing works, route signing and improvements at the junction with Rod El Farag and Shari Ecole de Mamalek.

5.16 On the radial routes of Shari El Gheish, Al Ahzar, El Qal'a which provide the main access from the east of Greater Cairo into the CBD via Ataba Square, improvements are proposed to surfacing, junctions and lane marking and route signing.

5.17 Tahrir Street is the main arterial route from Dokki to the Central Business District and improvements proposed for this route include the construction of Sudan Street from King

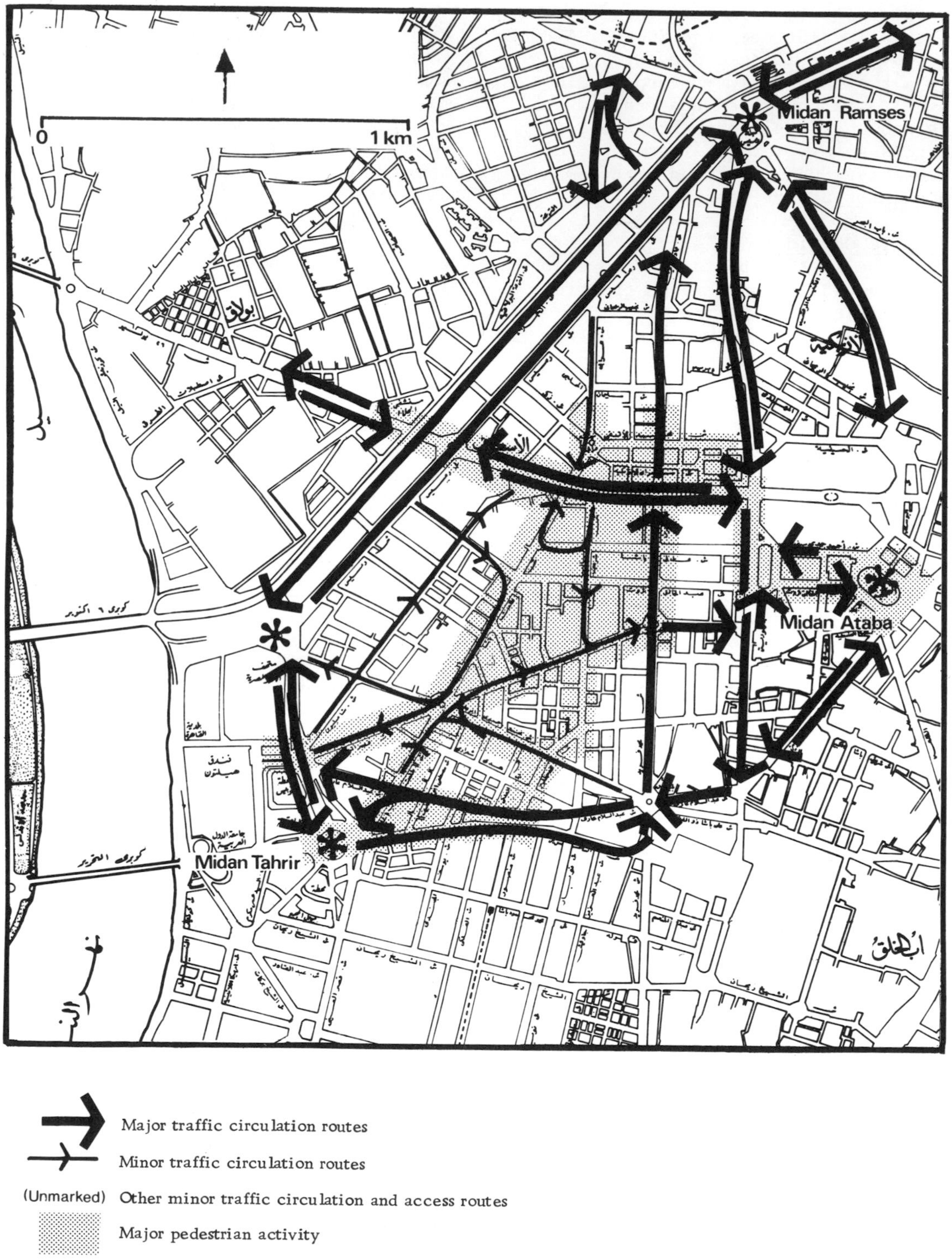

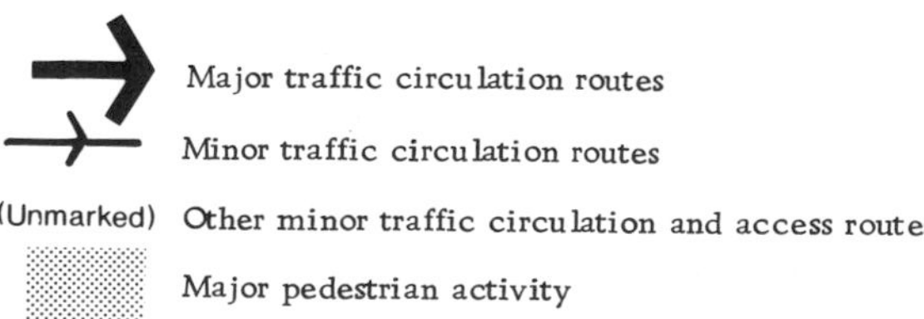

Fig. 4. CBD route strategy

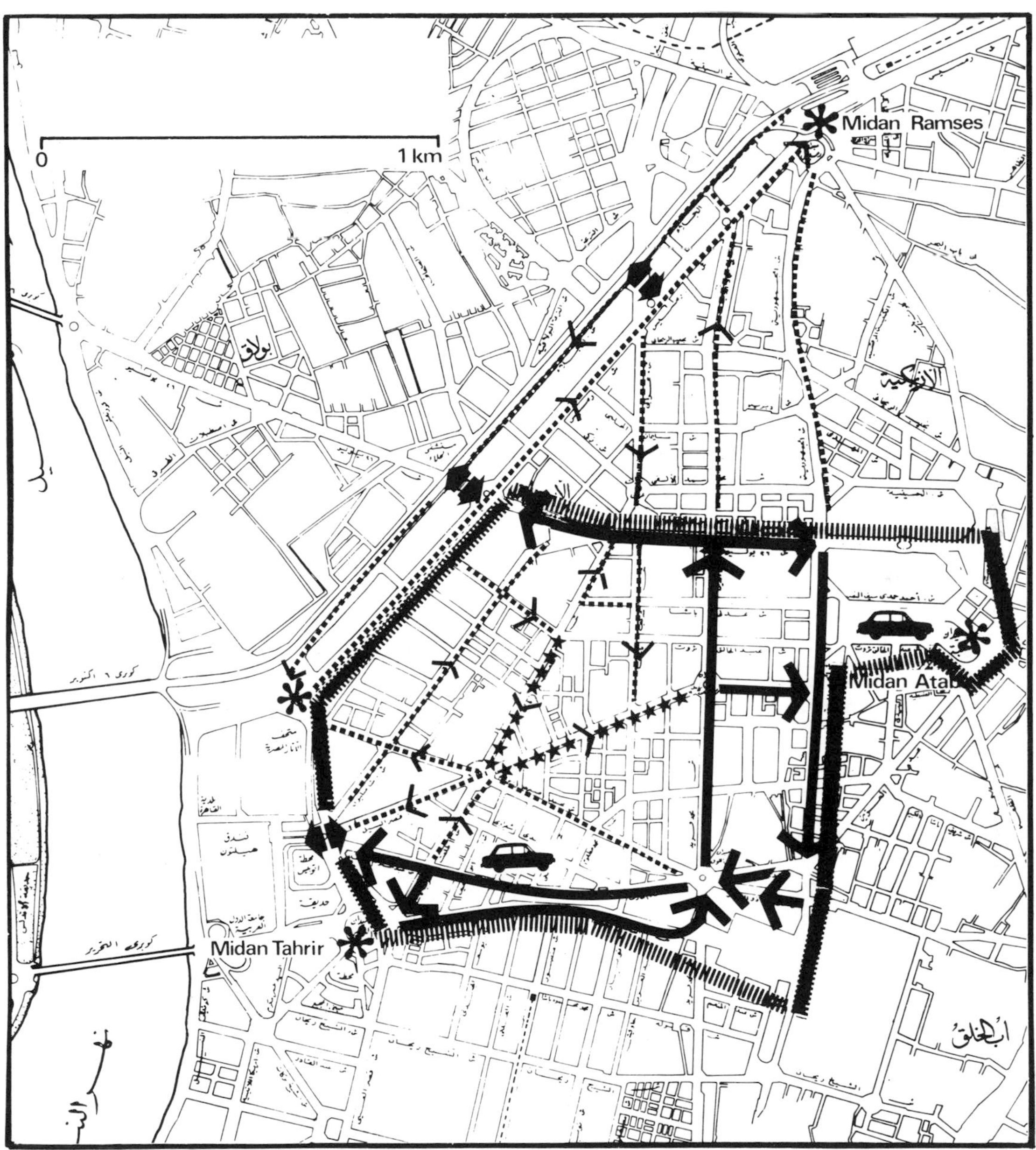

Fig. 5. CBD immediate action plan

Feisal Bridge to the junction with Tahrir Street, physical improvements to Tahrir Street and junction improvements at Galaa Square.

5.18 The construction of the Ramses Extension to Ghamra Bridge could reduce the need in the immediate future to increase further the capacity of the east-west highway system. However, local traffic would benefit considerably if another east-west distributor road was available between Ramses Street and the Ismailia Canal Road, a distance of just over 4 kilometres, as there are no complete east-west routes available. An incomplete route exists from the Corniche El Nile via Shari Ahmed Helmi on the western side of the Cairo-Alexandria railway and on the eastern side a route exists along Sekket El Waili connecting to Salah Salem and into Nasr City. The 'missing link' is a bridge crossing over the Egyptian Railway's main line and sidings and the benefits from constructing this could be considerable. It was recommended that a design feasibility study of the whole corridor should be undertaken.

6. CENTRAL BUSINESS DISTRICT

6.1 The existing conditions in the CBD require to be improved in order to reduce the considerable traffic, safety and environmental problems. The policy for the CBD is to introduce measures which provide:-

i) improvements to the environment for pedestrians.

ii) a reduction of the amount of on-street parking and the increase in the provision of off-street parking.

iii) development of charging for on-street and off-street parking.

v) encouragement to the diversion of non-essential cross-CBD traffic by improving the operation of routes around the CBD.

vi) an increase in the effectiveness of the enforcement.

6.2 The recommended strategy, Figure 4, restricts penetration through while re-organising accessibility within the CBD. It utilises the existing street system, with only minor changes to the existing directions of movement on some streets. Once established this plan allows progressive control of private vehicles and the level of imposed traffic restraint can be increased through time as demand increases and cannot be satisfied.

6.3 To achieve the plan, Figure 5, highway improvement works are required along seven major streets in the total circulation system to obtain orderly movement and control of vehicles throughout the CBD. The Sh. Tahrir/ Sh. El Bustane One-way System to the south, the Sh. Emad El Din/Sh. El Gumhuriya One-way System to the east, and Sh. 26th July through the CBD in an east-west axis form the basis of the traffic circulation system. Engineering and traffic management works are required, in particular, to increase their traffic throughput to accommodate some traffic diverted from streets within the core of the CBD.

6.4 Existing public transport does not penetrate the core of the CBD (except along 26th July Street) but utilises the perimeter streets with its routes linked to the transport terminals at Ramses Square, Tahrir Square and Ataba Square. Deeper penetration of public transport into the CBD is likely to necessitate the use of the shopping streets of Talat Harb or Qasr El Nil. In the compact street system of central Cairo it is doubtful whether the large buses should be allowed into a predominantly pedestrian environment. However, improved accessibility should be encouraged for small public transport vehicles - minibuses, shared taxis and taxis.

6.5 The CBD has over 21,000 parking spaces on-street (legal and illegal) and off-street (public and private). On average, the duration of stay is three hours and each space is occupied between three and four times per day. Parking is currently free or at very low charge and other than the difficulty of finding a space, few restraints are imposed. Controls on the amount and use of parking are required in order that there be a major impact on the traffic operations in the CBD and on the highway network in general. An area-wide policy and a Controlled Parking Zone needs to be implemented.

6.6 The constructions and operation of parking garages with over 1,000 spaces is now planned and with on-street and off-street controls on the parking stock and the introduction of an economic charging formula is proposed. The preferred locations for the first multi-storey garage developments are at Midan Opera and Sh. El Bustane. Integrating car parking with public transport terminal facilities is preferred and Midan Opera is a location which offers good accessibility to the CBD, especially for pedestrians, alternatively, the existing terminus at Midan Ataba could be developed. In addition, the provision of over 3,000 parking meters is planned.

7. AREA IMPROVEMENTS AND HIGHWAY MAINTENANCE

7.1 Within the majority of residential areas in Greater Cairo, and in particular within the low income areas, the physical conditions of the streets are poor and refuse collection and street cleaning is limited. It is proposed that zones should be selected by the Cairo Governorate and Giza Governorate within which essential improvements to the local street system are undertaken.

7.2 The design of the improvements are the responsibility of the Governorates, but it is suggested that the local communities should be involved in the design process. Through this involvement there is every possibility that the community will respond positively to the investments being made through the area improvements.

7.3 The general standard of distributor road and footway maintenance is also poor and this affects vehicle operating costs, safety, convenience and ease of movement. Maintenance is concentrated on main streets and in areas important to tourism and commerce. Although the responsibility for highway maintenance rests with the Governorates, the day-to-day mainten ance is carried out by the Districts which have limited equipment and where normal works consist of minor patching of roads and footways and emergency repairs.

7.4 Cairo Governorate own and operate a Barber Greene continuous mixer asphalt plant, but lack purpose designed control equipment and lorries for distributing asphalt throughout Cairo. The Governorate plan to extend their operations to include major surfacing and re-surfacing works and subsequently to programme for annual maintenance, say on a ten year cycle. To achieve this the Ghamra Plant is to be improved by up-to-date equipment and transport and the renovation of the offices and workshops.

7.5 Giza Governorate has some 1000 km of paved roads and 1000 km of unpaved roads and is totally dependent on contractors for asphalt for major and minor works. They propose to procure and install a 60-80 tons per hour output ashphalt plant and equipment plus the necessary laying plant and ancillary equipment.

7.6 Although the surfacing material is to the general specification of the Ministry of Transport, asphaltic concrete generally has a life of some 5 years in Cairo and the early failure results from poor mix proportioning. Improved quality control would extend the life of the material in service by a factor of two and this would have a dramatic effect on the whole maintenance programme. Technical and practical training is therefore planned to cover both the manufacture and laying of asphalt.

7.7 Approximately forty per cent of the roads in Greater Cairo are provided with street lighting. A planned maintenance programme of lamp replacement is to be initiated and supported by additional tower vehicles and light pick-up trucks.

8. TRAFFIC MANAGEMENT

8.1 The current traffic situation is typified by dense traffic flows on the main streets, low driving standards and driver behaviour, mechanical defects of many vehicles and flaunting of the road traffic laws, ignoring traffic signals and parking regulations.

8.2 The traffic management strategy aims to make the public aware of the disbenefits of the present situation, change the role of the traffic police and introduce physical improvements that will aid the move towards better conditions. The following policies have been identified as being appropriate:-

i) improvement of the control and management of traffic on the existing road system.

ii) increase in the training and equipment of the traffic police.

iii) pursuit of road safety for pedestrians and vehicles, and the provision of greater driver awareness of road safety.

iv) reduction of the environmental impact of traffic noise and pollution.

8.3 The principal objectives of the traffic police in Egypt are to promote the safe and orderly flow of traffic by providing a conspicuous presence, by setting an example of good driver behaviour to other road users, giving advice to them, and by enforcement of the traffic laws. The majority of traffic police are of soldier rank and most of them unable to drive. Soldiers and Constables do not have authority to investigate and prosecute, such authority is restricted to Officers. It is important therefore that the image of the traffic police, largely depicted by the soldier, is improved.

8.4 It is planned to centralise training to ensure standardisation of instruction, practice and enforcement. The Central Traffic Department, the main traffic agency for training of all police forces throughout Egypt, plan to develop a Traffic Training Academy. Priority will be given to training and re-training soldiers, constables and lower rank officers who are directly involved in traffic operations and control.

8.5 The development of specialised motorcycle units is also proposed, capable of dealing with extraordinary traffic problems, moving traffic control and supervising at high accident, high violation locations. The units would comprise officers and constables controlled by the Governorate's Traffic Police Department and by specialist accident investigators who would attend serious and fatal accidents and to investigate by scientific means the causes.

8.6 The implementation of the CBD traffic management proposals will require parking control units to enforce the extended traffic management regulations and to be equipped with tow trucks.

8.7 The Central Traffic Department are improving the current accident reporting system with advice from the Overseas Unit of the U.K. Transport Road Research Laboratory. The implementation of the accident analysis programme would relate more to the Governorates than to the Central Traffic Department, although close liaison would be essential.

8.8 Greater Cairo has thirteen vehicle

testing centres, each manned by a qualified traffic engineer and police officer. Private cars are tested every three years and trucks lorries and taxis are tested annually. It is proposed to improve safety by developing a more comprehensive vehicle test which would require improving the technical staff capability, improving the recording and testing procedure and purchasing equipment.

8.9 The existing vehicle records system is entirely manual, labour-intensive and time-consuming, affording no flexibility for recall of information and as the common practice of reporting traffic violations is by taking the registered number of the vehicle there is constant daily demand for files of registration numbers With the continuing growth in vehicle ownership and accidents there is a planned provision of a simple and rapid recording system.

8.10 The Governorate's driving schools play a significant role in training drivers and it is recommended that their influence is strengthened. It is proposed that an official Highway Code be published at the earliest possible time for distribution to the general public and to all driving licence applicants and driver training establishments. When the police driver training school has been established, driving instructors from Governorate Schools should attend courses on instructor training.

8.11 The present driving tests for both professional and private licences require only simple manoeuvres between traffic cones, on a traffic-free testing area off the public road. The test for each takes an average of 4 minutes to perform. It is recommended that a short, practical road test is introduced to correct this anomaly and a team of examiners trained in advanced driving and in assessing the learner standard is provided.

8.12 In recent years a form of traffic education in schools has been initiated by the inclusion of a small section on traffic training which has been written into the Social Science course of the third year, primary school syllabus. It is recommended that the first priority is now the training of Egyptian teachers who would specialise in a new Traffic Safety Programme.

8.13 All road safety publicity in Cairo is organised by the Central Public Relations Department, but to make the publicity campaign a more effective method of improving road behaviour and reducing road accidents it is recommended that a Road Safety Publicity Unit should be established within the Central Traffic Department and it should have its own budget.

REFERENCES

1. SOFRETU. Greater Cairo Transportation Study. Ministry of Transport, ARE, 1973.
2. PARSONS BRINCKERHOFT INTERNATIONAL INC. et al. Entrances to Cairo Study. General Office of Physical Planning, ARE, 1976.
3. CAIRO UNIVERSITY/MIT. Road Traffic Origin-Destination in Cairo, Ministry of Transport, ARE, 1976.
4. CAIRO UNIVERSITY/MIT. CTA Passengers Origin-Destination in Cairo, Ministry of Transport, ARE, 1977.
5. EL-REEDY T.Y. and EL-HAWARY M.A. Bus Lanes in Cairo. The Highway Engineer, 1981.
6. EL-HOSAINIS. Highway Traffic Management - Research and Practice in Egypt. The Highway engineer, 1979.
7. JAMIESON MACKAY AND PARTNERS. Cairo Urban Transport Management Project, Preparation Report. Ministry of Transport, Transport Planning Authority, ARE, 1980.

23

The costs of traffic accidents and the valuation of accident-prevention in developing countries§

P. J. HILLS, BSc(Eng), MSc, MICE, FIHE, MCIT, Professor of Transport Engineering and M. W. JONES-LEE, BEng, DPhil, Professor of Economics, University of Newcastle upon Tyne

THE NEED FOR AN ACCIDENT COSTING/VALUATION METHODOLOGY

1. Technical progress and economic development undoubtedly yield substantial benefits but they also involve inevitable costs. One of the most significant of these is the carnage and material damage caused by traffic accidents. While few people would deny the extent and importance of this problem, many transportation planners and politicians (in developed, as well as developing, countries) avoid the explicit definition and quantification of the costs of accidents and the value of accident-prevention on the grounds that, to do so, would be too difficult and indeed too controversial. Refusal to grasp such a potentially troublesome nettle is quite understandable but there are, nonetheless, very persuasive reasons for grasping it: the most significant of which derive from considerations of efficiency in the allocation of scarce resources.

2. Simply put, failure to associate explicit costs with traffic accidents (and values with their avoidance) will almost certainly result in the use of widely disparate criteria in the assessment of different projects that affect safety. In some cases, safety effects will merely be ignored while, in others, they will be informally weighted along with "intangibles" such as environmental effects; furthermore, the manner in which they are weighted will differ from one project to the next. Almost inevitably, this will result in a grossly inconsistent pattern of treatment of accident effects, yielding <u>implicit</u> costs and values of safety that range from zero right through to very substantial sums. Evidence from the UK and USA confirms this. As a consequence, it is extremely unlikely that the overall expenditure on safety in the transport sector will be in any sense "optimal" - in particular, if safety effects are widely ignored in transport planning and design, then there will probably be a severe <u>under-investment</u> in safety of travel. Even within a given transport budget, too much will be spent on some aspects of safety and too little on others; i.e. where a straightforward reallocation (at no extra cost) would reduce the overall level of accidents, the allocation of any given budget could be regarded as "inefficient".

3. Such allocative inefficiency would certainly be avoided by the consistent use of costs of accidents and values of accident-prevention in transportation planning. Notice that these problems will <u>not</u> be avoided by the application of uniform safety <u>standards</u> without any corresponding "cost-effectiveness" analysis. The former takes no account of the <u>cost</u> of achieving particular levels of safety and so, in general, leads to allocative inefficiency. Even cost-effectiveness analysis only tells one how best to allocate a predetermined overall safety budget and hence begs the vitally important question of determination of the optimum size of that budget.

4. The question then is how costs of accidents and values of accident-prevention are to be defined in principle and estimated in practice. The first and most fundamental point to appreciate is that the definition of costs or values depends crucially upon the <u>use</u> to which they are to be put. Costs and values may, for example, be intended solely for the assessment of the impact of accidents on national output or income per head. For this purpose, the "pain, grief and suffering" due to accidents, although of undeniable importance in general, may be more or less irrelevant. On the other hand, if costs and values are intended for use in investment planning and allocative decision-taking, then it may be that a wider view of the consequences of potential accidents is called for and, in a caring society, will almost certainly need to include components that take account of human suffering as well as material damage. In short, the selection of an appropriate accident costing and valuation method depends intimately on <u>the objectives</u> that are being pursued by the agency that will use the costs and values.

5. Four broad classes of objective have been identified which encompass the primary goals of the majority of planners in developing countries:

* <u>National Output Objectives</u> - such as maximisation of Gross National Product (GNP) or national income, maximisation of the rate of

§This paper is a summary of a report to the World Bank, presented in 1981.

Highway investment in developing countries. Thomas Telford Ltd, London, 1983, 189–196

growth of GNP or maximisation of the rate of growth of GNP per capita.

* Other Macro-economic Objectives - such as maximisation of the level of employment, minimisation of the rate of inflation or the maintenance of a stable balance of payments. In this category, we might also include the pursuit of "structural" economic goals (e.g. the development of an industrial sector or the establishment and maintenance of military and defence capabilities).

* Social Welfare Objectives - such as maximisation of some social welfare "index" that reflects the well-being of the individuals who comprise society. (This is the kind of goal that typically underpins conventional cost-benefit analysis). This category also includes the pursuit of general "quality of life" objectives, the most important of which, from a safety point of view, is the pursuit of "absolute" objectives such as the minimisation of the number of fatalities or injury-accidents in relation to traffic or the imposition and maintenance of pre-determined safety standards.

* Mixed Objectives - rarely, if ever, will governments and social decision-taking agencies pursue any one of the above objectives to the exclusion of all others. Rather, government decisions will typically reflect the simultaneous pursuit of some or all of the above (as well as other) objectives with the consequent and inevitable "trading off" of one objective against another.

6. To see that these objectives do intimately condition the definition (and hence the magnitude) of accident costs and the values of accident-prevention, one need merely ask oneself what are the relevant consequences of accidents, viewed from the differing perspectives of the various classes of objective. For example, while the "pain, grief and suffering" effects referred to earlier will certainly be directly relevant in the context of the third objective type, those relating to social welfare, they may well not be if the primary goals of allocative decision-taking are instead tied to the narrower type, economic output objectives. However, the relationship between objectives and the definition of accident costs and values can be more clearly seen by considering the various approaches that have been suggested or applied in the developed world and then assessing the relevance (or lack of it) of each of these approaches for the different categories of objective in the developing world.

THE DIFFERENT APPROACHES TO ACCIDENT COSTING/ VALUATION

7. It is possible to identify at least six different methods that have been proposed for placing a cost on accidents, in general (and traffic accidents, in particular) or a value on accident-prevention, any one of which may be "the best" depending on the choice of objective. While all of the methods are, with appropriate modification, applicable to non-fatal as well as fatal accidents, the discussion shall be confined, for the sake of clarity and simplicity, to accidents involving precisely one fatality. The six approaches to the costing and valuation of accidents are:

A. The "gross output" (or "human capital") approach - in which the cost of a traffic accident involving one fatality is treated as the sum of real resource costs (such as vehicle damage, medical and police costs) and the discounted present value of the victim's future output. The value of the prevention of an accident is correspondingly defined as the avoided cost. In some variants of this approach, a significant sum is added to the output loss and resource costs to reflect the "pain, grief and suffering" of the accident victim and those who care for him or her.

B. The "net output" approach - which differs from A only to the extent that the present value of the victim's future consumption is subtracted from the gross output figure.

C. The "life-insurance" approach - in which the cost of an accident or the value of accident prevention is defined as the sum of real resource costs and the amount for which "typical" individuals are willing to insure their own lives (or limbs).

D. The "court-award" approach - in which the sums awarded by the courts to the surviving dependents of those killed, as a result either of crime or of negligence, are treated as indicative of the cost that society associates with the fatality or the value that it would have placed on its prevention. Real resource costs are then added to this figure to obtain the cost of an accident.

E. The "implicit public sector valuation" approach - in which an attempt is made to determine the costs and values that are implicitly placed on accident-prevention in safety legislation or in public sector decisions taken either in favour or against investment programmes that affect safety. Thus, for example, if a government agency rejects a scheme costing $1 million that might have avoided 20 fatal accidents then the implicit value of accident-prevention must be less than $50,000 per fatality.

F. The "value of risk-change" approach - in which the premise is that a typical public sector investment in safety, in effect, provides each individual affected with a very small reduction in the risk of involvement in a fatal accident. The value of prevention of one accident involving one fatality is defined, therefore, as the amount in aggregate that all the affected individuals in society are willing to pay for these small (marginal) risk-reductions, both for themselves and for those about whom they care.

8. Not surprisingly, these six approaches generate substantially different costs and values for accidents involving one fatality. Typical figures derived from studies carried out in developed countries (UK, France and the USA) are summarised in Table 1.

9. There is much that can be said concerning the principles and specific details of the six approaches and their relevance for each of the four classes of objective discussed above (indeed, the full report devotes considerable attention to this). However, for the purpose of this summary, the report's findings concerning the relationship between costing/valuation methods and objectives may be paraphrased, as follows:

National Output Objectives

The gross output approach with no allowance for "pain, grief and suffering" gives a direct measure of the impact of the typical traffic accident on GNP and so is directly relevant. If court awards are determined primarily on the basis of output effects, then they too would be directly relevant. Implicit public sector valuation may reflect output effects and so also may have relevance. Gross output with pain, grief and suffering, net output and risk-change valuations on the other hand are not directly relevant; but, to the extent that they each contain components related to output effects, they have tangential relevance. Life-insurance based costs and values are definitely irrelevant, as these reflect only the purely financial impact of an individual's premature demise on his dependents.

Other Macroeconomic Objectives

Since none of the six accident costing and valuation methods makes specific allowance for the effects of accidents on employment, inflation etc., it is rather difficult to argue that any of the methods has other than superficial relevance to these goals. The essential difficulty is that the effect of an accident, viewed from the perspective of any of the macroeconomic goals, is highly specific to the characteristics and circumstances of the individual victim of the accident, whereas the costing and valuation methods all involve some sort of "averaging" across broad groups within the population.

Social Welfare Objectives

If accident costs and values are ultimately intended for use in a conventional cost-benefit analysis or in an assessment of effects

Table 1 Summary of costs/values of a statistical fatality in developed countries (*All costs and values are in US $ at 1979 prices. With the exception of the Abraham and Thedié figure, which is for France, all other figures are based on estimates for either the UK or the USA).

Valuation/Costing Method	Cost/Value* of an accident involving one fatality
A. Gross output approach:	
(a) including subjective component	$120,000
(b) including subjective component but increased by 50% and with reduced discount rate applied (Leitch, 1978)	$225,000
B. Net output approach:	
(a) excluding subjective component	$25,000
(b) including subjective component	$76,000
C. Life-insurance basis: Fromm (1965)	$930,000
D. Court-awards basis:	
(a) Abraham and Thedié (1960)	$83,000
(b) Shepherd (1974)	$1,100,000
E. Implicit public sector valuation	$3,000 - $60 M
F. Value of risk-change approach	$2,100,000

Table 2 Relevance of various accident costing/valuation methods for different decision-taking objectives

Valuation/ Costing Method	1. National Output Objectives	2. Other Macroeconomic Objectives	3. Social Welfare Objectives	4. Mixed Objectives
A. Gross Output	√√	(?)	X	√
Gross Output + Subjective Costs	X	X	√	√
B. Net Output	X	(?)	X	(?)
Net Output + Subjective Costs	X	X	(?)	(?)
C. Life-insurance	XX	X	X	X
D. Court-awards	√	X	(?)	√
E. Implicit public sector	(?)	√	(?)	√
F. Risk-change valuation	X	X	√√	√

√√ = definitely relevant (?) = relevance uncertain X = probably irrelevant
√ = possibly relevant XX = definitely irrelevant

on allocative efficiency, then the only costing/valuation method that is unabmiguously relevant for this type of goal is the value of risk-change approach (based upon aggregate "willingness to pay"), simply because this approach is explicitly and unequivocally designed for use in conventional cost-benefit analysis. Output-based measures, provided they include a "pain, grief and suffering" component, may be relevant but the extent to which they are depends upon how accurately they reflect individual's attitudes to the prospect of their involvement in accidents. The same can be said of court-awards and implicit public-sector valuations. The life insurance approach is again definitely irrelevant.

Mixed Objectives

In this case, any valuation method relevant to a single objective is also *a fortiori* relevant for a mixture that includes that one.

10. The overall position is summarised in Table 2. As to the costing and valuation methods actually employed by those countries in which explicit costs of accidents and values of accident-prevention are currently used, they are predominantly output-based and, indeed, largely assessed on an *ex post* basis; i.e.the output effects of past accidents are used. Costing/valuation methods and numerical magnitudes employed in a variety of countries - all expressed in US $ 1979 - are summarised in Table 3.

THE RELEVANCE AND FEASIBILITY OF ALTERNATIVE COSTING/VALUATION PROCEDURES FOR LDCs

11. The majority of governments and other social decision-taking agencies in developing countries will normally pursue many different objectives. However, it seems likely that these objectives will typically involve, as significant components, the maximisation of either national output or social welfare. It has been argued above that the only accident costing/valuation methods that appear to be directly relevant to these two objectives are: (a) the "gross output" method, for GNP maximisation; and (b) the "value of risk-change" method, for social welfare maximisation and cost-benefit analysis. Thus the inclination of the authors is to recommend the use of one or other of these methods of costing/valuation of accident-prevention in developing countries. Nonetheless, it may well be that wholesale adoption of one or other of these methods would, at least in the short run, be severely handicapped by data limitations and/or by practical and political problems. These problems would derive from the over-rapid transition from a situation in which accidents are effectively ignored in transportation planning to one in which they are viewed as involving substantial costs. Therefore, the recommendation should be tempered as follows:

(i) Since resource costs (such as vehicle damage and medical costs), together with net

Table 3 Summary of accident costing/valuation in various countries

Country	Method adopted *	Source/Report	Year	Average cost per accident (US $ 1979)				
				Fatal (1)	Serious injury (2)	Avge all injuries	Slight injuries (2)	Damage only
Australia	A	Troy & Butlin	1965/66	165,150	-	3,600	-	1,300
	A	Bureau of Transport Economics	1978	157,500	-	7,000	-	1,250
	A	Fox et al	1979	225,600	-	22,550	-	1,700
Canada	-	UK Dept. of Transport 'ACTRA' papers (1977)	1975	207,000	-	8,250	-	-
Denmark	(A)	NREA (1980)	1979	439,000	9,550	-	1,100	-
Finland	A	NREA (1980)	1975	442,950	-	14,400	-	1,900
	-	Tie-ja-vesira-kennus-halli-tus (1979)	1976	42,450	-	9,400	-	1,900
	(B)	" "	1976	388,150	-	-	-	-
France	A	UK Dept. of Transport 'ACTRA' papers (1977)	1973	147,350	-	9,800	-	-
Germany, W.	A	UK Dept. of Transport 'ACTRA' papers (1977)	1968	132,700	5,800	-	1,150	-
India	(B)	Srinivasan et al (1975)	1968	2,250	1,900	-	200	300
Kenya	B	Fouracre & Jacobs (1976)	1965	12,300	6,750	-	1,000	650
Netherlands	A	UK Dept. of Transport 'ACTRA' papers (1977)	1968	35,100	2,050	-	350	-
Norway	A	NREA (1980)	1976	206,100	10,650	-	600	-
Sweden	A	NREA (1980)	1975	-	-	90,200(3)	-	3,100
	A	NREA (1980)	1976	-	-	136,600(3)	-	4,700
Thailand	B	T.P.O'Sullivan & Partners (1968)	1964	1,550	-	300	-	100
United Kingdom	A	Sabey(1980)	1977	183,150	13,450	-	1,900	-
	A	Dept. of Transport(1979)	1978	179,400	13,450	9,100	1,850	800
	A	"	1978	257,600	13,750	11,000	1,850	800
United States of America	-	Roy Jorgensen & Associates (1975)	1953	14,350	-	2,350	-	550
	-		1955	10,150	-	3,500	-	850
	-		1959	24,950	-	4,050	-	500
	-		1964	21,450	-	5,500	-	1,650
	A		1966	208,150	-	3,800	-	850
	A		1971	428,650	-	20,400	-	900
	-		1972	144,800	-	6,000	-	850
	A	Faigin(1976)	1975	394,000	-	4,350		700

*see headings on Table 2. Notes: (1) definitions of fatality not necessarily consistent with UK (ie 30 days survival); (2) definitions of slight/serious injuries not consistent between countries; (3) includes fatalities.

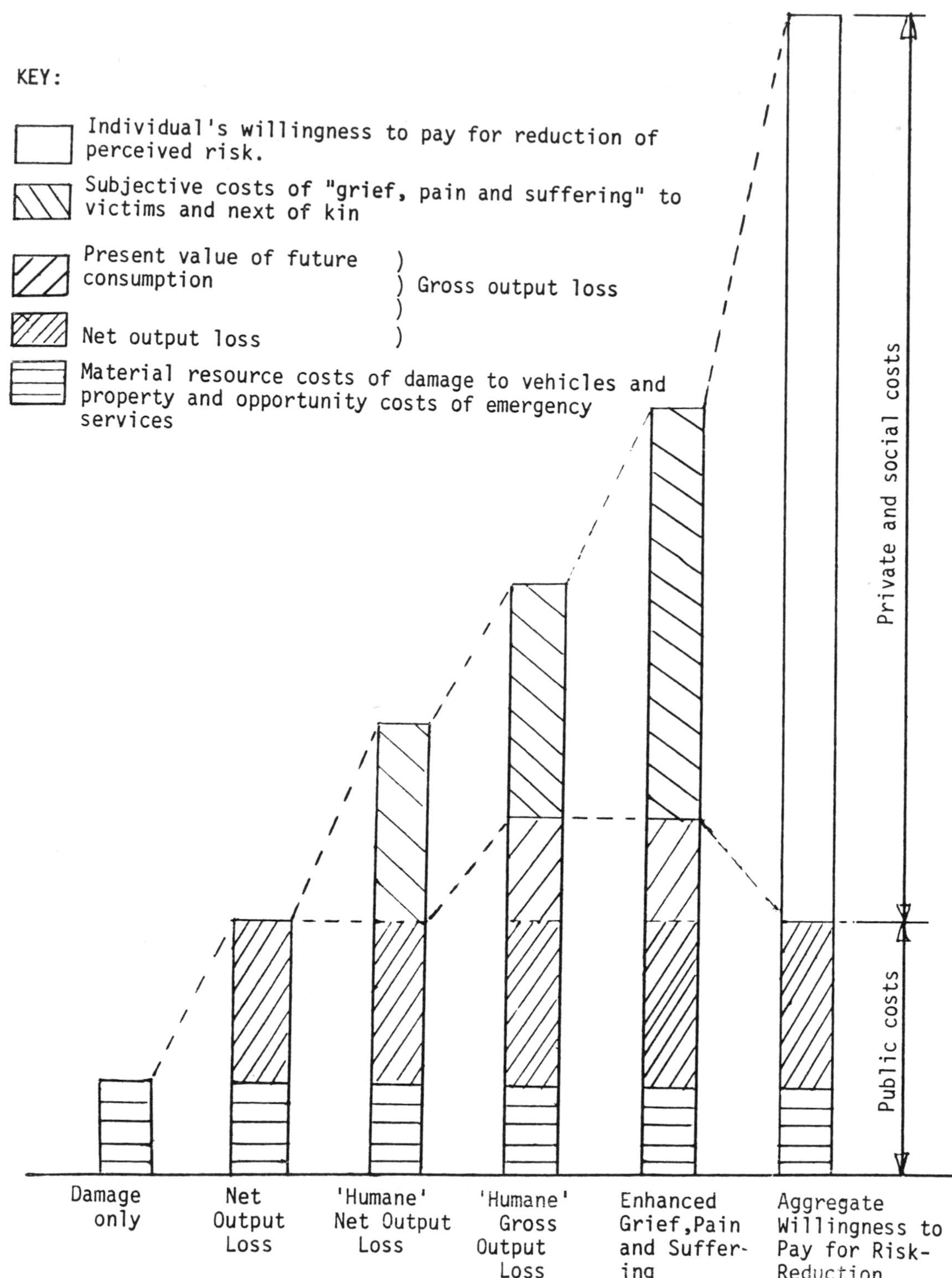

Fig. 1 Components of costs incorporated with alternative methods of evaluation accidents

output losses, form a common component of both gross output and value of risk-change and since they are relatively easy to estimate, these should be regarded as an absolute minimum cost of an accident or value of accident prevention.

12. To put the matter crudely, even if decision-takers appear disinterested in the continued survival or freedom from injury of their citizens *per se*, then the death or injury of any one member of society will still impose a cost upon the remainder of society, to the extent of the resource costs and net output losses that arise.

(ii) If the maximisation of GNP is (as we suspect) a primary goal for most decision-takers in developing countries, then a costing/valuation method based upon gross (rather than net) output should be used. Again, these should be regarded as minumum values, because the gross output measure takes no account of individual's and society's aversion to death and injury *per se*.

13. If, on the other hand, such matters are of concern, then developing countries may wish to add to gross output costs and values a component for "pain, grief and suffering". This would, incidentally, bring the accident-costing/valuation methodology into line with that employed in the majority of developed Western countries (Table 3).

14. However, it should be noted that at least some of the developed countries that currently use a gross output plus pain, grief and suffering approach now have doubts that this approach adequately captures the purely subjective effects of accidents on victims and those who care for them (Leitch Committee Report, 1978). In consequence, active investigation of the possibility of implementing a value-of-risk-change approach is under way.

15. Therefore, it is also recommended that:

(iii) If decision-takers in developing countries are seriously concerned for the quality of life and general well-being of their citizens, as expressed in typical social welfare-maximisation goals, then they should give serious consideration to estimating and using costs of accidents and values of accident-prevention based upon an aggregate willingness to pay for the reduction of risk plus the avoidance of resource-costs and net output losses.

16. In making the last of these recommendations, it should be made clear that such empirical work as has been done on the value of risk-change approach in developed countries suggests accident-costs and values substantially larger than those implied by output-loss approaches. Indeed, the figures revealed in Table 1 are of such enormity that their potential implication for allocative decision-making might serve to dissuade decision-takers in LDSs from giving serious consideration to this more sophisticated approach. However, it should be borne in mind that application of this approach to developing countries would almost certainly produce values substantially smaller in absolute magnitude and, quite possibly, such values would also represent a smaller multiple of corresponding output-based values. This would be the case if, for example, individuals' aversion to risk was found to be less marked in developing countries or if individuals' awareness of (or sensitivity to) risk were less acute.

17. Finally, it should again be stressed that the feasibility of any of the preferred methods will, at least in the short term, be conditioned by the availability of data of the appropriate kind and quality. In this respect, however, the authors would argue strongly that, in the evaluation of accidents, the choice of objectives and methods should always be viewed as the starting point with deficiencies of data being identified as a consequence, rather than vice-versa.

18. In summary, the main recommendation of the report is that the various preferred costing and valuation methods should be viewed as a "nesting hierarchy" which increases in both comprehensiveness and sophistication, from the hard and undeniable material costs of physical damage to vehicles and property, right through to the highly subjective values based on the willingness of individuals to pay for a reduction in their perceived risk of death or injury (Figure 1). The precise point at which decision-takers in a particular developing country choose to "plug in" to this hierarchy will, in part, depend on their choice of objectives and will also - at least in the short run - depend upon data-availability and the effectiveness of forecasting techniques. The great advantage of viewing the alternative costing and valuation methods in terms of a nesting elaboration is, of course, that data limitations will usually dictate the initial selection of more modest and straightforward methods than are ideally appropriate. The subsequent resolution to these data-problems will then permit a smooth and ordered progression to more sophisticated techniques of accident-costing and valuation.

SENSITIVITY TESTS

19. The report contains two distinct sensitivity exercises. The first is concerned with the impact of demographic characteristic and occupational risk - suggesting that it may well be important to identify carefully the demographic characteristics of the populations exposed to accident risks.

20. The second sensitivity exercise in the report examined the effect of variations in the magnitude of accident costs and values upon the ranking of projects in transport investment decision-making in developing

countries. This kind of sensitivity exercise is especially important in view of the very wide range of costs and values reported in Table 1 because, given this range, it is natural to ask whether such variations really matter and, if so, by how much. For example, if it turns out that substantial variation in the costs of accidents have little or no effect on project-ranking, then there is little point either in agonising over the appropriate costing/valuation method or of devoting scarce professional resources to the painstaking estimation of costs and values. In short, under such circumstances, while the contents of the report might be held to have some academic interest, they would for all practical purposes be largely irrelevant. Per contra, if variations of the extent reported do have a significant impact on project-ranking, then it would seem essential for decision-takers and planners in less developed countries to confront the accident costing/valuation issue explicitly. Their decision to opt for one or another approach in these circumstances will have a substantial effect on the allocation of scarce resources within the country concerned.

21. As a basis for this sensitivity test, a simplified version was used of the data for a group of mutually exclusive projects detailed in Adler's "Economic Appraisal of Transport Projects: A Manual with Case Studies. Very briefly, the alternative projects were, as follows:

Scheme A: "do nothing" to an existing two-lane stabilised gravel road, 190 kms in length, connecting two cities in India with populations of about 1 million and 400,000 respectively.
Scheme B: apply a bituminous surfacing over the existing road with only minor improvements to its alignment.
Scheme C: construct a bituminous pavement on the existing road with substantial improvements to its alignment, gradient, bridges and sight-lines.
Scheme D: construct an entirely new and shorter road, with the existing gravel road remaining in use for local traffic.

22. Following discussions with members of the World Bank with direct experience of such schemes, it has been assumed that Scheme B would raise accident-rates by 30%, Scheme C would have no net effect on accident-rates, while Scheme D (purpose-designed on a new alignment) would incorporate a number of safety features which together would serve to reduce accident-rates by 30%.

23. The effect upon project-rankings was then examined of varying the cost of a fatal accident from 0 to 400,000 Rupees and of varying the ratio of fatal to non-fatal accident-costs from five to twenty. It was found that, for all ratios of fatal to non-fatal accident-costs, variation in the cost of fatal accidents has a significant impact upon the net present value of those projects that alter accident

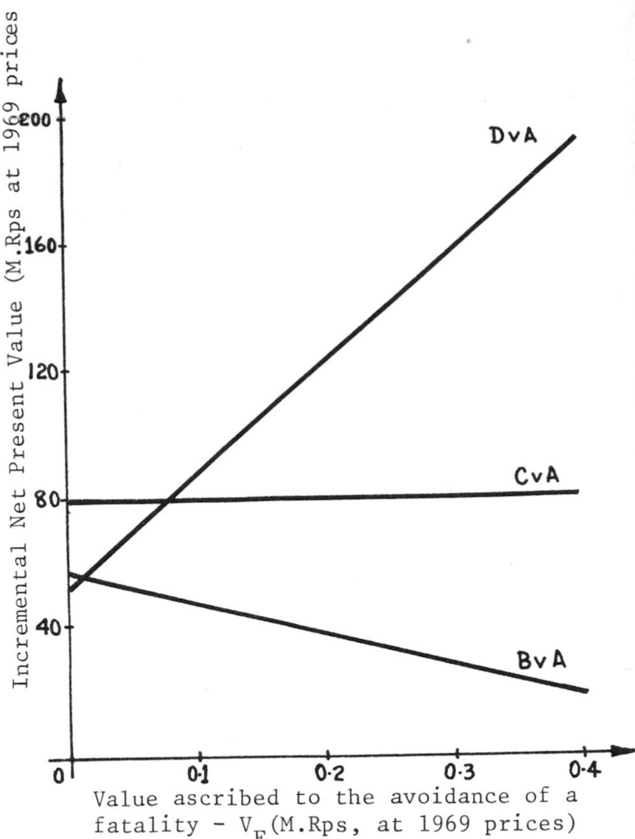

Fig. 2 The effect on overall net present value of accident-prevention (where a fatality is assumed to have a value 10 times that of an average injury)

rates and, more significantly, has a substantial effect upon project-rankings. The results for the intermediate fatal/non-fatal accident cost ratio are summarised in Figure 2. From this, it can be seen that an increase in the cost of a fatal accident from 0 to less than 100,000 Rupees serves to raise Scheme D from third to first place in the project-ranking.

24. In summary, the results of this sensitivity exercise indicate quite clearly that, far from being a matter of subsidiary importance, the size of accident-costs or values of accident-prevention can (and, in most cases, almost certainly will) have a marked effect both on the ranking of transport projects, in terms of net present value within mutually-exclusive groups, and on the magnitude of net benefits generated by any given project. In short, it would appear that the issue of the "appropriate" cost to associate with particular types of accidents - or values to place upon their avoidance - is not one that can legitimately be ignored on the grounds that accident-costs have little overall importance in project-appraisal: the simple message of this sensitivity test is that such costs are potentially very important indeed.

Discussion on Papers 21–23

Professor P.J. HILLS and Professor M.W. JONES-LEE *(Introduction to Paper 23)*: The report which we have presented to the World Bank - of which Paper 23 is a brief summary - is due to be produced in the Bank's Staff Working Paper Series (ref. 1). It goes into the questions in considerably more detail than can the Paper. By way of introduction however, we should establish precisely why it is that we should seek to place a value on 'saving a life' - or, more accurately, on the postponement of death. The distinction is, of course, crucial to understanding the proper basis of valuing health and safety as a resource.

Professional ambivalence

Engineers and planners often display considerable ambivalence in the way they treat safety, both in the design and evaluation of highways.

On the one hand, no-one would argue that traffic accidents are other than a tragic and wasteful consequence of a situation which could, in many cases, have been avoided by better visibility, fewer conflicts, higher levels of vehicle-maintenance or driver-training, more vigorous enforcement of speed-limits or blood-alcohol levels or whatever. In short, accidents are not only undesirable and costly but avoidable and (to some degree, at least) within our control.

On the other hand, many engineers and planners are not prepared to place a value on a life or a limb arguing that, to do so, would be inhuman, clinical or unfeeling; that it reduces human death and suffering to the level of mere commodities bought and sold in the market-place. Others argue that, since the identity of a future casualty is unknown, any value ascribed to the avoidance of his or her injury is bound to be arbitrary or, at best, averaged over a wide range of circumstance.

More sophisticated opponents of the explicit evaluation of traffic accidents point to the inherently random nature of some accidents which can bedevil attempts to estimate the frequency of their occurrence with any accuracy. Why agonize over the value of preventing an accident if the probability of doing so remains unknown?

Outwardly, these may seem to be powerful criticisms of the remarkably few and mostly fledgling attempts to research the subject of accident-valuation.

Design standards

The ambivalence however, reasserts itself strongly in the context of design. No-one wants to be accused of designing dangerous roads. Professional pride and integrity impels us to seek, in any new design, the highest possible safety standards. Indeed, we are prepared to commit often substantial amounts of our clients' resources to ensure that the scheme is adequately safe in operation. But how do we do this consistently as between one scheme and another? How do we know how safe is 'safe enough'? In setting design standards for safety, how do we compare their effects against other advantages such as time-saving or fuel-conservation and how do we weight the relative seriousness of different degrees of injury with that of death itself? These are the questions which cannot be avoided.

Methods in relation to objectives

Paper 23 emphasizes the main contention in the Report, namely that the choice of an appropriate method of evaluation must relate to the overall social or economic objectives being sought at national level. Thus, for instance, if the growth of GNP per head is seen as the goal for a particular country, then estimates of the costs due to pain/grief/suffering would have no relevance (since they do not detract from GNP) - whereas, if a welfare-maximization objective were being pursued, they clearly would.

The temptation at this point is to claim that, although accident-evaluation may be important from a humane standpoint, its effect in practice is trivial or at least dwarfed by comparison with other economic effects such as travel-time or fuel-consumption.

We can demonstrate that, in developed countries, this is certainly not so. Bearing in mind the nature of resources (both material and human) which are damaged and destroyed in traffic accidents, there is every reason to believe that they are as serious, relative to the size of the transport sector, in almost all countries. One thinks here of the foreign-exchange costs of repaired or replaced vehicles, the overstretching of scarce medical and hospital services and the death or debilitation of people who may be key to a country's development (accidents pay no respect to rank).

Paper 23 looks at the effects on a typical highway scheme of varying the assumptions on accident rate and accident value. It is clear from this that not only do accident-costs contribute significantly to Net Present Values of highway schemes but that, as values are increased, the rank-ordering of schemes can also change.

The significance of accident costs to NPV would be substantially greater, of course, where improvement in safety itself were the dominant theme rather than the improvement in travel-time or fuel-consumption. But, unless and until accidents are properly evaluated, safety will remain a secondary consideration expressed always in terms of a constraint and never as a design-objective in its own right.

Mr I.G. HEGGIE *(Freelance Consultant, Oxford)*: In Paragraph 19 of Paper 21, the Authors imply that private car users can be persuaded to use public transport by simply providing improved transit facilities. There is no evidence to support this contention. I published a paper in Policy & Politics (ref. 2) in 1976, reviewing a range of schemes which attempted to encourage car drivers to switch to public transport. They included greatly improved services (like the Stevenage Superbus), reduced or zero fares, innovations like Dial-a-Ride, and marginal improvements like bus lanes and priority intersections. The overall conclusion was that few, if any, car drivers were encouraged to change: the increased patronage - which in most cases was quite modest anyway - consisted almost entirely of new discretionary journeys. And in the case of the Stevenage Superbus this meant more, not less, motorized traffic was found entering the town centre after the scheme was introduced. It is not appropriate here to provide an exposition of why it should be difficult to persuade car drivers to switch to public transport. I would simply like to warn that one is generally dealing with exclusive markets - particularly for non-work journeys - and that new transit facilities will not automatically lead to a change in mode.

My second comment relates to the argument set out in paragraphs 25-30. In the first few paragraphs the Authors argue that private public transport operators tend to be more efficient than public ones and the Authors then go on in the later paragraphs to suggest threshold flows at which different urban technologies - varying from fairly conventional bus systems to full rail rapid transit - would be economic. I believe the analysis is too static. Firstly, the cases quoted to show that private operators are more efficient than public ones, do not in my view show that. They show that small, non-unionized undertakings are more flexible and more efficient than large bureaucratic ones. Secondly, one must view the technologies in a dynamic context in which unique thresholds have no real place.

If you look at their three cases, in reverse order, you have at the one extreme, a rigid rail technology combined with an inflexible bureaucratic management structure. At the other extreme, you have a highly flexible bus technology with no technologically-determined routes combined, if they are small, with highly flexible and demand-responsive operating units.

If you examine an urban area, you must accept that the rail solution will be inflexible and, as the urban area develops, the gap between what the system provides and what the customer wants will increase and productivity will fall. The bus solution, on the other hand, is organic: it can adapt and develop in response to changing urban structure and customer desires.

So, to turn my argument into a question, can the Authors say how robust are their threshold flows over a 15/20-year horizon?

Mr J.S. YERRELL *(Transport and Road Research Laboratory)*: I foresee turbulent times ahead for those concerned with road safety in developing countries. The reasons for this are contained in Tables 1 and 2 of Paper 23: in Table 2 we see that if we are concerned with maximizing Gross National Product then the gross output method is 'definitely relevant' and the risk-change valuation is 'probably irrelevant'. If, however, we are concerned with social-welfare objectives, the exact opposite is proposed i.e. the gross output method is 'probably irrelevant' and the risk-change valuation is 'definitely relevant'. This would not, in itself, be so serious except that Table 1 shows us that the developed-world valuations for fatal accidents for the two approaches are more than an order of magnitude apart: I cannot share the Authors' optimism that these large differences will be reduced significantly for developing countries. We are reminded that the selection of an appropriate method depends intimately on the objectives that are being pursued by the agencies that will use the costs and values, and this of course raises the question of which agency? Is it to be the donor - the bilateral or multilateral aid agency - or the recipient government? We must remember that road safety is often an emotional and politically sensitive topic within a particular country, and I think that we are going to need all our resources of objectivity and cold factual analysis in the debates ahead.

Even when the national planners have decided on mutually acceptable values for fatal-, injury- and damage-accidents, the real problems may be just beginning. They then have to know how many accidents of each kind they are considering, and - probably more difficult - what accident-reduction benefits are likely to flow from implementing various safety measures, be they in the fields of engineering, education or enforcement. The Overseas Unit of the Transport and Road Research Laboratory (TRRL) has been looking at road-safety problems in developing countries for many years, and our recent research is concentrating on the twin areas of devising appropriate accident-reporting, recording and analysis systems, and of evaluating the effectiveness of low-cost, appropriate remedial measures. Information on the latter is very difficult to collect, and I would like to make a plea for our colleagues to let us know of any schemes - particularly highway schemes - with which they may be concerned and where there might be identifiable safety features or measures (and a sufficiently reliable accident-recording system) to enable even a rudimentary evaluation of effectiveness to be made.

Mr J.B. COX *(N.D. Lea & Associates Ltd, Jakarta)*: The topic of this conference is criteria for planning highway investments in developing countries. It is therefore unfortunate and rather ironic that only in the last three Papers have equity considerations been considered in conjunction with efficiency considerations. It is ironic because the thrust into the feeder and rural roads in the mid-1970s was due to a change in development philosophy which emphasized basic human needs and equity considerations.

A minor mention of equity in Paper 23 prompts me to make a few points made previously in a paper to the First international conference on technology for development (ref. 3). The starting point is the realization that most rural areas in developing countries are undergoing a rate of increase per capita incomes of about only 30% of the national average increase. Conversely, the rate of increase of per capita incomes in urban areas is about 200% of the national average increase. Moreoever, we have to admit that most of our road development in developing countries is inequitable - as most of our road development occurs between urban centres and benefits this higher income, urban modern sector of the countries' population.

My approach, therefore, is to produce a social rate of return to determine priorities in highway investment, rather than an economic rate of return, this social rate of return being based on the criterion of equalizing the rate of increase of per-capita income throughout any one country. For example, transport savings accruing to a rural population which is presently receiving 30% of the national per-capita income increase would receive a weighting of three, whereas transport savings accruing to the modern sector would receive a weighting of only 0.5.

Now, how do we do this in practice? In the discussion on Papers 9-11, Mr Heggie mentioned that a World Bank rural roads evaluation programme in Indonesia was not successful in finding proxies for the internal rate of return. A further Asian Development Bank study (ref. 4) following on was, however, successful and found that the best proxy for internal rates of return was not any geographical or social indicators, but existing traffic on the rural roads. In Asia at least it is possible to measure traffic on links, easier in fact than many of the suggested proxies, as there is a quaternary road system below the classified primary, secondary and tertiary road system which is of equal size to the tertiary road system.

My proposal for taking account of equity considerations is therefore relatively simple and uses average daily traffic together with a knowledge of vehicle-type distributions. We know, for example, that trucks move goods between urban centres and benefit the modern segments of a country's population - so therefore let us place a weighting of 0.5 on road-user savings from this vehicle-type. Car road-user savings would also receive a weighting of 0.5 because these benefits accrue to upper classes. Similarly, as the dominant transport mode for the lower income rural segments are mini-buses, then I would weight these road-user savings by 3.0.

Social rates of return determined in this way would allow a ranking of projects on an equitable basis and highway planners could no longer be accused of being party to a development process which is efficient but inequitable.

Mr O. RENARD *(Bureau Central des Etudes Techniques, Ivory Coast)*: With regard to Paper 21, it is clear that the poor financial showing of public transport is often aggravated by the presence of private lines. During off-peak hours, large-capacity buses have a frequency of 15 minutes but are not able to pick up any passengers because the small-capacity vehicles come along every two to four minutes. The large-capacity buses (public transport) are left with the job of helping transport in the peak-hour demand only, which is very inefficient.

Predicting operating costs of public transport is very hazardous, especially when it is a new system in a developing country. In Abidjan, economic feasibility studies have shown that rail transit is worthwhile and would be profitable. Yet we are proceeding with a parallel project of an express bus on freeway because of the uncertainty on the costs of the rail system and the high capital expenditure for rolling stock at a time when resources are scarce.

Paper 22 shows that traffic management has a very high rate of return: more than 100%. Its problems are similar to those of maintenance: setting up an implementing agency, training and keeping qualified personnel, earmarking money for investment against investment allocated to more glorious projects.

Mr LEA and Mr ANDREWS *(Paper 21)*: Mr Heggie's comment is a useful addition to our brief reference to this topic. Care needs to be exercised in applying developed-country experience to less-developed country transit because the developing country may have a much lower standard of transit service and it may have the opportunity of influencing urban form and, as incomes increase, retaining a higher transit modal share which, in the long run, has the same effect as a diversion to transit.

Mr Heggie's second comment suggests some of the reasons why private transit systems tend to be more efficient than public ones. Flexibility and operator-motivation are certainly two important factors. The role of organization size is less clear.

Mr Renard raises the issue of skimming, which in many situations is important to a comparison between private and public bus operations. Where this condition is present, the more flexible operator provides service only on the more profitable routes, whereas the less flexible operator, who may be constrained by contract or government policy, must provide a service on the uneconomic routes and attempt to cross-subsidize it. Where such conditions exist, they require special consideration, but there is enough evidence from situations where skimming is not a serious factor, to suggest that for other reasons, private operators tend to be the more efficient.

Mr Cox raises the important question of weighting an economic analysis because of equity considerations, so that investments which are progressive in their income-distribution impact will show a higher 'social rate of return' than those with a regressive impact. The objective is laudable, but it may be preferable to take this equity consideration into account by some direct estimation of the income distributional effects of various options, rather than creating an unknown distortion to the economic rate of return.

Professor HILLS and Professor JONES-LEE *(Paper 23)*: We agree with much of what Mr Yerrell has to say. Tables 1 and 2 of Paper 23 do indeed pose a potentially serious dilemma for those involved in road planning and project appraisal for developing countries. It is also true that this dilemma can only be satisfactorily resolved by a clear specification of the economic, social and political objectives that underpin road planning in any particular developing country. This is one of the principal messages of the report on which our Paper is based (ref. 1) and if we have succeeded in getting the message across then we will be well-pleased. On the question of whose objectives are to count, we have to admit that there may be a difficulty if there is disagreement between the aid-agency and the recipient government over goals and priorities, but this kind of problem is not peculiar to safety and should in most cases be resolvable by discussion and debate between interested parties. For what it is worth, our view is that in the last analysis it is the objectives of the recipient developing country that should determine the selection of an appropriate accident costing/valuation methodology.

Mr Yerrell reasonably questions our suggestion that the ratio of 'willingness-to-pay' values of accident prevention to their 'gross output' counterparts for developed countries may be substantially reduced for developing countries. Again we must concede that we have no hard evidence for this assertion, but developed country data does indicate a significantly non-linear relationship between willingness-to-pay for safety and income which does suggest that willingness to pay in the poorer developing countries would be very much lower than in developed countries.

Finally we also agree that estimation of the effects of different safety measures is of vital importance. Without such data any accident-costing and valuation method - however clearly specified - is effectively useless. Equally, though, data concerning safety effects - however detailed and precise - are of little real use for project planning and appraisal without a well-articulated costing and valuation methodology. The two problems are vitally interrelated aspects of the overall appraisal of safety in road planning. Nonetheless, we would suggest that a prior settlement of the costing and valuation question might serve to direct data-collection efforts to their most effective use. For example, the requisite precision of estimates of vehicle damage might well depend on whether the accident-costing and valuation method was based on narrow output-related goals or on wider consideration of welfare and avoidance of pain and suffering.

REFERENCES

1. HILLS P.J. and JONES-LEE M.W. The costs of traffic accidents and the valuation of accident-prevention in less-developed countries. The World Bank, Washington. Staff Working Paper. To be published.
2. HEGGIE I. Consumer response to public transport, improvements and car restraint: some practical findings. Policy & Politics, 1976.
3. COX J.B. Efficiency and equity considerations in the redevelopment of ASEAN networks. 1st Int. Conf. Tech. Dev., Institution of Engineers, Canberra, 1980.
4. HOFF and OVERGAARD. Rural roads study, Central and East Java. The World Bank, Washington, 1981, Report to the Directorate General of Highways.

Closing address

Dr R. S. MILLARD, Highway Engineering Adviser, World Bank

Most of us suffer from an inability to remember funny jokes. Still worse is an aptitude for remembering jokes which are not funny. One such unfunny joke which haunts my memory is of a much married couple sitting at breakfast. He peers over his newspaper and pontificates: 'We are living in a period of grave transition' and the wife replies, '... and I've hardly a rag to put on'. That story is apt to our conference. We are living in times of great change and we at the conference are parading the rags we have to see us on our way through these changing times. And just as our lady undoubtedly has a wardrobe full of clothes, so our armoury of techniques is fairly formidable. The Institution of Civil Engineers and the Commission for European Communities are to be congratulated for promoting this forum at which we can parade and review our techniques of planning highway investment.

Firstly, to characterize this period of grave transition as it affects highways in the Third World: it is like trying to guess what is on the other side of a Black Hole. We can see where we have come from but we can only guess where we are going. There was in the Third World a period of phenomenal growth in motor traffic in the 1950s and 1960s, and this was a period of great activity in building up main road systems often using multilateral and bilateral aid. Emphasis started to change in the early 1970s towards secondary and minor roads increasingly as part of rural development, and also towards fostering road maintenance, this last usually with indifferent success so that many of the roads built 20, 10 even 5 years ago are now badly in need of rehabilitation. Aid donors, particularly the World Bank, continued in their resolve to demonstrate that wise choices were being made in allocating financial aid and thus they continued seeking to develop reliable and numerate means of determining project viability. On top of this came the fuel crisis, the first practical indication that our resources of fossil liquid fuels are finite, and with many consequences on the development plans of Third World countries.

Looking forward into the Black Hole, we start with the main conclusion from the Grieveson-Winpenny Paper (Paper 5) that the future seems bleak both for developing countries and for industrialized countries, with the corollary that we are moving away from building new roads into a period of consolidation where the main concern is to salvage, maintain and, where desirable, improve the road systems we now have. There are however signs that some countries are being tempted by aid donors to continue expanding their road networks beyond their capacity to provide adequate maintenance. There is a further corollary from the Grieveson-Winpenny Paper. The Authors refer to the need for cost-effective designs for staged improvement and proper maintenance. Cost-effective design is certainly needed but all too frequently we are not getting cost-effective production. Roads which are built with expected design lives of 10-20 years are failing much earlier than they should. Apart from the nonsense which this makes of the carefully worked-out economic appraisals, much genuine waste is involved. I am going to a Far-East country where a main road which has just been built is falling to pieces. The reasons are complex - they always are. But I mention this because planners and economists frequently believe that the science of highway engineering is more precise than in effect it is. Practice all too often falls far short of theory and one of the most important components of our efforts in this period of consolidation must be to make sure that we get value for money in the engineering of road construction, of road improvement, and of road maintenance.

The section of this volume on national and provincial roads has produced two complementary sets of papers: one set dealing with the elaborate and detailed studies of the transport infrastructure in what is happily called the Southern Cone of Latin America - the overall planners' approach; the other exploring the use of road-user charges and self-regulating systems. These two themes run parallel through these proceedings. At the extreme they represent opposite political poles, one a devotion to a centrally controlled highly planned economy and the other reaching out towards Adam Smith, Milton Friedman, Mrs Thatcher and President Reagan, in the belief that all will be right in the world if we can allow economic cause and effect to operate. Those of us who are empiricists believe we need a bit of both. Certainly it is good to see the dedication to overall transport planning in the Southern Cone of Latin America and probably for the first time in the world, working methodologies are beginning to emerge which provide a numerate basis for planning transport investment and determining road-user taxation policies.

On road-user charges, it seems that toll

roads have rather a limited application in the Third World, but a good picture of the uses of road-user taxation is given in Paper 4 with, in the discussion, some notice of that prevailing problem: an inability in Third World countries to control the size and weight of vehicles in the interests of overall road-transport economy. By this time another prevailing theme had emerged - the general neglect of road maintainance in Third World countries. There are clearly large problems remaining in both the funding and the execution of road maintenance. Paper 17 indicated one trend which is clearly useful, the increasing use of contractors to undertake routine and periodic maintenance. This raises yet another theme first emerging in the Grieveson-Winpenny Paper and also mentioned in Paper 13, that our situation of grave transition is calling for a greater flexibility in arrangements for awarding contracts in situations where it is not easy to specify with precision the nature and quantities of work that will be required.

The section on feeder and rural roads produced the same dichotomy between the planners and the Friedmanites, and again interest in the discussions concentrated on the planning aspects, with another conflict lying just under the surface between those seeking to simplify the planning process in the light of the very crude planning data available and those who, recognizing the complexity of the development process, want to give due consideration to all recognizable facets of the process. It is good to see ex-post studies being reported in Papers from Malaysia and Ghana indicating that at village level at least the demand from transport facilities can be somewhat different than might be expected. In this respect it was somewhat surprising to find institutional development being challenged as a passing gimmick. Perhaps this reflects a cynicism about the trend followers who jump on every new bandwagon. But institutional development is more than a passing trend. It is vital that rural people themselves be brought to play a responsible part in the development process. No matter who pays for the roads in the beginning, the local people must be brought to realize that the local roads and tracks are theirs to look after. They must take part in the planning process and they must contribute to their construction and subsequent maintenance - an essential part of the development of responsible local government. It must not be forgotten that there is great scope for the use of tracks carrying two-wheeled vehicles to provide the link between the farm and the road.

There is a further general point to be made about the participation of local communities and local contractors. One distinction between rail and road is that roads do provide the opportunity to develop local community responsibilities and local entrepreneurship - both of which are vital in nurturing development.

Papers 12, 18 and 19 are concerned with pavement strengthening. On this subject there is a principle to be recognized - that our methods of pavement engineering imply that pavements need to be strengthened from time to time to safeguard the original investment. This pavement strengthening may be classified as road maintenance, but what we are discussing is developing an approach to planning the orderly strengthening of road pavements in a country network to meet traffic needs at minimum total cost. These three Papers are concerned with producing such programmes of orderly strengthening. In a sense the Papers from Jamaica and from the Ivory Coast present different extremes: the pragmatic and the theoretical. As the discussion showed, particularly the contribution from Mr John Cox, current theory on pavement deterioration is not yet adequate to explain the behaviour of lightly constructed pavements in hot climates. The theories developed on pavement stiffness vital for airfields and heavily trafficked roads in cool climates have been misleading us. We should be thinking more of critical strain rather than critical stress, of flexibility rather than stiffness, and of total transport costs rather than maintenance costs.

Clearly there is a problem here in defining how road pavements deteriorate in Third World countries and more work is needed to resolve it.

Paper 21 points out how public transport, indeed how the growth of cities themselves, can be distorted by ill-considered subsidies. The outstanding feature of Dr Latchford's Paper (Paper 22) is the courage to take on the terrifying traffic problems of metropolitan Cairo. May I implore him to come back in, say, five years' time and tell us how well the different features of the plan are working.

Paper 23 on the costs of road accidents has been included in the section on urban roads. But in almost every developing country road accidents present a countrywide problem with accidents and fatalities at an appalling level. Until recently this has been regarded as one of the inevitabilities of rapidly increasing road traffic. But many governments are stiffening in their resolve to reduce the toll of road accidents and this Paper provides a valuable review of the options in costing road accidents and road-safety measures. For planning wise investment in road safety we need much more data than are at present available on accident costs and on the efficacy of different road-safety measures. Developing countries often have an advantage in that they can adopt unconventional road safety measures which are both cheap and effective. I conclude this review by giving two examples. In Argentina if you are caught speeding you are likely to be hauled out of your car and made to sit at the curbside for 10 minutes for each kilometre per hour that you were exceeding the speed limit. In Lagos if you are caught jay-walking you are likely to be compelled to spend ten minutes running on the spot.